198
Topics in Current Chemistry

Springer-Verlag Berlin Heidelberg GmbH

Design of Organic Solids

Volume Editor: E. Weber

With contributions by
Y. Aoyama, M. R. Caira, G. R. Desiraju, J. P. Glusker,
A. D. Hamilton, R. E. Meléndez, A. Nangia

Springer

This series presents critical reviews of the present position and future trends in modern chemical research. It is addressed to all research and industrial chemists who wish to keep abreast of advances in the topics covered.

As a rule, contributions are specially commissioned. The editors and publishers will, however, always be pleased to receive suggestions and supplementary information. Papers are accepted for "Topics in Current Chemistry" in English.

In references Topics in Current Chemistry is abbreviated Top. Curr. Chem. and is cited as a journal.

Springer WWW home page: http://www.springer.de
Visit the TCC home page at http://www.springer.de/

ISSN 0340-1022
ISBN 978-3-642-08427-0 ISBN 978-3-540-69178-5 (eBook)
DOI 10.1007/978-3-540-69178-5

Library of Congress Catalog Card Number 74-644622

Originally published by Springer-Verlag Berlin Heidelberg New York in 1998
Softcover reprint of the hardcover 1st edition 1998

Cover design: Friedhelm Steinen-Broo, Barcelona; MEDIO, Berlin
Typesetting: Fotosatz-Service Köhler GmbH, 97084 Würzburg

SPIN: 10552164 66/3020 – 5 4 3 2 1 0 – Printed on acid-free paper

Volume Editors

Prof. Dr. Edwin Weber
Institut für Organische Chemie
Technische Universität Bergakademie Freiberg
Leipziger Straße 29
09596 Freiberg/Sachsen, Germany
E-mail: weber@tu-freiberg.de

Editorial Board

Preface

Considering the high level of our knowledge concerning covalent bond formation in the organic chemistry of molecules, our understanding of the principles involved in organic solid design is almost in its infancy. While chemists today are able to synthesize organic molecules of very high complexity using sophisticated methods of preparation, they lack general approaches enabling them to reliably predict organic crystalline or solid structures from molecular descriptors – no matter how simple they are. On the other hand, nearly all the organic matter surrounding us is not in the single-molecule state but aggregated and condensed to form liquid or solid molecular assemblages and structural arrays giving rise to the appearances and properties of organic compounds we usually observe. Obviously, the electrical, optical or magnetic properties of solid organic materials that are important requirements for future technologies and high-tech applications, as well as the stability and solubility behavior of a medicament depend on the structure of the molecule and the intramolecular forces, but even more decisively on the intermolecular forces, i.e. the packing structure of the molecules to which a general approach is lacking. This situation concerned J. Maddox some years ago to such a degree that he described it as "one of the continuing scandals in the physical sciences" [see (1998) Nature 335:201; see also Ball, P. (1996) Nature 381:648].

The problem of predicting organic solid and crystal structures is very difficult. The solid state and crystalline arrangements adopted by organic molecules depend on a subtle balance of intermolecular interactions which can be achieved for a given conformation in a particular packing arrangement giving rise to what is called polymorphism. Any approach to using theoretical methods to help predict organic crystalline and solid structures must therefore take into account this problematic balance between the intra- and intermolecular interaction energies present – strictly speaking, a point currently unsolvable.

Nevertheless, with the beginning of the supramolecular concepts in chemistry [see (1993) Top. Curr. Chem. Vol. 165 and (1995) Top. Curr. Chem. Vol. 175] about 25 years ago the situation has continued to improve little by little. In keeping with the new way of thinking in supramolecular dimensions, i.e. beyond the molecule, structural chemists and crystallographers have had little difficulty in recognizing a molecular crystal as the supermolecule par excellence and an organic solid not just as a collection of molecules but as a defined network structure. Indeed, the crystallization or solid formation process itself is an impressive display of supramolecular self-assembly, involving specific molecular recognition at an amazing level of precision. Taking up the words of one of the authors (G.R. Desiraju), crystals and solids constitute one end of the supramolecular continuum and may be viewed as "hard" supermolecules in contrast to the "softer" supramolecular aggregates which exist in solution. Consequently, supramolecular chemistry today encompasses the study of molecular crystals with all the applications and ramifications that such a study implies in the field of solid-state chemistry, crystal-engineering and materials science both from organic and inorganic viewpoints.

Many of the prerequisites for such an improved understanding of solid-state supramolecular chemistry but being focused on organic or essentially organic solids and their potential design are discussed in the present book, beginning in the contribution of J. P. Glusker (Chapter 1) in which she considers the *directional aspects of intermolecular interactions* and the directional preferences of binding of functional groups. In Chapter 2, A. Nangia and G. R. Desiraju attempt to show the importance of *pattern recognition* using *supramolecular synthons* in organic crystal-engineering and referring also to the implications of such ideas in related areas. R. E. Meléndez and A. D. Hamilton, in Chapter 3, take up this topic in a more specific way and present advances in this field, directing their focus on the design of organic solid structures based on the *hydrogen-bonded tape, ribbon and sheet motifs* involving complementary intermolecular interactions. There is much current interest in organic hosts, whose guest binding properties are reminescent of traditional inorganic zeolites but are composed of organic and metal-ion building blocks to form controlled solid network structures by making use of directional interactions such as hydrogen bonding and coordination. The present stage of these *organic zeolite analogs* and prospects associated therewith are discussed from both static and dynamic viewpoints in Chapter 4 by Y. Aoyama. In the closing report of M. R. Caira (Chapter 5), a summary of the thermodynamic, kinetic and structural considerations of *crystalline polymorphism* which is encountered in all areas of research involving organic solid substances, giving rise to unique difficulties in solid materials design with facing of current facts, is presented.

I hope that this book will stimulate new work on the design of organic solids which is a truly promising topic relating to many important areas of research and industry . To conclude these words of introduction, I wish to express my heartfelt appreciation to all the contributors who have made this book possible.

Edwin Weber Freiberg, July 1998

Contents

Directional Aspects of Intermolecular Interactions
J. P. Glusker 1

Supramolecular Synthons and Pattern Recognition
A. Nangia, G. R. Desiraju 57

Hydrogen-Bonded Ribbons, Tapes and Sheets as Motifs for Crystal Engineering
R. E. Meléndez, A. D. Hamilton 97

Functional Organic Zeolite Analogues
Y. Aoyama 131

Crystalline Polymorphism of Organic Compounds
M. R. Caira 163

Author Index Volumes 151 – 198 209

Contents of Volume 196

Carbon Rich Compounds I

Volume Editor: A. de Meijere
ISBN 3-540-64110-6

Design of Novel Aromatics Using the Loschmidt Replacement on Graphs
Y. Sritana-Anant, T. J. Seiders, J. S. Siegel

Modern Routes to Extended Aromatic Compounds
S. Hagen, H. Hopf

Carbon Rich Cyclophanes with Unusual Properties – an Update
B. König

Unsaturated Oligoquinanes and Related Systems
R. Haag, A. de Meijere

The Centropolyindanes and Related Centro-Fused Polycyclic Organic Compounds
D. Kuck

Contents of Volume 197

Dendrimers

Volume Editor: F. Vögtle
ISBN 3-540-64112-1

Iterative Synthesis in Organic Chemistry
N. Feuerbacher, F. Vögtle

Supramolecular Chemistry Within Dendritic Structures
V. V. Narayanan, G. R. Newkome

Divergent Approaches to Phosphorus-Containing Dendrimers and Their Functionalization
J.-P. Majoral, A.-M. Caminade

Chiral Dendrimers
D. Seebach, P. B. Rheiner, G. Greiveldinger, T. Butz, H. Sellner

Dendrimers with Polymeric Core: Towards Nanocylinders
A.-D. Schlüter

Electrochemical and Photochemical Properties of Metal-Containing Dendrimers
M. Venturi, S. Serroni, A. Juris, S. Campagna, V. Balzani

Directional Aspects of Intermolecular Interactions

Jenny P. Glusker

The Institute for Cancer Research, The Fox Chase Cancer Center, 7701 Burholme Avenue, Philadelphia, PA 19111, USA. *E-mail: jp_glusker@fccc.edu*

Directional intermolecular interactions can be found by statistical analyses of the surroundings of functional groups in all crystal structures containing them. The most notable such interaction is the hydrogen bond, which, if strong as between OH or NH groups and oxygen or nitrogen acceptors, is approximately linear. This provides an important means of aligning molecules together. Analogous but weaker interactions described here are C-H···O, F···H and H···π, and also those between C-S-C groups and electrophiles or nucleophiles. Some directionality can also be identified in aromatic-aromatic interactions. Metal ion coordination can, in certain instances, also have a directional component, particularly if the coordination geometry is inflexible, as for the octahedral binding of divalent magnesium. The geometries of interactions of metal ions with various functional groups in proteins are described, and in many cases they are more rigid than the analogous interaction involving a hydrogen bond. The emerging use of crystal surfaces as probes of molecular recognition is then discussed. Finally some examples of molecular recognition in biological macromolecules are given; these stress the importance of pattern recognition in hydrogen bonding, together with the significance of weaker interactions.

Keywords: Recognition, Molecular, metal ion coordination, Hydrogen bonding, Intermolecular interactions, Directionality, Crystal surface recognition.

1 **Introduction** 3

2 **Methods of Analysis** 4

2.1 Sources of Crystallographic Data 6
2.2 Spectroscopic Methods of Studying Intermolecular Interactions . . . 7

3 **Types of Intermolecular Forces** 7

4 **Directed Organic Interactions** 8

4.1 Intermolecular Interactions in Hydrocarbons 8
4.2 Nucleophile and Electrophile Interactions with Sulfur and Halogen Atoms 12
4.3 Hydrogen Bonding 15
4.4 C-H···O Interactions 20
4.5 F···H Interactions 22
4.6 H···π Interactions 24

5 Interactions of Planar Functional Groups in Proteins 26

5.1 Surroundings of Carboxylate Groups 26
5.2 Surroundings of Imidazole and Histidine Groups 29
5.3 Surroundings of Arginine Side Chains in Proteins 30
5.4 Surroundings of α-Hydroxycarboxylate Groups 31

6 Metal Ion-Based Directional Interactions 31

6.1 Metal Ion Coordination 33
6.2 Magnesium Binding 33
6.3 Zinc Binding 34
6.4 Calcium Binding 35
6.5 Ionophores and Other Metal-Binding Molecules 37
6.6 Metal-Binding Sites in Proteins 38

7 Networks of Interactions 39

7.1 Descriptions of Hydrogen Bonding Networks 40

8 Crystal Surface Recognition and Chirality 40

9 Applications to Macromolecular Structures 42

9.1 Binding of Drugs to Dihydrofolate Reductase 42
9.2 Binding of Drugs to Nucleic Acids 44
9.3 Protein-Nucleic Acid Interactions 50

10 Incipient Chemical Reaction Pathways 52

11 Conclusions 53

12 References 53

List of Abbreviations and Symbols

A	adenine
A · T, G · C	base pairs
C	cytosine
CSD	Cambridge Structural Database
DNA	deoxyribonucleic acid
G	guanine
ICSD	Inorganic Crystal Structure Database
NADPH	nicotinamide adenine dinucleotide phosphate
PAH	polycyclic aromatic hydrocarbon
PDB	Protein Data Bank
T	thymine

1 Introduction

Intermolecular interactions are essential to life. They are what hold us together, determine how the molecules within us interact with each other, and control how we relate to our surroundings. They are also at the heart of all the amenities for living that we value so highly, such as the construction in buildings, the printing and binding of books, and countless other advantages of modern-day life. Therefore it is essential that we understand the way that molecules bind to each other in terms of their constituent functional groups – how and why these interactions take place. The characteristics of intermolecular interactions that will be reviewed here are their geometries, their strengths, and their angular dependencies [1]. With such information we can then hope to be able to devise new materials with required physical properties.

When two molecules interact they may either bind to each other or they may undergo a chemical reaction. The two initially interacting molecules tend to approach each other in directions that are governed by the distribution of electronic charge on their respective surfaces, and any geometrical effects (such as steric hindrance) that modify the availability of sites for binding. If the interaction between the two molecules is only binding (as discussed in this article), and no chemical reaction takes place, the resulting complex has an atomic arrangement of relatively low energy and therefore reasonable likelihood. It contains atoms poised at equilibrium distances because they attract each other at long distances, but repel each other at very short distances [2]. Generally the positively charged group on one molecule will attract a negatively charged group on another molecule, and aggregation will result.

What types of intermolecular interactions occur and which biophysical techniques can be used to characterize them? An example of a strong interaction, and one that is very directional, is the hydrogen bond. It serves to align molecules because it involves small bond and activation energies [3]. At present the best experimental way to determine the directionalities of intermolecular binding is to investigate how molecules pack with each other in the crystalline state, since high-resolution geometrical information is available from structure determinations by X-ray and neutron diffraction methods. A crystal is defined as a solid within which the arrangement of molecules or ions within it can be considered in terms of a unit cell, the contents of which are repeated in a regular manner in three dimensions. Intermolecular interactions found in a crystal structure lead to a minimization of the free energy of the entire atomic arrangement throughout the crystal. This overall arrangement is determined by the forces between atoms, expressed by the sizes, shapes, charges, dipoles, and hydrophobicities of the individual molecules or ions. In order to start to form a crystal, two molecules must interact with each other in some way. Solvent may participate if the interaction occurs in solution. As the crystal develops, more and more molecules bind to each other, generally, with the same types of interaction that lead to the initial formation of a dimer. Therefore the atomic arrangement in the crystal, in which the total energy is minimized, contains information on the forces between groups that lead to the initial aggregation.

Perturbations to the directionality of interactions may result from steric interactions between different neighboring groups around a functional group; they necessitate a statistical analysis of many different crystal structures containing the interaction of interest. This type of analysis is made possible by the availability of computer-based crystallographic databases: the Cambridge Crystallographic Database [4] contains three-dimensional coordinates of crystal structure analyses of small molecules containing at least one carbon atom, the Protein Data Bank contains coordinates of protein and nucleic-acid structures [5, 6], and the Inorganic Crystal Structure Database [7] contains three-dimensional coordinates for atoms in crystals of inorganic compounds.

A variety of methods can now be used to probe intermolecular interactions. The structural information on intermolecular interactions obtained from X-ray and neutron diffraction studies can be compared with gas-phase experimental data from pure rotational or rotation-vibrational spectra [1] and the energies obtained from ab initio molecular orbital calculations. It is found that each of these methods generally gives essentially the same result. While most X-ray diffraction studies are on crystals of small molecules, comparisons with the lower-resolution results of protein crystallographic studies give information on interactions in an environment that consists of about 50% water by volume [8].

Several types of directional intermolecular interactions will be considered in this article. The main categories are the hydrogen bond (and weaker interactions that could be considered a subset of this), and metal ion coordination (which can, in certain cases, be directional in terms of ligand disposition). Examples of each as they occur in small molecules and proteins will be described and discussed. Essentially one is looking for rules that govern how molecules interact with each other and how they bind to each other, so that predictions will be possible.

2 Methods of Analysis

The orientations of binding between different functional groups on molecules or ions in the solid state can readily be obtained from three-dimensional crystal structure coordinates. Instead of calculating the distances between bonded atoms (as is usual), it is necessary to consider the relationships between non-bonded atoms, invoking the space-group symmetry where necessary, and then to calculate their interatomic distances and the directions in which they lie with respect to some chosen axial system in the crystal. Similar data from solutions lack directionality because of averaging of the orientations of interactions throughout the entire solution. Therefore, at present, it is necessary to substitute averaged data in solution by statistical analyses of interactions as they occur in the solid state. Interactions in one crystal structure might be biased by packing requirements of all of the functional groups contained in the particular molecule that has been crystallized. A statistical analysis, however, which will involve many crystal structure determinations made for a wide variety of reasons and on a very diverse sampling of molecular types, should average out any perceived problems with solid-state distortions.

The nonbonded surroundings of a functional group of a molecule in a crystal can be analyzed in the following way to assess any preferred directions of interaction of the group with its neighbors [9, 10]. A functional group, such as a carboxylate group or an epoxide group, is selected. Three-dimensional coordinates of atoms in crystal structures containing this group are obtainable from the Cambridge Crystallographic Database (CSD) [4]. This chosen functional group in each crystal structure is moved to a standardized position and orientation, and then the positions of atoms packed around it are superimposed to give a scatterplot, as diagrammed in Fig. 1. Points in a scatterplot are sometimes hard to analyze objectively; outlying points may deflect the attention of the observer, thus giving them more significance than is warranted. To avoid this problem the scatterplot can be contoured [10] by putting a Gaussian-like peak on each point, and contouring the result (as for an electron-density map). In this way the probability plot of the three-dimensional positions of points on the scatterplot is smeared and any directional preference of binding is, therefore, highlighted as areas of high density of points in the scatterplot.

a Search the database for crystal structures containing a metal ion and a carboxylate group.

b Lay each carboxyl group in a defined orientation and position and mark where the metal ion lies with respect to this.

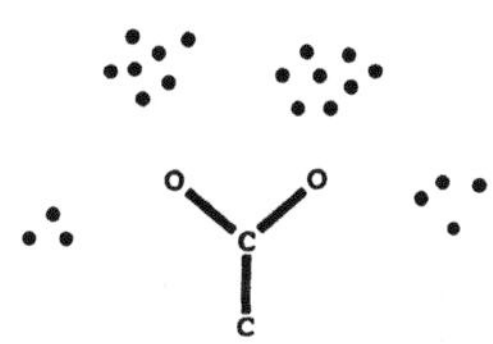

c Put a Gaussian function on each point on the scatterplot.

d Contour the result. This gives directional preferences of binding of a metal ion to a carboxylate group.

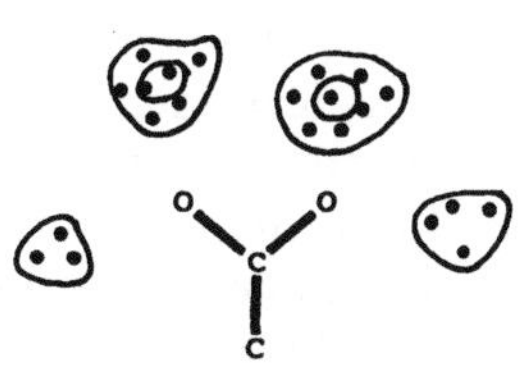

Fig. 1 a–d. Construction of a scatterplot of intermolecular interactions around a functional group [10]

2.1 Sources of Crystallographic Data

Two major requirements for meaningful comparisons of three-dimensional structures for statistical analysis are:

1. ready access to the three-dimensional coordinates of all the appropriate crystal structure determinations that contain the features chosen for comparison, and
2. effective and easy-to-use analysis techniques.

Fortunately, efficient computer-based crystallographic databases containing three-dimensional coordinates of atoms in crystal structures reported in the scientific literature are now available to the scientific community. Therefore it is no longer necessary to type large sets of numbers into a computer, because they can be accessed in the correct format from the databases. In addition, there are many excellent computer-graphics, geometrical, and statistical programs available for application to these three-dimensional coordinates obtained from the databases. This ensures that it is possible to compare structures. It is prudent, however, for the investigator to build molecular models, as required, of the ball-and-stick or space-filling variety, in order to obtain chemical or biochemical insight from any comparisons that have been made.

The Cambridge Structural Database (CSD) [4] contains unit-cell dimensions information on approximately 170,000 three-dimensional crystal structure determinations that have been studied by X-ray or neutron diffraction. Each crystal structure is identified by a unique six-letter code, called its REFCODE. Duplicate structures and remeasurements of the same crystal structure are identified by an additional two digits after the REFCODE. The CSD may be searched in several ways. It is possible to find a list of bibliographic references to reported data on all compounds with given chemical characteristics such as steroids or peptides. In particular, however, the computer software provided with the CSD allows one to search for a small group of atoms (either a full molecule or a fragment of a chemical structure) with bonding that is precisely defined by the codes that are provided by the user as input to the search. When groups of atoms that meet the required specifications have been extracted from the CSD, tables of required geometrical data may be generated and statistical methods applied to the results. In analyses of hydrogen bonding the information on crystal structures obtained by neutron diffraction (in which hydrogen atoms are located with more precision than is possible for X-ray diffraction) are important.

The Protein Data Bank (PDB) [5] is a computerized archive for the three-dimensional structural data on biological macromolecules – proteins and nucleic acids. Each protein structure reported has an identifying code (IDCODE), a header record containing useful information on the protein such as the name and source of the protein and the resolution of the structure, together with a series of references to published articles on the protein. Data are included on the refinement methods used, such as the programs used, the R value, the number of Bragg reflections, the root-mean-square deviations of the bond lengths and angles from ideality, and the number of water molecules that have been located. Then follows a description of the protein, its amino acid sequence, including an analysis of which parts of the backbone fold as helix, sheet, or turns. The main entry is a list of atomic coordinates (ATOM) and information on three-dimensional coordinates of metal ions, substrates, and inhibitors bound to the protein (HETATM). These coordinates can readily be extracted in a suitable form for use with a computer-graphics system.

The Inorganic Crystal Structure Database (ICSD) [7] contains information on all compounds containing at least one nonmetallic element but no C-C or C-H bonds (because these are covered by the CSD). Each reported crystal structure has a separate entry. Information provided in the database includes the chemical name, phase designation, unit cell dimensions, density, space group, and the oxidation state of the elements. Atomic information includes three-dimensional coordinates. Also listed are the *R* value, temperature, pressure, method of measurement and the full journal reference.

2.2 Spectroscopic Methods of Studying Intermolecular Interactions

Spectroscopic methods of analysis of intermolecular interactions involve studies of pure rotational and high-resolution vibration-rotation spectra of weakly bound heterodimers (that is, dimers composed of two different species). The two compounds are cooled until heterodimers form, sometimes by seeding with argon and expanding the mixture adiabatically through a nozzle [1]. The rotational constants measured give the principal moments of inertia from which information on the geometry of the dimer can usually be deduced. The absolute intensity of a rotational transition will give a measure of the intermolecular stretching force constant and the dissociation energy of the dimer. The centrifugal distortion constant of a rotating molecule, obtainable from the rotational spectrum, also gives information on the stretching force constant. These measurements make it possible to measure the strength of the intermolecular interactions.

3 Types of Intermolecular Forces

There are several types of intermolecular interactions, each of which involves electrostatic forces of some kind or other. An example is provided by the ion-ion interactions between cations and anions. Pure electrostatic (Coulombic) interactions are long-range and many, such as hydrogen bonding, are directional. Molecules can, however, be distorted by the electric fields of surrounding molecules, even if the molecules themselves are electrically neutral.

If molecules are polar, that is, if they have a dipole moment, they may interact with each other in a head-to-tail arrangement (a dipole-dipole interaction). If one molecule is polar (with a dipole moment) and the other is nonpolar but polarizable, the polar molecule may induce a dipole in the nonpolar molecule (a dipole-induced dipole interaction). These two dipoles, one permanent and the other induced, will interact in the same way as two dipoles. The strength of this interaction depends on the magnitude of the permanent dipole moment of the polar molecule, and on the polarizability of the second molecule. Even if the two molecules are nonpolar, there can be attractive, low-energy molecular interactions between them. These are induced dipole-induced dipole interactions, also called London dispersion forces, in which a nonpolar molecule induces a small instantaneous dipole in another nearby polar molecule. The force F (in dynes) between two charges q and q′ (in electrostatic units) is expressed by Coulomb's equation:

$$F = /q\,q')/(\varepsilon r^2)$$

where r is the distance (in cm) between the charges and ε is the dielectric constant of a medium (such as solvent). The higher the dielectric constant of the medium, the more the force between the two groups interacting with each other is reduced. For example, hydrogen chloride exists as H^+ and Cl^- in water, which has a high dielectric constant of 80, while in vacuo, where the dielectric constant is unity, the two components combine directly to give HCl. Thus water is an excellent solvent because Coulombic interactions in it are sufficiently weak to allow ions to remain separated.

4 Directed Organic Interactions

Analyses of packing in crystals have, in many cases, shown that there are directional preferences of binding. The nature of this directionality appears to depend on the partial charges developed on the interacting atoms. Some examples will now be described.

4.1 Intermolecular Interactions in Hydrocarbons

The nature of intermolecular interactions is exemplified by the crystal structures of polycyclic aromatic hydrocarbons (PAHs) which contain only carbon and hydrogen atoms. Analyses of crystal structures have led to the derivation of numerical constants describing the forces between pairs of atoms [11, 12]. Thus the potential energy expression involves an equation of the form

$$V = \Sigma_{j,k}[-A_{jk}r_{jk}^{-6} + B_{jk}\exp(-C_{jk}r_{jk}) + q_j q_k r_{jk}^{-1}]\,. \qquad (1)$$

In this equation r_{jk} is a nonbonded interatomic distance between atoms j and k, q is the point electrostatic charge on an atom, and A_{jk} B_{jk} and C_{jk} are adjustable parameters that have been obtained from experimental measurements of unit cell dimensions, interatomic distances, and packing arrangements in crystal structures. A_{jk} represents the coefficient of the London dispersion attraction term between atoms j and k, while B_{jk} and C_{jk} are short-range repulsive energy terms. The summation is over all interatomic interactions (between all j atoms and all k atoms). For PAHs the terms in Eq. (1) represent forces between pairs of carbon atoms, pairs of hydrogen atoms, and individual pairs of carbon and hydrogen atoms. The dominant intermolecular interactions in crystals of the smaller PAHs are between hydrogen atoms and carbon atoms.

Delocalized electrons on the carbon atoms of benzene, and slight residual positive charges on the hydrogen atoms, result in the packing arrangement shown in Fig. 2 for crystalline benzene. The hydrogen atoms are attracted to the more electron-rich carbon atoms to give a herring-bone arrangement of molecules. As the PAH becomes larger, the carbon-to-hydrogen ratio increases and intermolecular carbon-carbon interactions become relatively more important. The result is that large PAH molecules stack one above the other [12, 13]. Equation (1) makes it possible to predict packing arrangements for PAHs. In particular, unit cell dimensions can be predicted for a crystalline PAH and their values can be compared with experimental values [11].

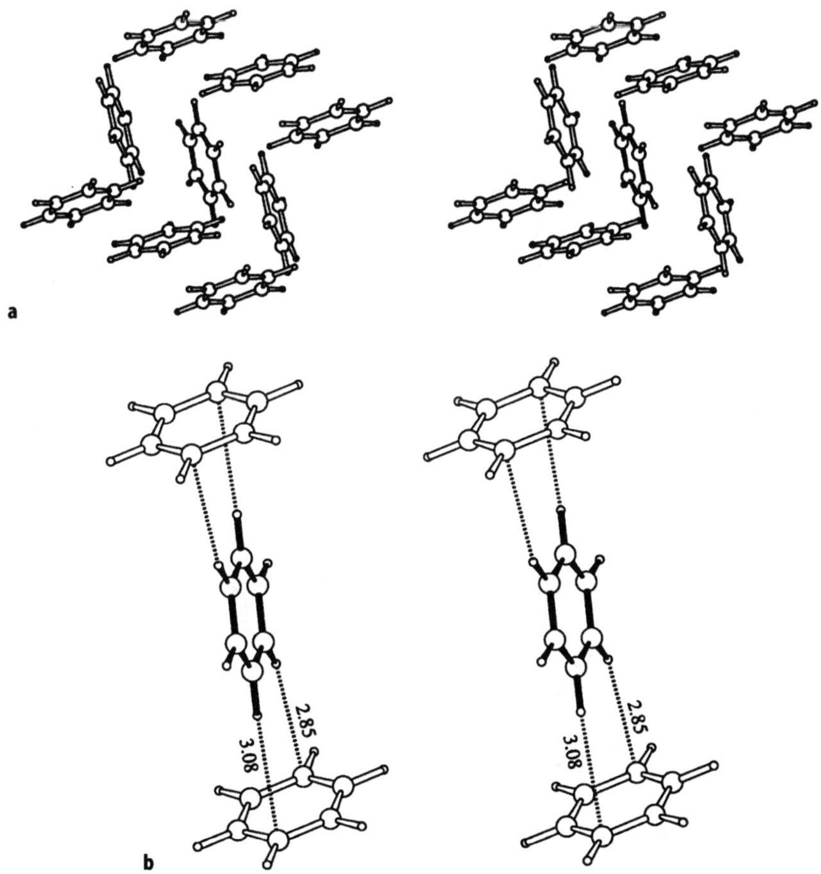

Fig. 2a, b. Packing of molecules in crystalline benzene, showing: **a** a stereoview of the overall structure; **b** the shortest C···H distances in Å (representative of an H···π interactions) [109]

The forces here have some directionality, evidenced by the retention of the PAH packing arrangement in both the crystal structure of a PAH alkylating agent and that of its adduct with a tripeptide, shown in Fig. 3. In spite of the presence of hydrophilic peptide groups in the adduct, the aromatic rings pack in essentially the same manner. Similarly in the cinnamic acids, close approaches of carbon-carbon double bonds in the crystalline state can, on photoactivation, lead to cyclobutane products, as diagrammed in Fig. 4 [14]. These examples suggest that there can be some directionality in C···C and C···H interactions in crystals. The forces between molecules containing oxygen in addition to carbon and hydrogen, however, are much more complicated, especially if hydrogen bonding is possible. It is therefore still necessary to study the results of crystal-

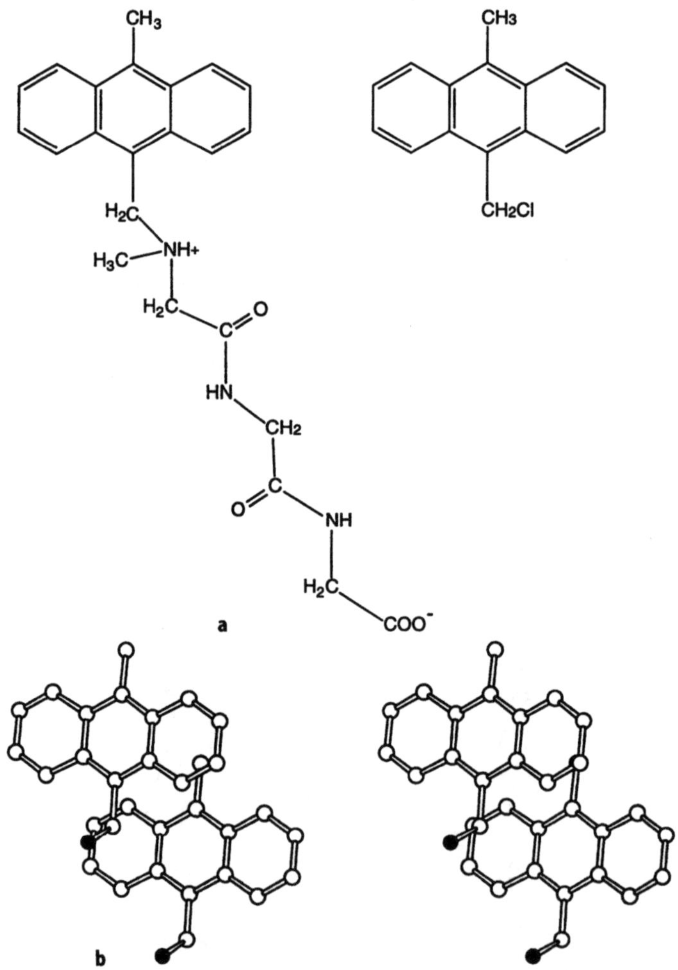

Fig. 3 a, b. Crystal packing of aromatic groups in an anthracene derivative and in a tripeptide alkylated by this derivative [14]: **a** chemical formulae of (*left*) the alkylated peptide and (*right*) the alkylating agent; **b** stereoview of packing in chloromethylanthracene. (H omitted)

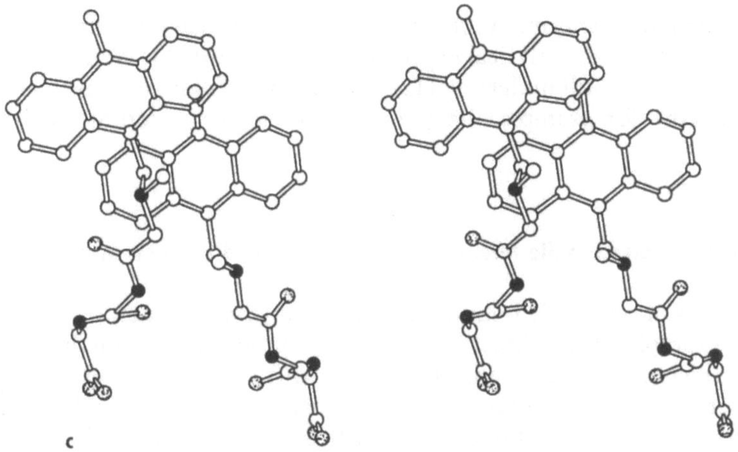

Fig. 3c. Stereoview of packing in the alkylated sarcosylglycylglycine. In this and all subsequent figures oxygen atoms are *stippled* and nitrogen atoms *filled circles*

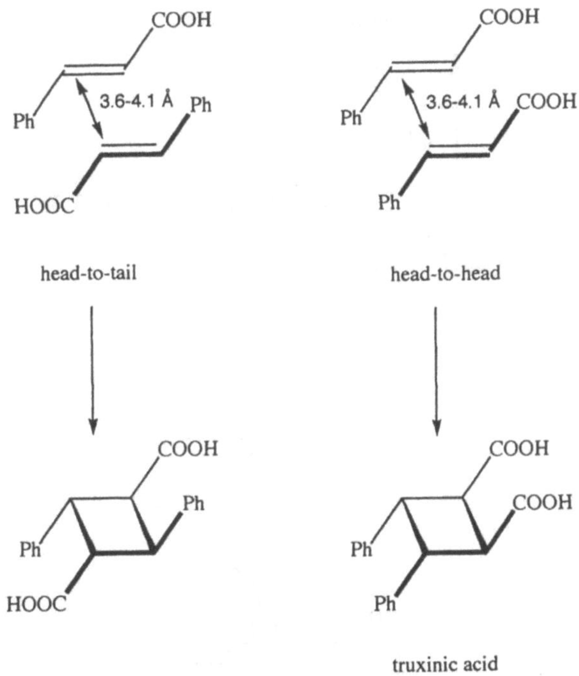

Fig. 4. Effects of proximity of carbon-carbon double bonds in a crystal. Photoactivation of a cinnamic acid derivative to give a cyclobutane [110]. The double bonds lie 3.6–4.1 Å from each other in the crystal. Note that the identity of the product depends on the alignment of the side chains in the crystal structure

lographic structure determinations in order to derive analytical expressions (in the same manner that analytical expressions were derived for PAHs), and these types of studies are still underway [12, 13]. The force fields so derived are used to predict how, for example, drugs bind to biological macromolecules [13, 15–17].

4.2 Nucleophile and Electrophile Interactions with Sulfur and Halogen Atoms

Surroundings of covalently bound sulfur atoms in crystal structures provide an excellent example of the directionality of intermolecular interactions. Directional preferences of binding around divalent sulfur bonded to two other atoms (XSY where X and Y are commonly carbon) were studied in a wide variety of crystal structures. This was one of the early analyses of intermolecular interactions in crystal structures, before the CSD was generally available [18]. The analysis was made on nonbonded contacts, up to 4.2 Å from the sulfur atoms; it showed, as indicated in Fig. 5, that the locations of functional groups around the CSC group depend on their (partial) charges.

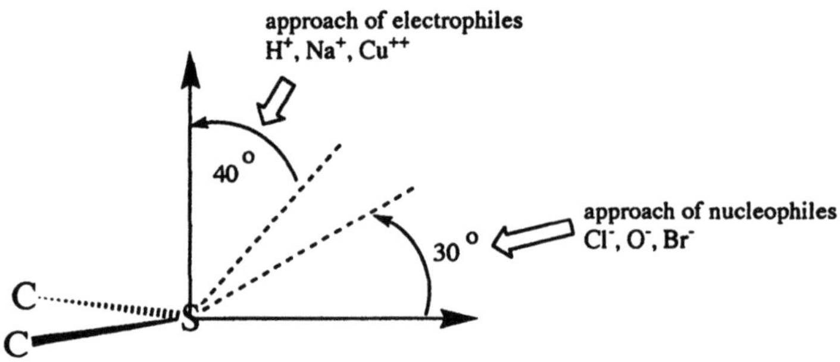

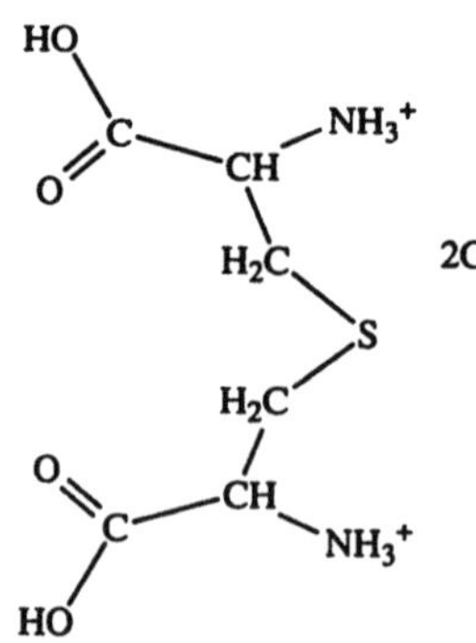

Fig. 5a–c. Directional preferences of binding of X-S-Y groups: **a** diagram of the directions in which charged groups approach the sulfur atom; **b** chemical formula of mesolanthionine dihydrochloride

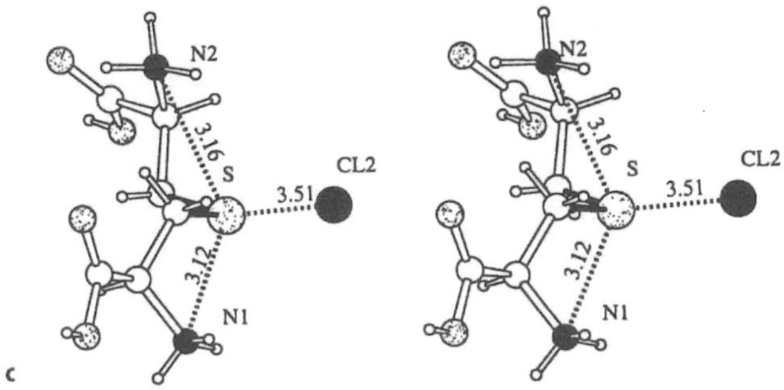

Fig. 5c. Stereoview of the surroundings of C-S-C (filled bonds) in mesolanthionine dihydrochloride [111]. Note Cl^- in C-S-C plane, N^+ perpendicular to it

Electrophiles such as metal cations or hydrogen bond donors, with at least a partial positive charge, approach in a direction that is from 50 to 90° from the CSC plane. Frontier orbital theory [19] would suggest that these electrophiles (electron acceptors) are interacting with the highest occupied molecular orbital of the sulfur atom. This is a lone-pair orbital nearly perpendicular to the C-S-C plane (an *sp* hybrid orbital, while the S-C bonds are pure 3*p* orbitals). On the other hand, nucleophiles (electron donors), such as negatively charged groups or ions, approach the sulfur within 30° of the C-S-C plane, and tend to lie along the extension of one of the C-S bonds, the direction predicted for the lowest unoccupied molecular orbital (able to accept electrons). Therefore electrophilic attack of divalent sulfur would be expected to occur in a direction perpendicular to the sulfide plane, while nucleophilic attack would be expected to occur in the direction of an S-C bond. This study was then extended to S···S contacts.

An analysis of crystal structures showed that if one of the sulfur atoms acts as an electrophile, it approaches the other which thus has nucleophilic characteristics [20]. Thus the mutual arrangement of the two nonbonded XSY groups is such that one sulfur atom (S_1, the electrophile) approaches another sulfur atom (S_2, the nucleophile) in the $X_2S_2Y_2$ plane in such a way that S_2 lies in a direction perpendicular to the $X_1S_1Y_1$ plane (Fig. 6).

An analysis by Chakrabarti [21] of protein structures in the PDB showed that metal ions approach the sulfur of methionine at about 38(5)° from the perpendicular to the C-S-C group. This is similar to values in the range found, as just described, for small-molecule crystal structures in which the metal ion is presumed to interact with a sulfur lone-pair orbital [18]. It was also found that metal ions approach cysteine residues such that the M···S-C-C torsion angle is ±90° or 180°, and that the conformation of the cysteine side chain is generally affected by the metal ion. The metal ions that readily bind to sulfur in proteins are copper, iron, mercury, and zinc. The geometry of binding of metal ions to methionine or cysteine did not appear to depend on the identity of the individual metal.

In a similar analysis of the surroundings of C-Cl, C-Br, and C-I groups in crystal structures it was found that electrophiles approach the carbon-halogen bonds at an angle of about 100° (nearly perpendicularly), while nucleophiles

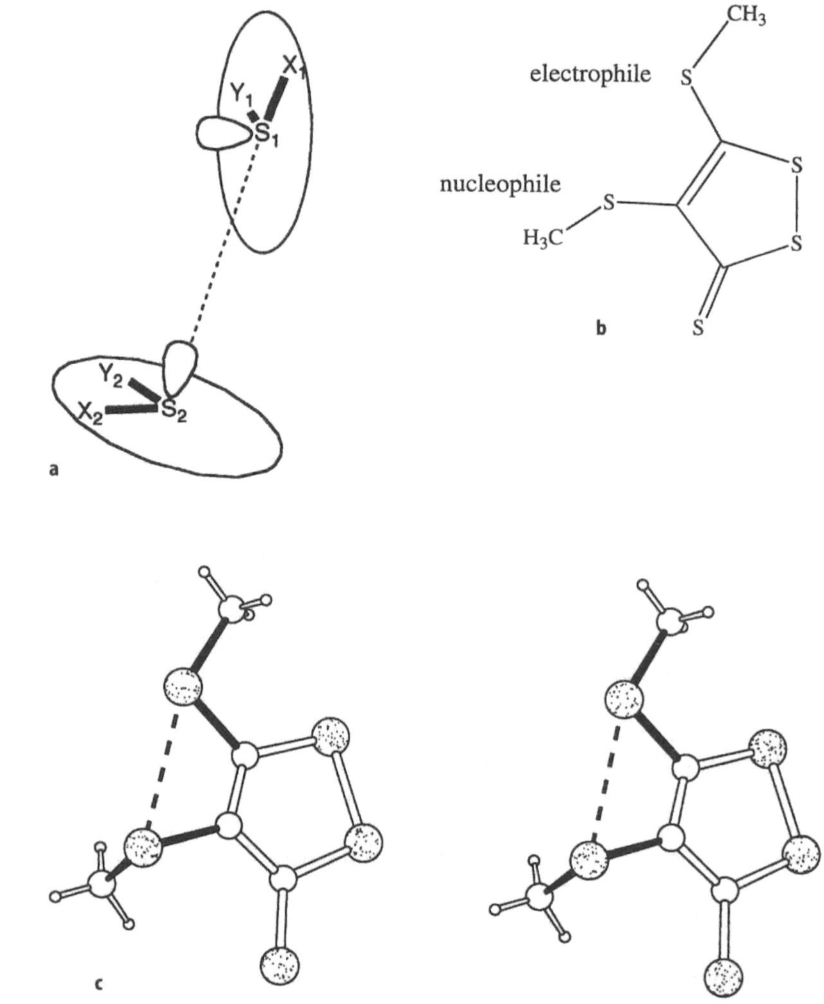

Fig. 6a–c. Directional preferences of binding of disulfide (-S-S- groups): **a** diagram of the interactions (X_1, X_2, Y_1, Y_2 = any atom); **b** chemical formula of 4,5-bis(methylsulfanyl)-1,2-dithiole-3-thione; **c** stereoview of the S···S interactions in 4,5-bis(methylsulfanyl)-1,2-dithiole-3-thione [112]

approach at about 165° (almost directly) [22, 23]. This is analogous to the situation around sulfur, just described. It was found that C-Cl···O contact distances are shorter in the C-Cl bond direction than perpendicular to it. On the other hand, C-C···H distances do not show this effect. Distances involving bromine or iodine show even greater asymmetry. This is described as "polar flattening" [24]. In addition, this anisotropy can lead to a quadrupole at the nucleus of the halogen atom, with lower electron density in the carbon-halogen direction. This will facilitate an attraction of electronegative oxygen atoms.

4.3 Hydrogen Bonding

Hydrogen bonds are among the strongest and most directional intermolecular interactions. They play a crucial role in aligning molecular components of biological systems [3, 25, 26]. The hydrogen bond, usually characterized as *A*-H···*B*, is an interaction between a hydrogen atom covalently bonded to an electronegative atom *A* and another atom *B* that has lone pairs of electrons or polarizable π electrons, and is fairly electronegative; the proton (H) donor is *A* and the proton (H) acceptor is *B*. The hydrogen bond is formed when the electronegativity of *A* relative to H in the covalent *A*H bond is such as to withdraw electrons and leave the proton partially unshielded. The acceptor *B* then interacts with the exposed proton [3, 26, 27].

The hydrogen bond (*A*-H···*B*) is mainly electrostatic in character. Its strength appears to be a complicated balance of various factors [28] which include:

1. the interactions between fractional charges that have developed on *A*, *B*, and H,
2. the deformability (polarizability) of the electron cloud around the acceptor atom *B* so that it can make its lone pairs available to the proton (the softness of *B*),
3. the transfer of electronic charge from *B* to H (σ-bond transfer),
4. the electronic repulsion between *A* and *B*,
5. how readily a hydrogen-bond donor atom *A* will lose its covalently bound hydrogen atom as H^+ (related to the electronegativity of *A* and the strength of the *A*-H bond), and
6. how readily the hydrogen bond acceptor *B* can accept the H^+ (the electronegativity of *B*).

Of these, the important attractive forces are electrostatic (1), polarization effects (2), and charge-transfer interactions (3). The electrostatic component (in 1) falls off less rapidly as a function of distance than do the others, and therefore at longer distances it is the most important. These energy components are balanced by repulsive forces (4) which become important at short distances.

Hydrogen bonds will be formed wherever possible [3, 27, 29]. If the numbers of potential hydrogen bond donors and acceptors in a crystal are not equal, water, which can donate two and accept either one or two hydrogen bonds, can help settle the balance of donors and acceptors [30]. Hydrophobic groups in a molecule tend to pack near hydrophobic groups of other molecules in the unit cell. If a molecule contains both hydrophobic and hydrophilic areas, these will pack in separate areas in a crystal, as far apart as possible.

The strength of a hydrogen bond lies somewhere between that of a weak covalent bond and a van der Waals interaction; 48 kcal mol^{-1} for strong O-H···N hydrogen bonds as in arginine···aspartate hydrogen bonds in proteins, and 2–7 kcal mol^{-1} for weak C-H···O hydrogen bonds. Dunitz [31] pointed out that the energy required to extend a covalent bond by 0.2 Å is of the order of 50 kcal mol^{-1}, while that to extend a hydrogen bond the same amount is only 1.2 kcal mol^{-1}; this illustrates the difference in strengths on the two types of bonds. The more readily the hydrogen is removed as H^+ from atom *A*, the stronger is the hydrogen bond H···*B* [26]. Moderate hydrogen bonds are formed to neutral atoms, and weak hydrogen bonds are formed when the hydrogen is attached to a neutral atom such as carbon (rather than oxygen or nitrogen as in the strong hydrogen bonds), or when the acceptor group *B* has a π electron system rather than a lone pair of electrons.

It has been usual to take the sum of the van der Waals radii as a criterion for hydrogen bonding; if the distance between *A* and *B* is less than the sum of the van der Waals radii, the interaction is presumed to be a hydrogen bond, provided a hydrogen atom is available to form such an interaction. The *A*-H···*B* angle defines the directionality of the hydrogen bond, lying near 180° for high directionality; it may, however, be bent to as low a value as 120°. The hydrogen bond is, however, primarily electrostatic and far-ranging in effect. As a result, sometimes the hydrogen atom may be attracted by two electronegative atoms and a three-center hydrogen bond may be formed, even though distances from the hydrogen atom may be longer than the sum of their van der Waals radii [3]. In such a case, all four atoms involved are in approximately the same plane, and the distances between the hydrogen atom and the two acceptor atoms are often not the same. This arrangement of atoms occurs about 25% of all O-H···O and N-H···O hydrogen bonds [3]. On the other hand, it can be argued that the directionality of hydrogen bonds excludes its description as a purely electrostatic interaction.

The strongest hydrogen bond is believed to be the $[F\cdots H\cdots F]^-$ bond in the hydrogen bifluoride ion (Fig. 7). It has an energy near 50 kcal mol^{-1} [2] and an F···F distance of 2.27(6) Å in potassium hydrogen bifluoride (KHF_2). The hydrogen atom appears to be symmetrically located between the two fluorines [32]. Another very strong hydrogen bond is the O-H···O hydrogen bond in certain carboxylates and organic hydrogen anions. The distances are also very short for O···H but the O-H bond is lengthened so that both O-H and O···H distances lie in the range 1.1–1.3 Å, with O···O distances in the range 2.35–2.5 Å. The O-H···O angle is in the range 170–180°. Normal O-H–O and N-H···O hydrogen bonds, with angles of 165(5)° are powerful aligners of molecules, and are well known for their great contributions to macromolecular folding and function [3, 25]. H···O distances are generally in the range 1.8–2.0 Å. Such hydrogen bonds account for the formation of α helices and β sheets in proteins, base pairing in nucleic acids, and many protein-nucleic acid interactions. Examples of such hydrogen bonds in sulfuric acid and its hydrate [33] and in citric acid monohydrate [34] are shown in Fig. 8 and 9. These illustrate the cooperativity of hydrogen bonding so that a given oxygen atom both gives and receives a hydrogen bond, so that such interactions continue throughout the crystal.

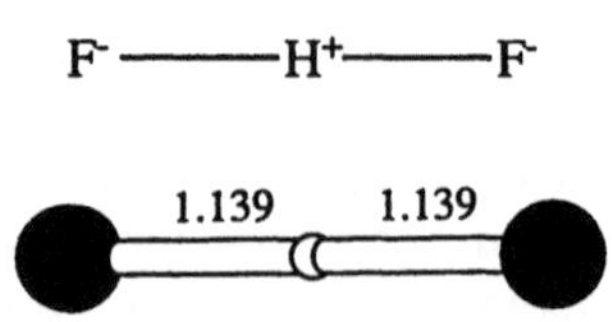

Fig. 7. The bifluoride ion in the crystal structure of its potassium salt [32]

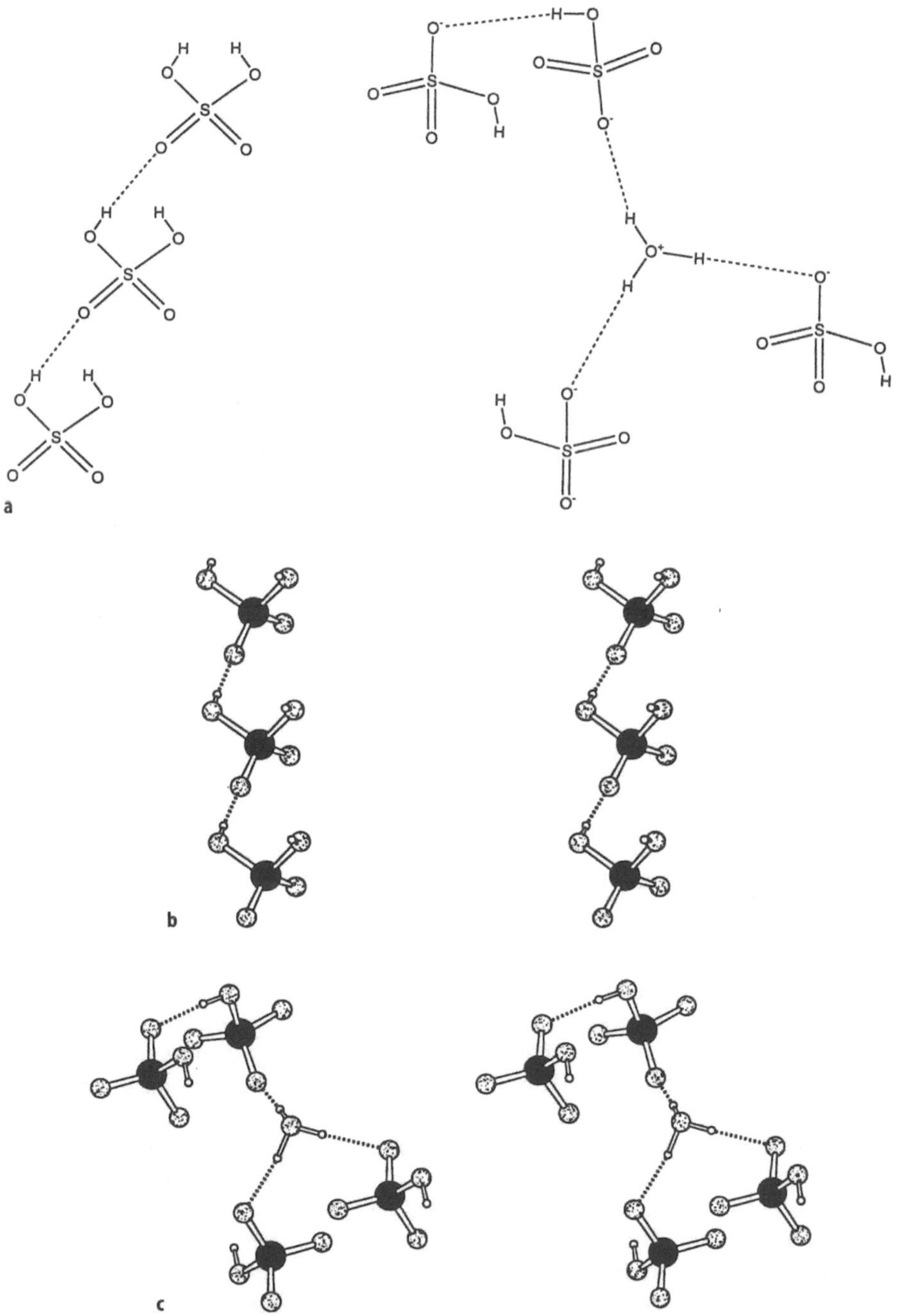

Fig. 8a–c. Hydrogen bonding in sulfuric acid and its hydrate [33] from neutron diffraction data: **a** chemical formulae of the stereoviews in (b) and (c); **b** sulfuric acid, showing hydrogen bonding that extends through the crystal; **c** sulfuric acid monohydrate, showing hydrogen bonding and a hydrated hydrogen ion

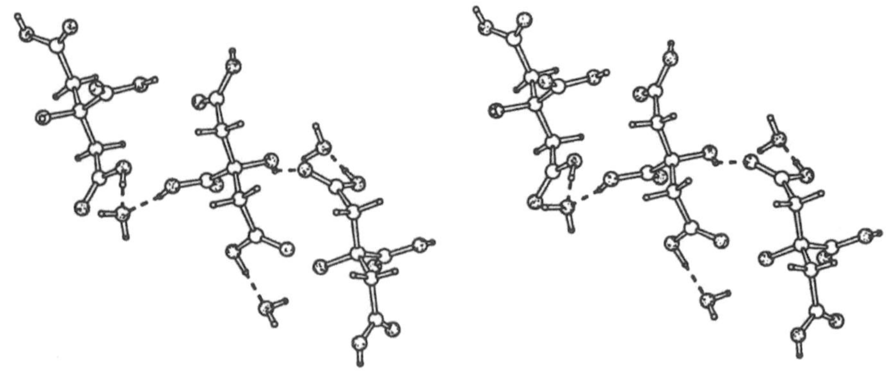

Fig. 9. Citric acid monohydrate [34]. This stereoview shows hydrogen bonding that is cooperative, extending through the crystal in such a way that oxygen atoms both give and receive hydrogen bonds

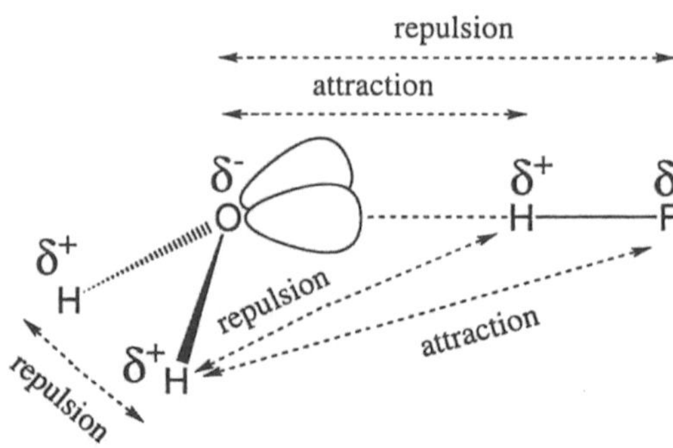

Fig. 10. Repulsions and attractions in the $H_2O\cdots HF$ complex

Spectroscopic studies [1] have shown that the $H_2O\cdots HF$ interaction is such that the HF molecule points its hydrogen atom directly towards the oxygen atom, and that the axis of the HF molecule lies in the direction expected for a tetrahedral arrangement of nonbonding lone pairs of electrons on the oxygen atom. This implies that the HF molecule is recognizing an area of high electron density and orienting itself in order to minimize the electrostatic energy of the system. However, as shown in Fig. 10, it is necessary to consider not only the electron density on the lone pair of electrons on H_2O, but also the charges on any other atoms in the immediate neighborhood. For example, a point positive charge near a water molecule gives a potential energy curve with two minima, symmetrically placed at about ±30°. The repulsive portion of the potential energy equation has lowered the height of the barrier and moved the minima to smaller angles.

The directions from which hydrogen bonds [9] approach *sp*^{2}- and *sp*^{3}-hybridized oxygen atoms in ethers, ketones, esters, and epoxides have been studied.

The geometry around the hydrogen bond acceptor geometry is considered. Contoured scatterplots showed that the largest concentration of hydrogen-bonded OH or NH groups interacting with these oxygen atoms lie in the directions of the lone pairs of electrons of the oxygen atoms. For the oxygen atom in the carbonyl groups of ketones, enones, and esters, the contoured scatterplots showed maxima in the plane of the carbonyl group at about 130° to the C=O bond, with a range from 90–180°. The two maxima in the contoured scatterplots, shown in Fig. 11, were found to lie in a plane perpendicular to that of the C-O-C group in epoxides and ethers. There is a narrower distribution of hydrogen bond donors to ether oxygen atoms than to the oxygen atoms in ketones or epoxides. These imply that the lone-pair electrons show directional properties, albeit weak and diffuse [3]. Studies of hydrogen bonding of such functional groups to alkanol hydroxyl groups [35] gave similar results. It was noted that (*E*)-ester oxygen atoms form very few hydrogen bonds, presumably because of competition with the adjacent carbonyl group. Hydrogen bonds to (*Z*)-ester oxygen atoms are considered to be destabilized by the repulsive electrostatic interaction with the carbonyl group (see Fig. 12) [35]. Main-chain carbonyl groups in proteins bind metal ions in the

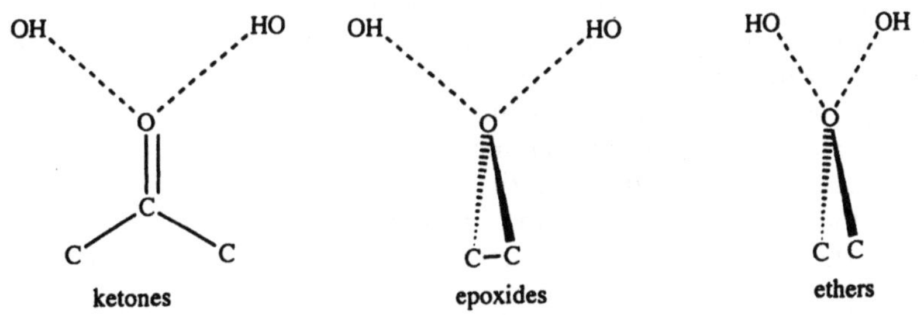

Fig. 11. Scatterplot around C=O [9]. Hydrogen bonds are indicated by *broken lines*. Note that the distribution of hydrogen bond donors is similar for ketones (in the plane of the carbonyl group), and epoxides (perpendicular to the epoxide ring plane). These are the directions of the lone pair of electrons in both cases. The directions of hydrogen bonding around ethers are closer together, as diagrammed

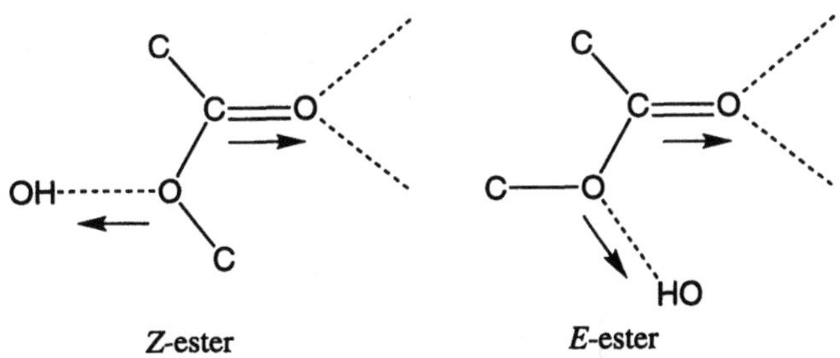

Fig. 12. Interactions around esters [39]. Hydrogen-bond directions are indicated by broken lines. The directions of OH and C=O dipoles are indicated by arrows. Note the difference between (*Z*)- and (*E*)-esters

peptide plane and approximately near the C=O bond direction, so that the M···O=C angle is 140–170° for unidentate binding and 110–130° for bidentate binding. The metal ion is rarely more than 35° from the peptide plane [36].

4.4
C-H···O Interactions

The directionality of a hydrogen bond is partially a function of its strength – the shorter the hydrogen bond the more likely the *A*-H···*B* angle is to be near 180°. Weaker interactions, such as C-H···O or C-H···N, described by Sutor [37], are found when the hydrogen atom is attached to multiple bonds or strong electron withdrawing groups (favoring the formation of H^+). These interactions, if short or the order of the sum of the van der Waals radii, are generally attractive rather than repulsive, and have come to be regarded as very weak hydrogen bonds.

The interaction between a C-H group and a neighboring oxygen atom has about one third the energy of an O-H···O hydrogen bond. If, however, there is a large number of such interactions, their contributions to the total energy of a system may be substantial. For example, the importance of C-H···O interactions in aligning molecules has been shown by the design, by Desiraju and co-workers, of dibenzylidene ketones (Fig. 13) that interact by way of C-H···O interactions with 1,3,5-trinitrobenzene [38]. Use was made of the great acidity of the hydrogen atoms in 1,3,5-trinitrobenzene, and the propensity of nitro groups to interact with hydrogen atoms bound to carbon atoms. It was shown by X-ray crystallographic studies that there is a similarity of binding of dibenzylidene acetone, cyclopentanone, and cyclohexanone. These complexes are analogous to the designed complexes containing O-H···O and N-H-O hydrogen bonds (Fig. 13c), but in the complex with the dibenzylidene ketone described here only C-H···O interactions could be formed.

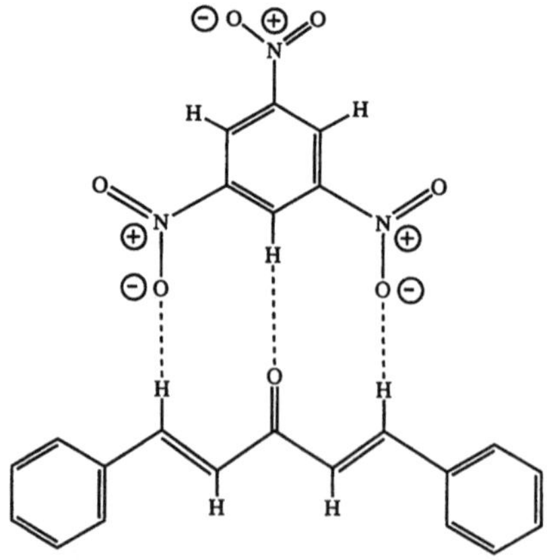

a

Fig. 13a–c. The complex of a dibenzylidene ketone with 1,3,5-trinitrobenzene, showing C-H···O hydrogen bonds [38]: **a** chemical formula

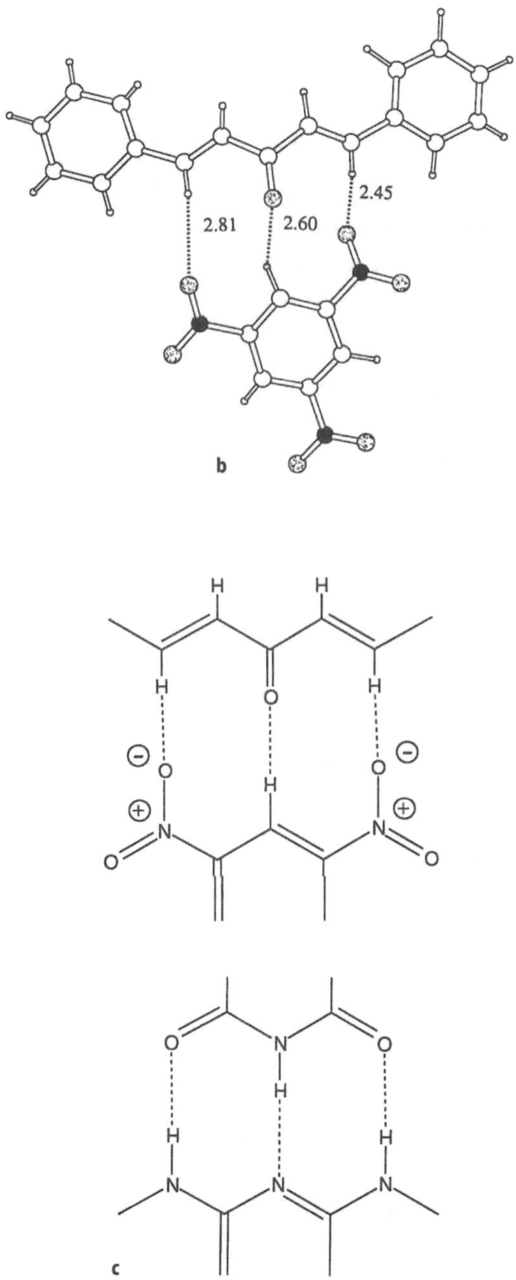

Fig. 13b, c. **b** view in crystal structure; **c** comparison of the interactions in (**b**) and those in which conventional O-H···N and N-H···N hydrogen bonds are formed

4.5
F···H Interactions

When fluorine is covalently bound to a carbon atom (a C-F bond) it is found that the fluorine atoms approach hydroxyl groups with minimum F···H distances of 2.3 Å, much longer than O···H distances of 1.4–1.7 Å in strong O···HO hydrogen bonds [39–42]. The probability of O···H-O hydrogen bond formation is significantly higher than that for the formation of a hydrogen bond involving a C-F group; thus O···H-O hydrogen bonds will form wherever possible; rather than C-F···H hydrogen bonds. Fluorine is the most electronegative element [26], and therefore it is interesting that C-F bonds rarely form hydrogen bonds, especially since the H···F⁻ interaction (involving a fluoride ion rather than organic fluorine) forms the shortest hydrogen bonds measured (H···F⁻ = 1.14 Å) [29]. Some C-H···F interactions in a fluoro PAH are shown in Fig. 14.

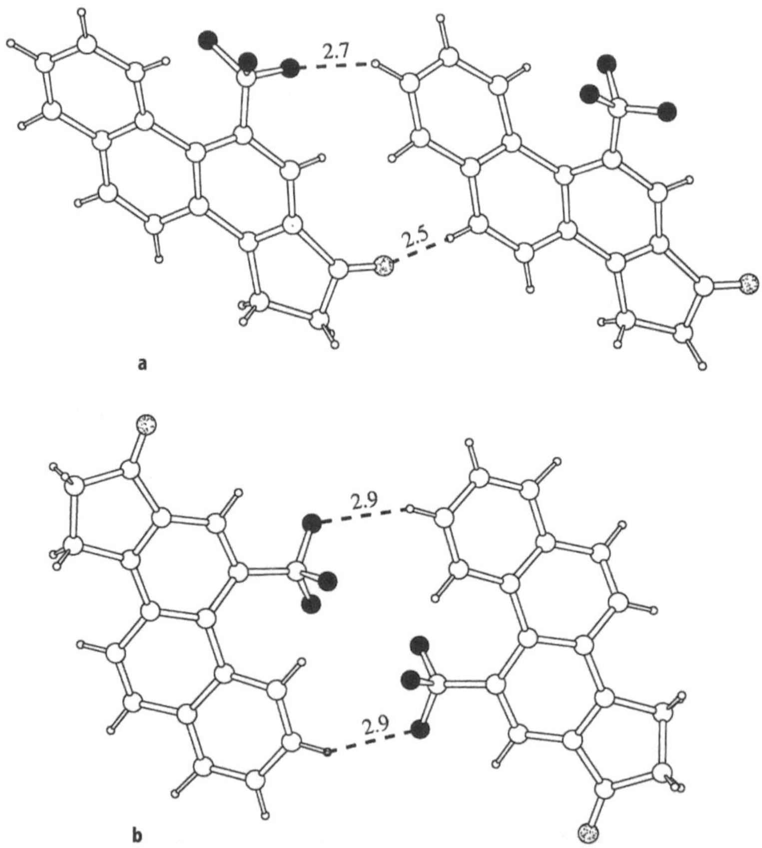

Fig. 14 a, b. A fluorinated PAH showing C-H···F interactions [113]. The compound is 11-(trifluoromethyl)-15,16-dihydrocyclopenta[*a*]phenanthren-17-one. Two examples of interactions between molecules are shown in (**a**) and (**b**). They both involve C-H···F interactions. H···F and H···O nonbonded distances are given in Å

The fluorine atom in a C-F bond can only act as a proton acceptor in a hydrogen bond, while the oxygen atom in the C-OH bond can act both as a proton donor and acceptor. The result for oxygen is a cooperative effect – oxygen (as O-H) can both give and accept hydrogen bonds, as shown in Fig. 9, while fluorine can only accept them.

A three-center hydrogen bond from N-H to O and F is shown in Fig. 15. The H···O distance, however, is shorter than the H···F distance. The fluorine atom in a C-F group does not, apparently, develop much negative charge, and, in addition, it is not readily polarized. Therefore the C-F group only forms very weak hydrogen bonds (as the hydrogen-bond acceptor). While fluorine is more electronegative than oxygen, oxygen is more polarizable. On the other hand, the reason that the fluoride ion can form a short H···F^- hydrogen bond in hydrofluoric acid, is explained as the result of a proton resting between two negatively charged ions, a powerful electrostatic effect.

As pointed out by Dunitz and Taylor [42] the hydrogen bond *A*-H···*B* can be thought of as orbitals on *A* and *B* competing for a proton, and therefore the basicities of *A* and *B* must not be too different. This is the case if *A* and *B* are O and N, but not if *A* is O or N and *B* is F. In addition, since fluorine forms only single bonds, it cannot readily take part in a delocalization scheme for electrons and therefore does not build up as large a partial negative charge as does oxygen or nitrogen. These studies show that examination of crystal structures of fluo-

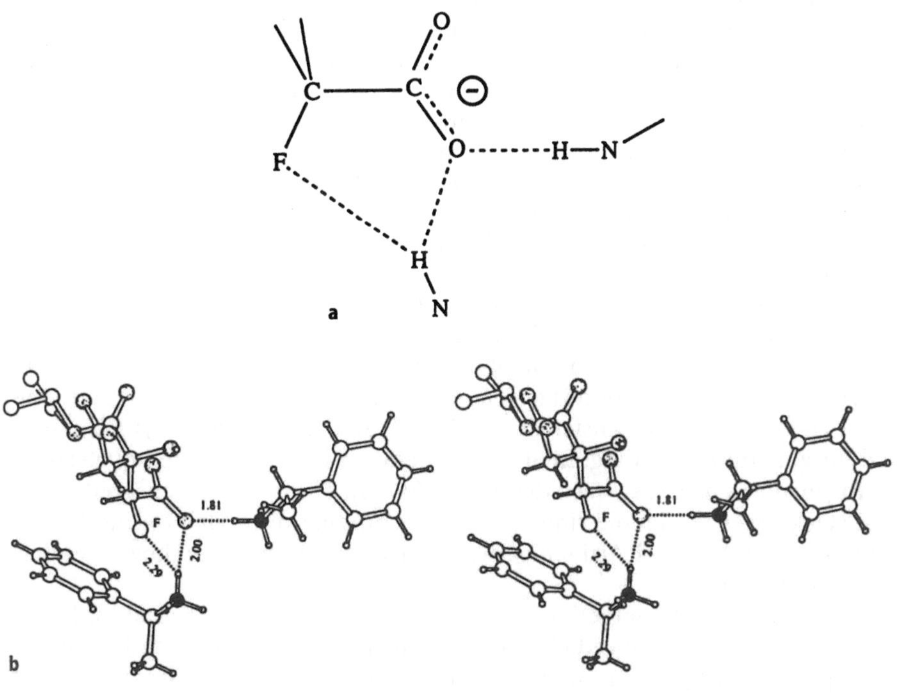

Fig. 15 a, b. Three center hydrogen bonding from NH to O and F [39]: **a** interaction diagrammed; **b** interaction shown as a stereoview rine-containing compounds leads to a better understanding of hydrogen bonding in general.

An interaction between a C-F bond and a metal ion (C-F$\cdots$M^{n+}) has been observed so far by X-ray diffraction studies only when the metal ion is an alkali metal ion [40]. The coordination of carbon-bound oxygen and fluorine atoms around the metal ion can be analyzed in terms of the requirement that the sum of the bond valences should be near 1.0 for a monovalent cation [43]. This sum is only achieved if the M^{n+}$\cdots$F-C interaction is taken into account as well as those involving oxygen atoms. The fluorine atom contributes to the local neutralization of the charge of the cation, and is, indeed, a significant member of the first coordination sphere of the metal ion.

4.6 H$\cdots\pi$ Interactions

Strong O-H$\cdots$O and N-H$\cdots$O hydrogen bonds can be used in the design of structures because they generally dominate packing if they can be formed. Other types of interactions, however, need also to be taken into account. An excellent example of this is provided by a neutron diffraction study of 2-, 3-, and 4-aminophenols [44–46]. In these crystal structures, studied by neutron diffraction, there are significant N-H$\cdots\pi$ interactions in which the hydrogen atom points to the electron-rich aromatic ring, implying the importance of herring-bone interactions of the type described earlier for benzene and PAHs. Each nitrogen and oxygen atom in these structures is surrounded by three groups, which, including the bond to the carbon atom of the molecule, leads to a tetrahedral bonding situation – but not one that would have been readily predicted. This is shown in Fig. 16. Thus there are two important types of interactions involving delocalized π-electron systems, as described earlier. These are the C-H$\cdots\pi$ or C-H$\cdots$C interactions (where π may be any part of an electron-rich aromatic system, a delocalized double bond or a triple carbon-carbon bond) which give rise to a "herring-bone" arrangement of molecules, as shown for benzene in Fig. 2.

O-H$\cdots\pi$ interactions are also seen in 2- and 3-aminophenol crystals in 2-ethynyladamantan-2-ol [47], and in several protein structures in the PDB. In one (*O*-glycosyl hydrolase, with bound 4-acetylamino-3-amino benzoic acid, 1IVE, [48]) an O-H$\cdots\pi$ interaction is envisioned between Tyr 406 and Asp 151. In the other (μ-glutathione *S*-transferase, 1GST, [49]) an O-H-π bond to Tyr 6 is essential for glutathione binding. Interactions between indole rings and quaternary nitrogen atoms have also been described [24, 50]. The positively charged nitrogen stacks over the indole ring system. The binding of tacrine (tetrahydro-9-aminoacridine), an anti-Alzheimer drug, to acetylcholinesterase shows a similar interaction [51].

The other type of interaction, a π-π interaction, is found in crystals of PAHs; these are rich in delocalized electrons and mainly consist of sp^2 carbon atoms with peripheral hydrogen atoms attached. Their structures resemble that of graphite in which planar hexagonally disposed layers of carbon atoms are separated by about 4 Å. Large PAHs also form molecular complexes with other flat

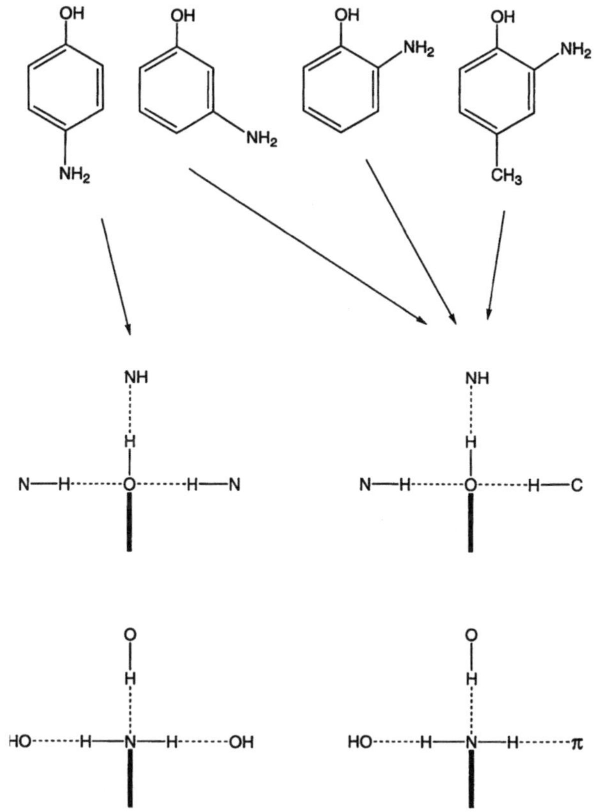

Fig. 16. The different styles of packing in the crystalline state of some aminophenols [44–46]. It is shown that *p*-aminophenol utilizes O-H···N and N-H···O hydrogen bonds in crystal packing, while the *o*- and *m*-compounds and 2-amino-4-methylphenol also utilize C-H···O and N-H···π interactions

molecules containing electron-poor ring systems. The resulting complexes, which involve interactions between the π-electron systems of the two types of molecules, are generally called π-complexes. They are characterized by short intermolecular distances perpendicular to the stacking direction, like graphite. Therefore it is common to find crystals growing as needles, elongated along the stacking direction. Examples of these types of complexes are provided by the trinitrobenzene complexes of PAHs, diagrammed in Fig. 17 [52, 53].

The two molecules forming such a π-complex consist of a donor molecule with a low ionization potential so that an electron can be readily lost (a delocalized π-electron of the PAH) and an acceptor molecule with a high affinity for electrons (the aromatic ring of trinitrobenzene which is electron-poor because the three nitro groups pull electrons out of it). As a result, stacks of alternating donor and acceptor molecules are found in the crystal. The relative orientations within the parallel planes of these donor and acceptor molecules are determined, to a considerable extent, by the orientations of charge distributions of the

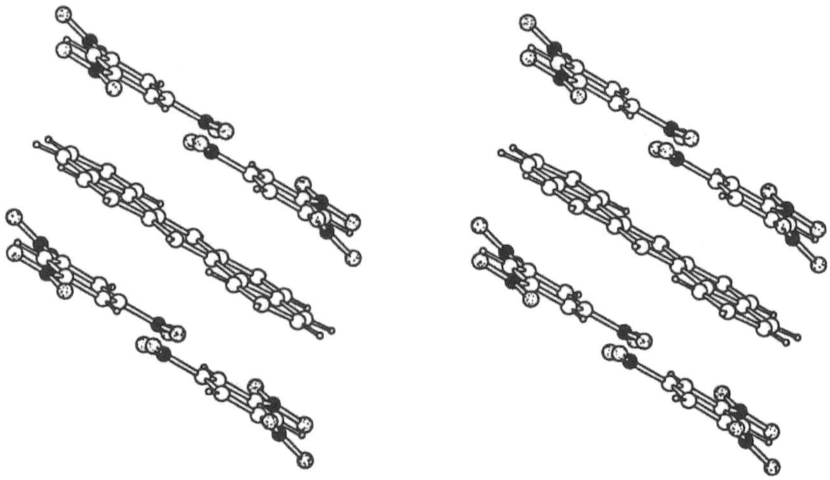

Fig. 17. The packing of molecules in crystalline trinitrobenzene complexes of PAHs [52, 53, 114]. The PAH is sandwiched between trinitrobenzene molecules

highest occupied molecular orbital of the polycyclic aromatic hydrocarbon from which an electron will come, and the lowest unoccupied molecular orbital of the trinitrobenzene, to which the electron will go. From such considerations it is possible to predict how the molecules will be aligned in alternate stacks, implying a directionality in the interactions.

In some complexes of this type, charge transfer can occur between the two molecular components. This is evident if there is a special spectral band corresponding to the charge transfer. In such compounds, however, it is also found that the interplanar spacing of flat molecules is smaller than usual. Thus, while the normal spacing is 3.4 Å, it can decrease to 3.1 Å in a charge-transfer complex.

5 Interactions of Planar Functional Groups in Proteins

The geometry of the interactions of metal ions with selected side-chain groups found in proteins will now be considered; this provides a promising avenue for the design of supramolecular metal ion-complexing agents.

5.1 Surroundings of Carboxylate Groups

The relationship of a metal ion to a selected functional group, such as a carboxylate group, has been studied. Carboxylate ions have one delocalized negative charge, and each oxygen atom has two lone pairs disposed at 120° to the C-O bond. The entire carboxylate group and its lone pairs of electrons lie in a plane. The directionalities of these oxygen lone pairs are important for metal-ion binding. It is found in various crystal structures that a given carboxylate group may bind up to six cations and that it may, in certain circumstances, share the metal cation between both oxygen atoms of the carboxylate group in a bidentate fashion, as has been found for calcium ions [54].

The lone-pair electrons in a carboxylate group are designated *syn* or *anti*, as shown in Fig. 18. The proton on another molecule can approach in either of these directions. In the *syn* conformation (*Z*-form) the proton is on the same side of the C-O bond as the other C-O bond; this is the conformation found when carboxyl groups dimerize by forming two hydrogen bonds. On the other hand, in the *anti* conformation (*E*-form) the proton is on the opposite side of the C-O bond from the other C-O bond. Ab initio quantum chemical studies of formic acid indicate that the *syn* (*Z*) conformation is more stable than the *anti* (*E*) conformation by about 4.5 kcal mol^{-1}, implying that the *syn* lone pairs are more basic (and therefore bind metal ions more readily) than do the *anti* lone pairs [55]. Carboxylates in active sites of enzymes generally seem to employ the more basic *syn* lone pairs for metal chelation [56], and it has been estimated that *syn* protonation is 10^4-fold more favorable than *anti* protonation (since 1.4 kcal mol^{-1} corresponds approximately to a tenfold increase in rate). The carboxylate ion is therefore a weaker base when constrained to accept a proton in the *anti* (*E*) direction.

An analysis of the directions in which metal ions approach a carboxyl group in crystal structures [57] showed that alkali metal and some alkaline-earth metal ions, which form strong bases, have less specific directions of binding, and show extensive out-of-plane interactions. By contrast, most other metal cations bind in or near the plane of the carboxylate group. Direct or bidentate binding, in which the metal ion is equidistant from the two oxygen atoms of the carboxylate ion, appears to be preferred when the metal-oxygen distance lies in the range 2.3–2.6 Å; otherwise *syn* binding is generally preferred. These results are a function of the carboxylate "bite" size (2.2 Å) and of a need to keep $O\cdots M^{n+}\cdots O$ angles larger than approximately 60°.

These results are also found in protein crystal structures [58]. Here carboxylate groups are found in aspartate and glutamate side chains. When ligands

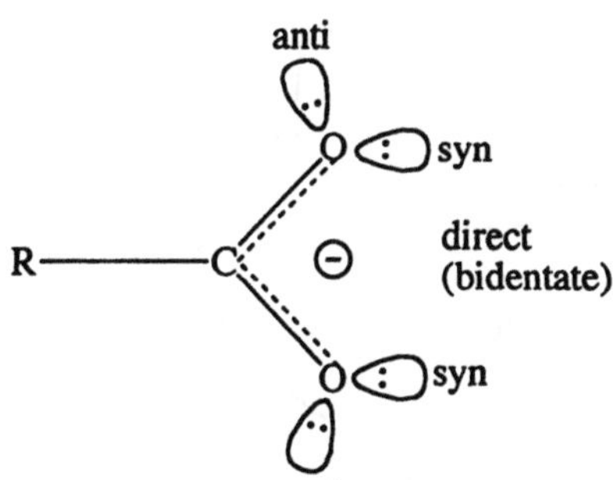

Fig. 18. Geometry of a carboxylate group showing that it has four lone pairs of electrons on its oxygen atoms [57]. These lie in the plane of the carboxylate group and provide optimal directions for hydrogen bonding and the binding of metal ions

bind to a metal ion there are steric interactions between them. As a result, calcium ions can only bind near the end of an α helix. Metal ions with a smaller coordination number, such as zinc ions, do not have this constraint.

When the carboxyl group is not ionized it is found, from detailed neutron diffraction studies, that the hydroxyl portion of the carboxyl group, while a powerful hydrogen-bond donor, does not accept hydrogen bonds [59]. This information can be used to assign ionization states in protein crystal structures in which hydrogen atoms are not located. The results for lysozyme, shown in Fig. 19, were in line with those deduced by other methods.

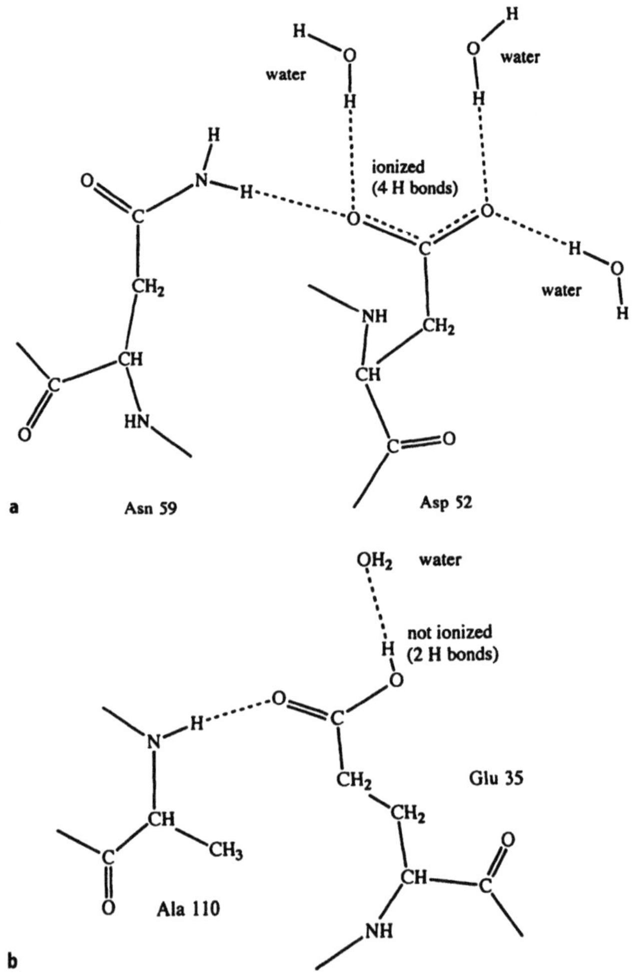

Fig. 19 a, b. Different surroundings of carboxylate and carboxylic acid groups [59]: **a** ionized carboxylate group which can form several hydrogen bonds (a maximum of six in general); **b** carboxylic acid group which only forms two hydrogen bonds. This information can be used to differentiate between ionization states when a protein structure determination does not have hydrogen atoms located

5.2 Surroundings of Imidazole and Histidine Groups

Histidine residues are common in the active sites of enzymes and for binding metal ions, especially divalent zinc and copper ions [60]. Imidazole groups in histidyl side chains bind in proteins to metal ions, but the metal cation nearly always lies in or very near the plane of the imidazole group along the lone-pair direction of the nitrogen atom. This has been verified by studies of crystal structures in the CSD [61]. One imidazole group, since it has two nitrogen atoms, can bind one or two metal ions, Zn^{2+}, Fe^{2+}, and Cu^{2+} being common binders. In proteins [62], the metal ion shows a preference for binding to the nitrogen atom furthest from the C_{α} of the histidine, as shown in Fig. 20; the identity of the metal ion does not appear to affect the overall geometry. The histidine group may be precisely oriented by additional hydrogen bonding, as shown in Fig. 21.

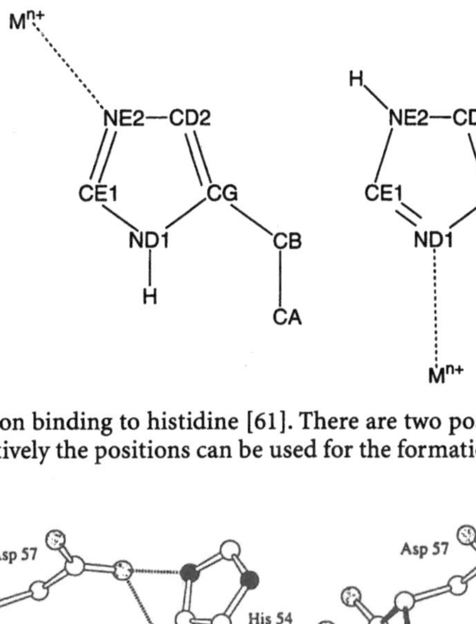

Fig. 20. A metal ion binding to histidine [61]. There are two positions available for metal ion binding. Alternatively the positions can be used for the formation of hydrogen bonds

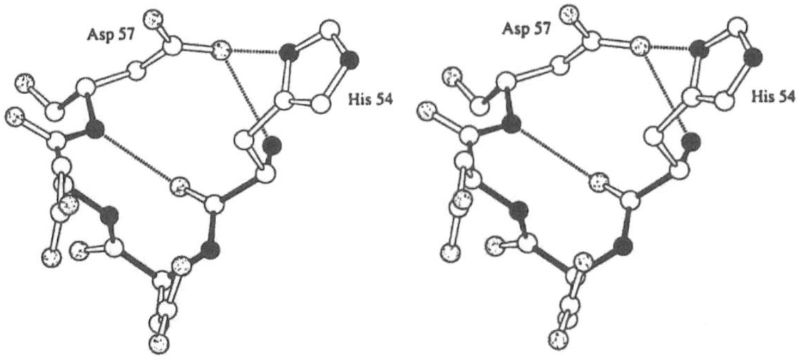

Fig. 21. Constraints of hydrogen bonding on the orientation of a histidine side chain in D-xylose isomerase [81]. The backbone of the protein is indicated by *filled bonds*. Note that the aspartate side chain is constrained in position by the backbone -NH- group of the histidine, and that the orientation of the histidine ring system is constrained by this constrained aspartate group

As for carboxylate groups it is found that transition-metal ions lie closer to the plane of the imidazole than do hydrogen bonding groups [63]. Thus metal ion binding is more precisely directional.

5.3
Surroundings of Arginine Side Chains in Proteins

Arginine side chains in proteins provide an excellent means of aligning carboxylate and phosphate groups so that an O-C-O or O-P-O group lies in the plane of the arginine. This multidentate protein side chain is used with success in protein-nucleic interactions, as shown in Fig. 22 [64].

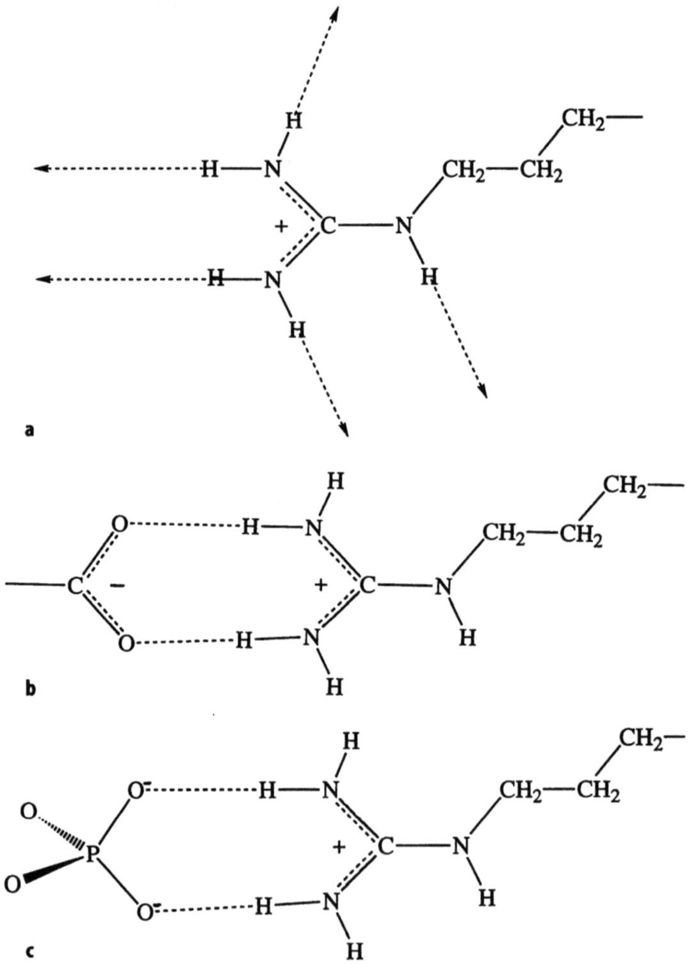

Fig. 22 a–c. Arginine interactions [64]: **a** with indicated hydrogen bond directions; **b** the binding of a carboxylate group; **c** the binding of a phosphate group

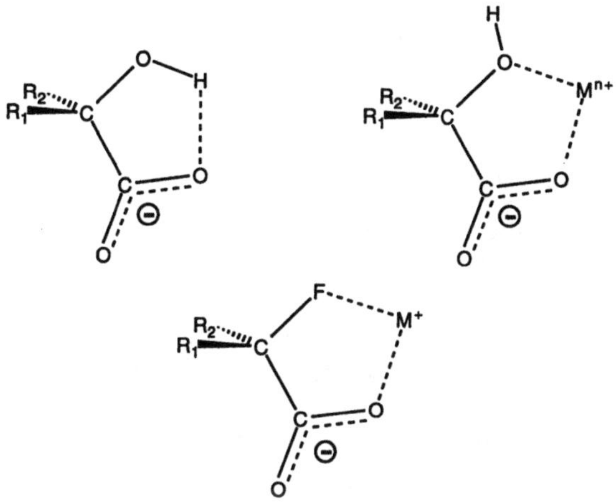

Fig. 23 a–c. The α-hydroxycarboxylate group [66]: **a** ionized group; **b** the effect of binding a metal ion; **c** the effect of replacing the hydroxyl group by fluorine

5.4
Surroundings of α-Hydroxycarboxylate Groups

Chelating groups that bind metals include hydroxamates, some of which are powerful enough to extract iron from stainless steel, and the α-hydroxycarboxylate groups, HO-CR_2-COO^- (R=any group), found in many important biochemical compounds such as citrates and malates [65]. X-ray crystallographic studies indicate that, if the cation is of a suitable size, the entire chelating group is approximately planar, even though the O···O distance is shorter than the sum of the van der Waals radii for oxygen atoms. A study of one form of potassium citrate revealed, surprisingly, that the potassium ion did not chelate the α-hydroxycarboxylate, but that the hydroxyl hydrogen atom formed an internal hydrogen bond. This was verified by a neutron diffraction study [66]. When the central hydroxyl group of citric acid is replaced by fluorine, the α-hydroxycarboxylate group is still a good chelating group, but no hydrogen atom is available to form an internal hydrogen bond. Therefore, in dipotassium 3-fluorodeoxycitrate, a potassium cation is chelated by the α-fluorocarboxylate group. This is shown in Fig. 23.

6
Metal Ion-Based Directional Interactions

The role of hydrogen bonding as a means of binding molecules together in a directional manner has been described. The effect of replacing the hydrogen atom by a metal ion, which is much larger, will now be explored. As will be shown, a metal ion can bring molecules together in a manner analogous to that of a hydrogen bond. A specific example is provided by the magnesium-binding site of the protein integrin [67], diagrammed in Fig. 24. The role of metal ions in

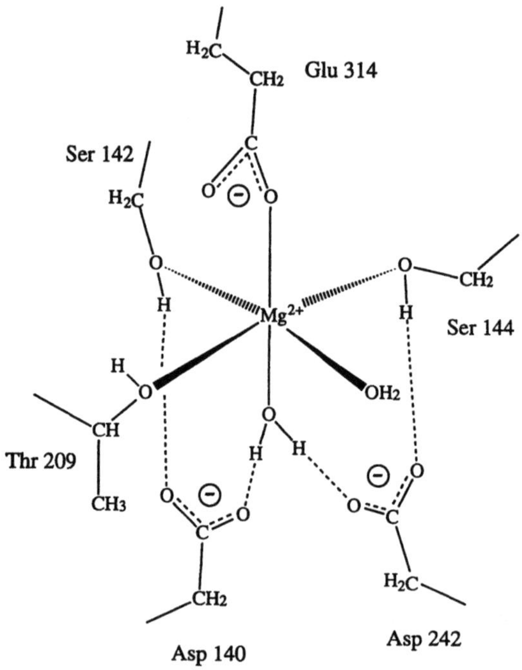

Fig. 24. The magnesium-binding site in integrin [67]. Note that, two carboxylate groups, two hydroxy groups and one water molecule bind in such a manner that two ring motifs (each involving eight atoms) are formed. This coordination of magnesium ions serves to bind two portions of the molecule together

protein folding is similar; while the interaction in this case may be termed "intramolecular," the components on the protein may be removed in sequence from each other. Positively charged metal ions act as electrophiles, seeking electron density on other atoms or ions. Therefore, in order to achieve local electrical neutrality, metal cations attract around them those atoms that have some negative charge. The interatomic distances between cations and anions will depend on the strength of this charge-charge interaction. Thus the shorter a metal ion-oxygen distance, the higher the negative charge on the oxygen. Assuming that the cations and liganding anion atoms are hard spheres, the shortest distance between them can be presumed to be the sum of their ionic radii. Ionic or atomic radii are measured as half of the closest distance that two identical nonbonded atoms or ions approach each other in crystal structures. Additional ionic radii can be deduced from minimum distances between different ions.

The number of ions or atoms that pack around a metal ion in a crystal is referred to as its coordination number of the metal ion. Its value is related to the ratio of the radius of the cation to that of the anion that packs around it [26]. Small, highly charged cations, such as Be^{2+} and Al^{3+} have low coordination numbers, although they are usually higher in crystals than in gases or liquids. Larger cations have higher coordination numbers, 6 being the most common value found, but values may reach 12. Ionic radii vary with the coordination number [68]. For example, a cation with a tetrahedral coordination of four anions has 93–95% the radius of the same cation with a coordination number of 6, but if the coordination number is 8, the radius is 103% that for coordination number 6. Similarly, radii for the oxide anion vary from 1.35 Å (coordination number 2), to 1.40 Å (coordination number 6), to 1.42 Å (coordination number 8).

The directionality of metal-ion coordination can be considered in two ways – from the point of view of a coordinating ligand such as a carboxylate or imidazole group in a protein side chain, and from the point of view of the metal ion, that is, how directional the bonds in its coordination sphere are. These two aspects of metal binding are discussed here. They are relevant to the manner in which metal ions mediate intermolecular as well as intramolecular interactions.

6.1 Metal Ion Coordination

Metal ion coordination is generally considered to consist of anions arranged around the metal in order to fill space. Directional components of the interactions involved are not usually taken into consideration. If, however, the cation has a rigid coordination sphere – a regular tetrahedron for coordination number 4 or a regular octahedron for coordination number 6, and if at least two coordination positions are filled (as, for example, by a bidentate ligand) – then the directions of binding of the other coordination positions are well fixed. Thus cation coordination can, in certain cases, have a strong directional component.

6.2 Magnesium Binding

Some metal ions have fairly rigid coordination spheres. A prime example is the divalent magnesium ion which shows a great preference for six oxygen atoms (such as from water or carboxylate groups) in its inner coordination sphere [69]. The coordination sphere can be considered rigid and if the orientation of two ligands with respect to the magnesium ion are fixed, then the directionalities of the other ligands are also fixed (see Fig. 25). This has been verified by studies of

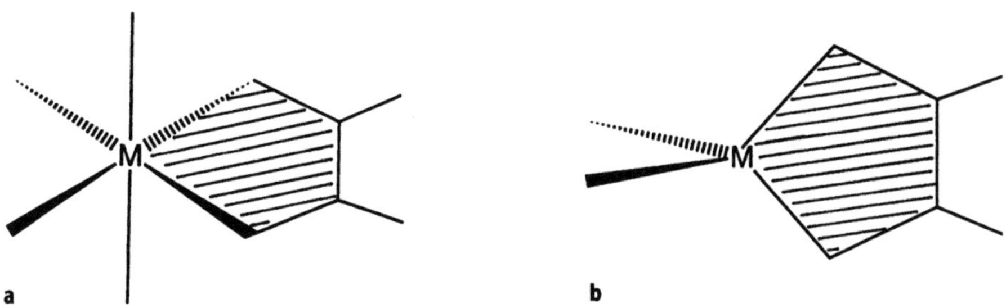

Fig. 25 a, b. If two locations on a rigid coordination polyhedron are fixed, the directions of binding of the other ligands is also fixed: **a** 2 $Mg^{2+}\cdots O$ interactions fixed in an octahedral coordination scheme; **b** 2 $Zn^{2+}\cdots S$ interactions fixed in a tetrahedral coordination scheme structures of magnesium-containing molecules in the CSD and PDB and by ab initio molecular orbital calculations. The magnesium ion maintains water in its inner coordination sphere in an orientation in which the $Mg^{2+}\cdots OH$ angle is 120–130°. This then orients groups in the second coordination sphere that are bound to these inner sphere water molecules. This means that magnesium ions are involved in precise orientations of functional groups such as phosphate groups in proteins. In a motif involving an additional water molecule (see Fig. 26) [58], this orientational capability is further accentuated, as shown in Fig. 27.

6.3 Zinc Binding

Zinc ions have approximately the same size and charge as magnesium ions, but different chemical characteristics because zinc is a transition element and magnesium is not. This difference is evident in its coordination behavior. Unlike magnesium, zinc exhibits coordination numbers of 4, 5, and 6 with almost no energy consequences for changes between these [70]. This is in line with the way in which these metal ions are used in biological systems; magnesium ions generally play a structural role, while zinc ions are more likely to be directly involved

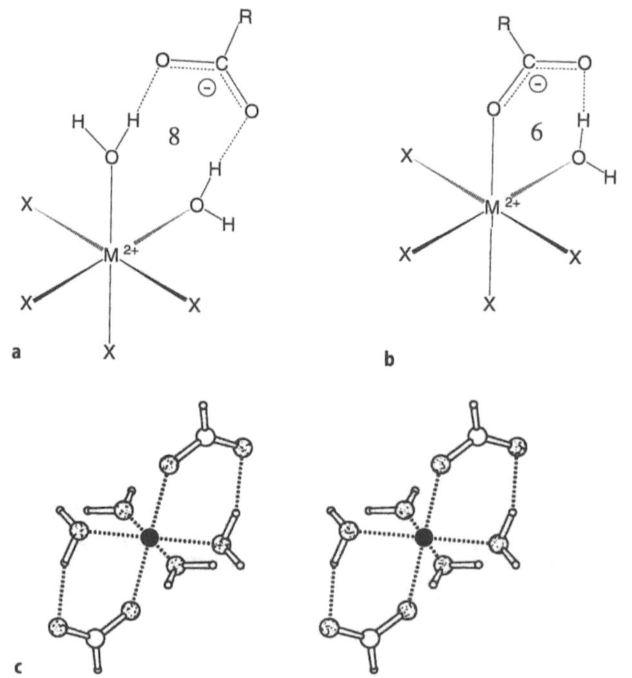

Fig. 26a–c. Motifs found in magnesium-binding proteins: **a** the ring noted in Fig. 24; **b** an aquated carboxylate motif; **c** stereoview of the motif found in cobalt formate [63]. These motifs are also used by other metal ions that have similar sizes and ligand preferences

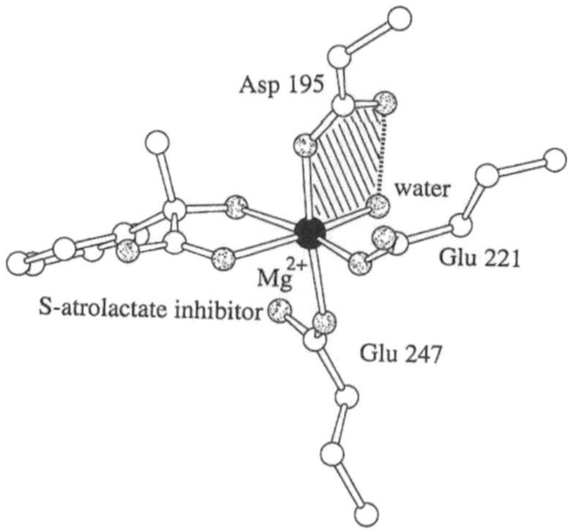

Fig. 27. Use of a motif to orient the magnesium octahedron in a protein (mandelate racemase) to facilitate a reaction [115]. The motif in Fig. 26b serves to orient the direction in which substrate (modeled in this crystal structure by an inhibitor) binds

in catalytic processes, with possible changes in coordination number during reaction. When the coordination number is low, zinc binds nitrogen and sulfur, preferring oxygen for the coordination number 6. As a result, coordination of 6 ligands to zinc does not show the firm directionality found for magnesium coordination. Ab initio molecular orbital calculations show that the energy cost of changing the coordination number of magnesium away from 6 is high, whereas it takes almost no energy for zinc ions to change their coordination number between 4, 5, and 6. Tetrahedral coordination of zinc can be used to control protein folding.

An interesting metal-binding motif in proteins is the zinc finger, which is a helix-turn-helix motif that binds a zinc ion via four sulfur atoms of cysteine residues, or via two sulfur and two nitrogen atoms (from histidine). This motif, shown in Fig. 28, was first identified by NMR studies [71], and has now been found in the crystal structures of several protein-nucleic acid complexes [72].

6.4
Calcium Binding

Calcium ions, which are larger than magnesium ions, bind oxygen but their coordination number can vary between 6, 7, and 8. The energy penalty for a change among these three coordination numbers is low [73]. There are some directional qualities to calcium binding in proteins. Calcium tends to bind carboxylate groups in a bidentate manner, particularly at coordination numbers higher than six [73].

One calcium-binding motif found in proteins is the helixloophelix motif, also known as an *EF* hand [74, 75]. A comparison, diagrammed in Fig. 29, is made

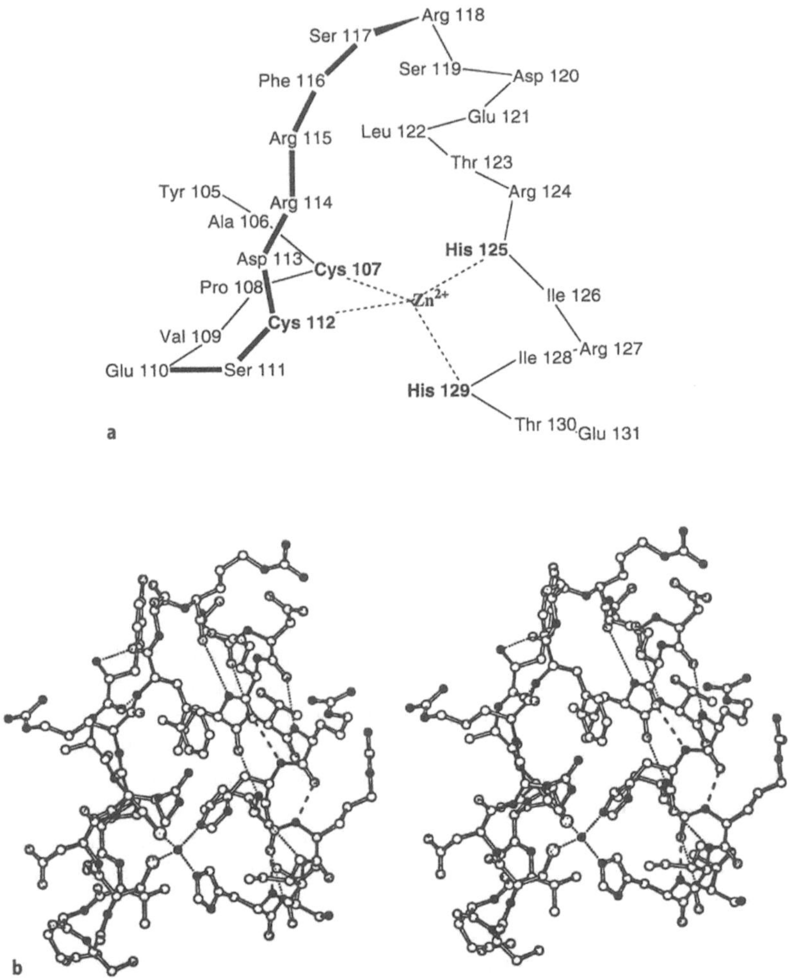

Fig. 28 a, b. A zinc finger [72]: **a** diagram of the protein folding and the tetrahedral binding of a zinc ion; **b** stereoview of one of the zinc fingers in the zif268 protein-nucleic acid complex [72]

with a right hand with an index finger (the *E* helix), a curled second finger (the loop), and a thumb (the *F* helix). This metal-binding motif has a strong affinity for calcium ions. The motif is generally considered as an octahedral arrangement around the calcium ion with one site shared by both carboxylate oxygen atoms of a glutamate side chain to give a coordination number of 7. The affinity of this arrangement for calcium ions is highest when there is a pair of contiguous *EF* hands. For each hand the affinity is related to the net ligand charge of the loop, and the hydrophobicity of various residues in the *EF*-hand pair [76]. Nature has utilized this calcium-specific binding motif in several proteins, such as troponin C, calmodulin, parvalbumin, and lactalbumin.

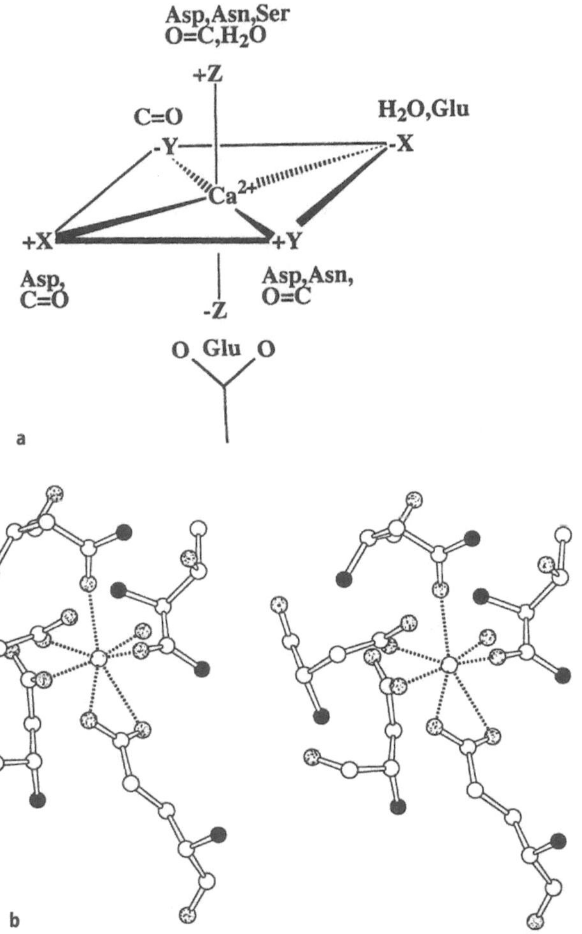

Fig. 29 a, b. An *EF* hand [74]: **a** diagram of ligands commonly found around calcium ions in this motif; **b** stereoview of the binding in calmodulin in a similar orientation to that in (a)

6.5
Ionophores and Other Metal-Binding Molecules

Ionophores are designed by nature to carry small ions across lipid membrane barriers. They provide a torus to capture the required ions. The first to be characterized was valinomycin, a cyclic dodecadepsipeptide with a ring of 36 atoms. Its crystal structure [77], illustrated in Fig. 30, shows carbonyl oxygen atoms 2.7–2.8 Å from the potassium cation that it is specifically engineered to bind. Specific hydrogen bonding in the valinomycin cage prevents it from contracting to bind the smaller sodium ion, and therefore this ionophore has a high K^+:Na^+ selectivity (favoring potassium). Crown ethers, such as dibenzo-18-crown-6 (Fig. 31) hold the metal ion (K^+) approximately in the plane of the oxy-

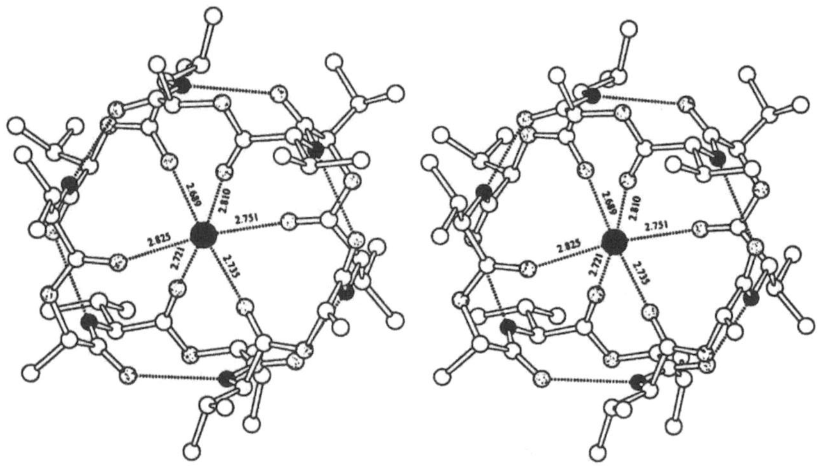

Fig. 30. Complexation of potassium by valinomycin [77]. Hydrogen bonding in the complexing agent helps assure a torus-like structure for metal ion binding

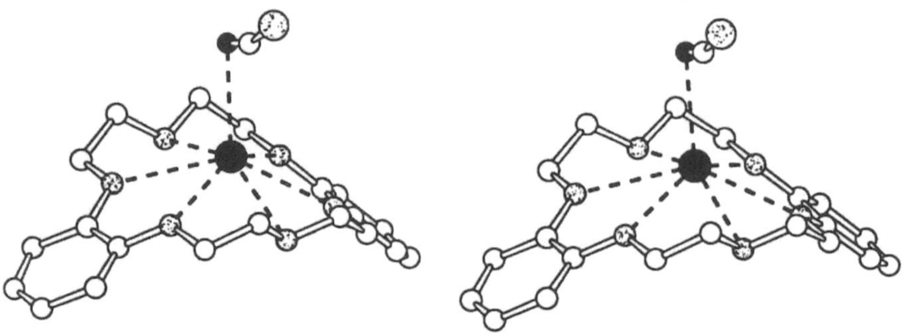

Fig. 31. Complexation of a metal ion (K^+ of KSCN) by dibenzo-18-crown-6 [78]. This stereoview shows that the metal ion is not exactly in the plane of the complexing oxygen atoms

gen atoms [78]. Cram [79] and Lehn [80] made more complex molecules, designing them to perform certain functions with the specific aims of mimicking selective transport across biological membranes and certain aspects of enzyme chemistry.

6.6 Metal-Binding Sites in Proteins

The experimental observation that there is a high probability that the metal ion lies in the plane of a neighboring carboxylate ion provides a mechanism for searching for metal-binding sites in proteins. The method was used successfully to locate two metal sites in the enzyme xylose isomerase from *Streptomyces rubiginosus* [81]. One metal-binding site involves three carboxylates (aspartate and glutamate), one histidine, and water, and the other involves four carboxylate groups and water. This test is complementary to that suggested by David Eisenberg and co-workers [82], in which the existence of hydrophilic areas in the protein (negatively-charged oxygen atoms) with a hydrophobic area immediately behind them (carbon atoms of the carboxylate group) are identified.

7 Networks of Interactions

When large molecules interact they recognize motifs containing appropriate functional groups. These motifs may lie in one molecule or may be formed as two or more aggregates. Therefore networks of interacting groups are of interest in studies of binding of large molecules to each other. As a result the types of interactions on a multimolecular scale are now being studied, having been characterized as "supramolecular synthons" by Desiraju [83] (see also the article by A. Nangia and G.R. Desiraju in this volume). Such synthons are structural units that can be assembled by intermolecular interactions. Each of the types of interactions described here can take part in such synthons. Some examples are shown in Fig. 32. Major participants in supramolecular synthons are hydrogen bonds and therefore they are discussed here. But any other types of intermolecular interactions, such as metal coordination or C-S-C interactions, could be involved.

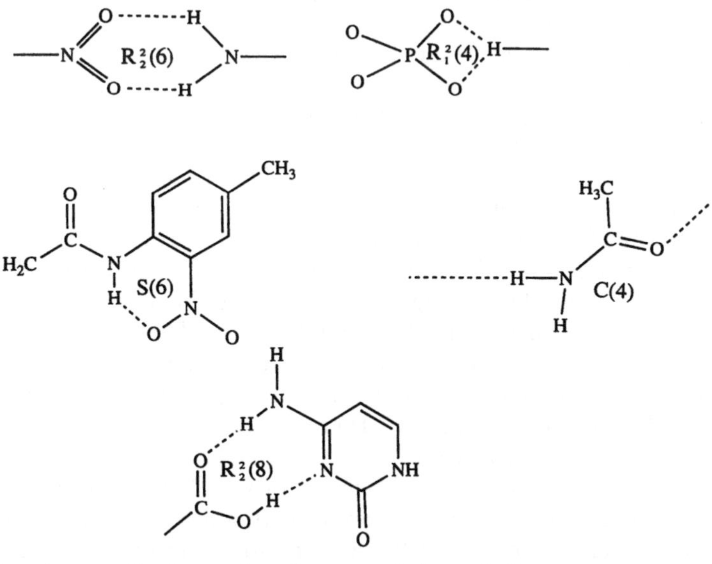

Fig. 32. Descriptions of hydrogen-bonding networks [30, 84]. Shown are some examples with their graph-set designations (R=ring, C=chain, S=intramolecular)

7.1 Descriptions of Hydrogen-Bonding Networks

Extended networks of hydrogen bonds are found in many crystal structures and, as they run continuously through the crystal, they act cooperatively [3] with lower energy than expected from the sum of the energies of the individual hydrogen bonds. Hydrogen-bonding schemes in crystal structures are readily analyzed, provided all the hydrogen atoms have been included in the refined crystal structure. It is simply a matter of determining the interatomic distances and angles between hydrogen-bond donors and acceptors, keeping in mind the periodicity and space-group symmetry of the crystal structure.

Graph theory has been used [30, 84] as a method for describing such hydrogen bonding. The pattern is described as G, where G describes the type (S for self or intramolecular, D for noncyclic finite pattern such as a dimer, R for rings, and C for chains). The sub- and superscripts d and a refer to the number of donors and acceptors, and r is the size of the pattern. This scheme highlights extended systems of hydrogen bonds and similarities and differences of patterns of hydrogen bonding in crystals of related compounds. Some examples are given in Fig. 32 (see also the article by M. R. Caira in this volume).

8 Crystal Surface Recognition and Chirality

The effects of certain impurities on the habit of a growing crystal provide another directional aspect of intermolecular recognition. If an impurity is similar (but not identical) to the molecules on the surface as well as in the bulk of the crystal, it is adsorbed on a specific crystal face. Presumably the surface has recognized that part of the impurity molecule that most resembles the molecules composing the crystal. The impurity, however, presents to the exterior a side onto which further molecules of host material cannot bind, as diagrammed in Fig. 33. Therefore the growth of that face will be inhibited, and will be prominent in the habit of the resulting crystal. For example, benzamide forms platelike elongated crystals. If, however, some benzoic acid is added to the solution, the resulting crystals will be elongated along a different direction.

Directional binding of molecules to crystal faces provides an interesting means of controlling crystal habit and resolving optically active molecules [85–87]. It provides food for thought with respect to the design of agents with specific recognition requirements. An example is provided by the crystal structure of glycine (which lacks an asymmetric carbon atom). This crystallizes as bipyramids [86] in a monoclinic, centrosymmetric space group. The packing of molecules and the *pro*chirality of the C-H groups they present on the surface are different for the 010 and $0\bar{1}0$ faces of the crystal. On addition of resolved amino acids, pyramidal crystals with a basal face are formed. This is because the *pro-R* and *pro*-S-hydrogen atoms of glycine can be distinguished in the crystal structure of glycine, one lying on the (010) faces and the other on the $(0\bar{1}1)$ faces, as shown in Fig. 34. Only D-amino acids [(*R*)-amino acids] can readily replace glycine at growing (010) faces while only L-amino acids [(*S*)-amino acids] can grow

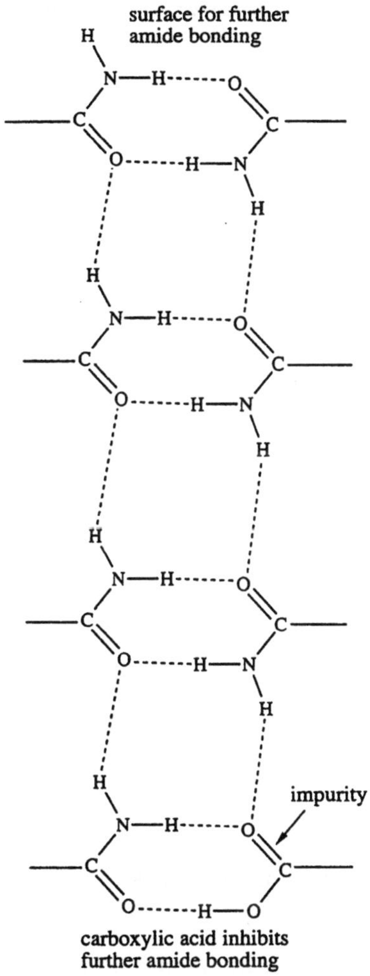

Fig. 33. Effect of an impurity on the surface of a growing crystal [85]. Once this impurity has bound no more crystal growth can occur in that direction

on $(0\bar{1}0)$ faces. Thus, if glycine is grown in aqueous solution and forms a plate, as it does in the presence of resolved amino acids such as L-leucine[(*S*)-leucine], it will float with its (010) face exposed to air and will grow on the other side [on the $(0\bar{1}0)$ face]. Thus, in a solution of a DL-mixture of any amino acid, the crystal will bind (*R*)-amino acids on the (010) faces and will enrich the solution in (*S*)-amino acids, and vice versa (see also the article by M.R. Caira in this volume).

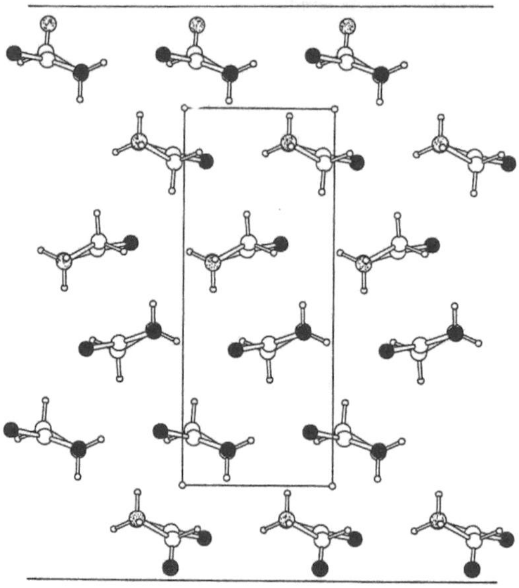

Fig. 34. Glycine crystals binding optically active amino acids [86, 116]. Shown is the binding of D-alanine to the 010 face and L-alanine to the 0$\bar{1}$0 face. The methyl groups of alanine are *stippled circles* for the former, and *filled circles* for the latter

9
Applications to Macromolecular Structures

Macromolecular recognition often involves many points of interaction, whether it is subunit assembly, substrate or inhibitor binding or the operation of control factors. The binding of drugs to biological macromolecules involves a fit between the drug and its receptor that depends not only on their respective shapes, but also on the distribution of electronic charge on the surfaces of both molecules so that a complementary fit may be obtained. Molecules with similar, but not identical, formulae may bind to the same receptor in the same or different ways, depending on the nature of the functional groups in each.

9.1
Binding of Drugs to Dihydrofolate Reductase

An extensively studied enzyme-inhibitor system involves the protein dihydrofolate reductase [88]. Crystallographic results demonstrate an important feature of drug-enzyme interactions; an inhibitor drug may not bind in the same way as substrate even though both have similar chemical formulae. This enzyme catalyzes the reduction of 7,8-dihydrofolate to 5,6,7,8-tetrahydrofolate, an essential coenzyme used in the synthesis of thymidylate, inosinate, and methionine. Methotrexate is an antimetabolite that is a folate analogue, and is used as a cancer chemotherapeutic agent. It inhibits dihydrofolate reductase by binding tightly and thereby causing a cellular deficiency of thymidylate ("thymine-less death") so preventing DNA synthesis. Folate and methotrexate differ by the substitution of the pterin ring carbonyl in folate by an amino group, and by methylation of an exocyclic folate NH-group, as shown in Fig. 35. Thus the hydrogen

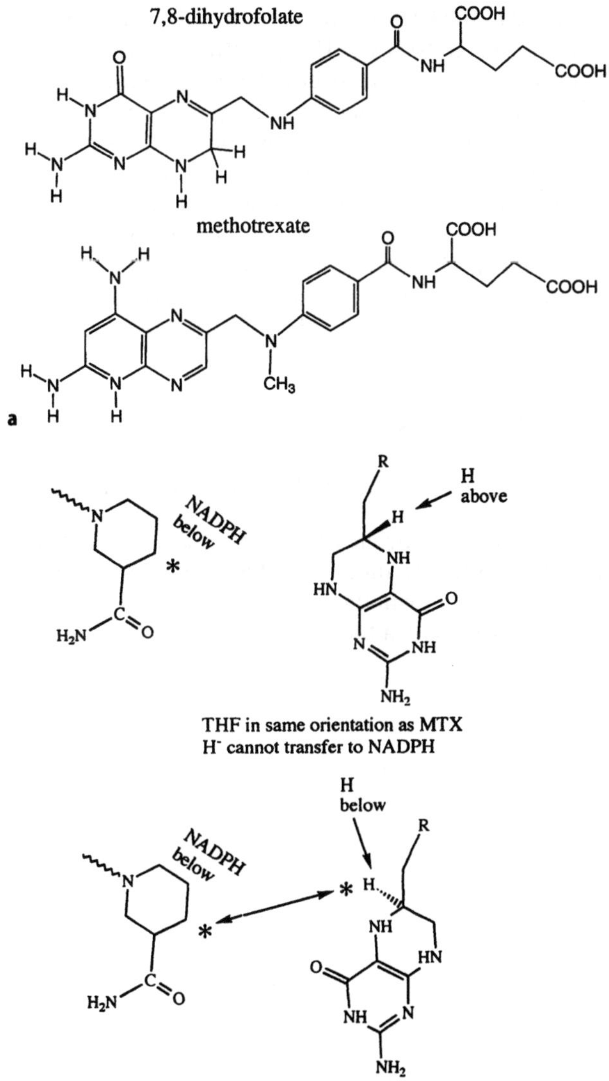

Fig. 35 a, b. Action of dihydrofolate reductase [89, 90]: **a** formulae of dihydrofolate and methotrexate; **b** stereochemistry of the enzyme action (THF = tetrahydrofolate, MTX = methotrexate)

bonding capabilities of the two compounds are different. Methotrexate is found in a cavity 15 Å deep in the *E. coli* enzyme complex. The entire inhibitor molecule is bound by a system of hydrogen bond and hydrophobic interactions. A similar binding is found for the *E. coli* and *L. casei* enzyme complexes.

The enzyme promotes hydride transfer from NADPH to the pterin ring. This is facilitated by the development of a charge on C6 and by protonation of N5. The absolute configuration of tetrahydrofolate has been established [89, 90] to be *S* at C6, but if dihydrofolate is bound in the same orientation as methotrexate, the configuration should be *R*, as shown in Fig. 35. When dihydrofolate (or tetrahydrofolate) is bound in the enzyme in the same way as methotrexate, the hydrogen atom that is transferred to NADPH is pointing in the wrong direction. Thus dihydrofolate and methotrexate must bind to the enzyme with different orientations as established later by X-ray crystallographic studies. Studies of additional bound substrate analogues and inhibitors [91, 92] confirm that binding in the active site of the enzyme is governed mainly by hydrogen bonding, and not simply by molecular shape. These situations are illustrated in Fig. 36. They show that substrate and inhibitor have different and unique binding preferences and therefore the fact that their modes of binding are different could have been predicted from our understanding of donor and acceptor roles in hydrogen bonding (see also the article by A. Naugia and G. R. Desiraju in this volume).

9.2 Binding of Drugs to Nucleic Acids

Antitumor agents that interact with DNA do so in one of three ways.

1. Intercalation between the nucleic acid bases of DNA. This mode of binding is found for anthracyclines, actinomycin D and acridines. The DNA backbone flexes so that the bases, normally separated by 3.4 Å, are moved to approximately 7 Å apart. As a result, a flat aromatic molecule can slip between them.

a

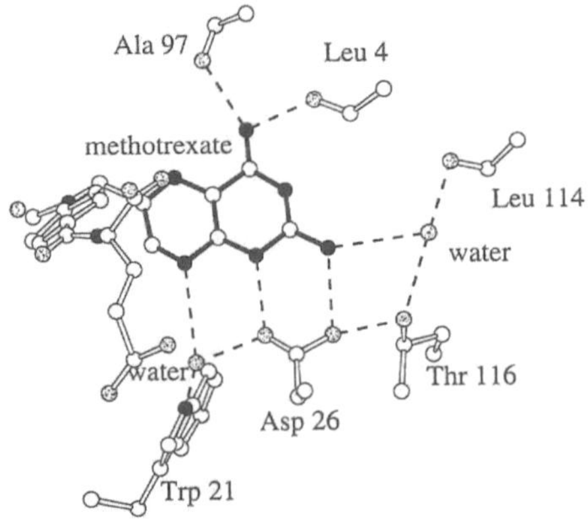

Fig. 36 a–c. Binding in dihydrofolate reductase [88]: **a** binding of the pterin ring of methotrexate in crystals of the *Lactobacillus casei* enzyme

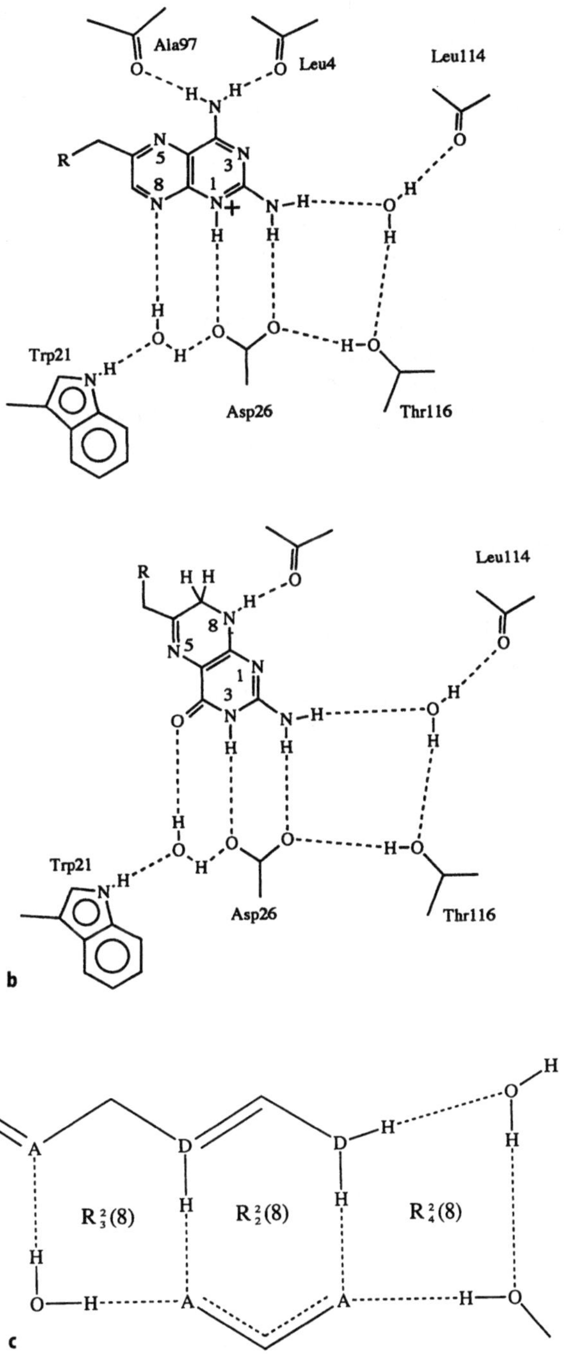

Fig. 36 b, c. **b** More detailed diagram that compares the binding of methotrexate (*upper diagram*) and folate (*lower diagram*); **c** motif for recognition of substrate or inhibitor inferred from these crystallographic studies, together with their graph-set designations

2. Non-intercalative groove binding. The drug lies in one of the grooves in DNA, following the helical contour of DNA as it binds. Netropsin and distamycin bind in this way.
3. Covalent bond formation. This occurs with mitomycin C.

The anthracyclines, such as daunomycin and its hydroxy analogue, adriamycin, are naturally-occurring tetracyclic aminoglycosidic derivatives that contain an anthraquinone chromophore. The main target of these drugs appears to be DNA, probably by formation of an intercalation complex. Daunomycin and adriamycin have a planar chromophore with an amino sugar extending out of this plane. In the crystal structure of the complex of adriamycin with a self-complementary hexadeoxynucleotide, the oligonucleotide forms a right-handed helix with two molecules of adriamycin intercalated in the d(CpG) sequences at each end, as shown in Fig. 37 [93–95]. Adriamycin and daunomycin are intercalated in a similar manner with their long axes perpendicular to those of the base pairs; the 14-hydroxyl group does not significantly affect the orientations of the chromophore or the amino sugar. The hydroxyl group forms two hydrogen bonds to an adjacent guanine. The amino sugar lies in the minor groove of the distorted helix, but does not form bonds to the nucleic acid. Thus ring A anchors adriamycin in DNA, and the nucleic-acid backbone is distorted by the interactions formed. Glycosylation of C5 of a thymine residue in an oligonucleotide does not prevent intercalation [96].

Recently, a bis intercalating derivative of daunomycin has been studied [97]. In this compound the amino groups are joined by a xylyl group. This daunomy-

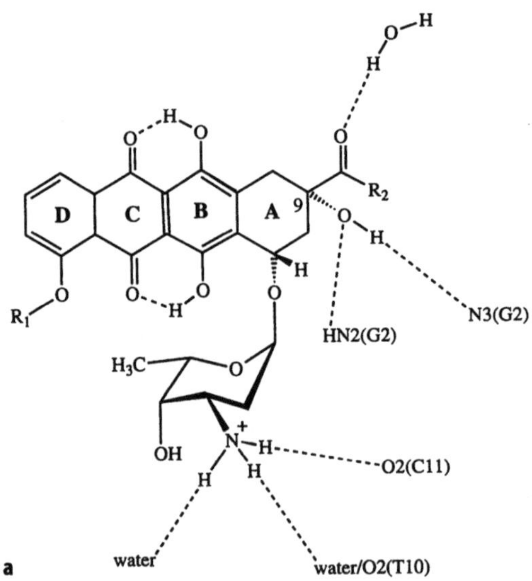

a

Fig. 37 a–c. Binding of daunomycin to an oligonucleotide [95]: **a** chemical formula of daunomycin, showing some interactions in the crystalline state. $R_1 = CH_3$, $R_2 = CH_3$ for daunomycin, CH_2OH for adriamycin

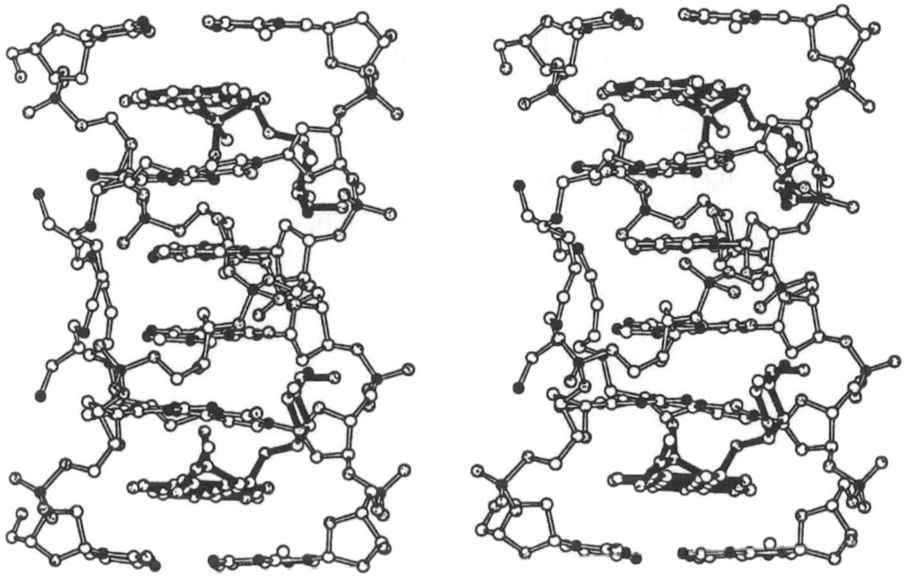

b

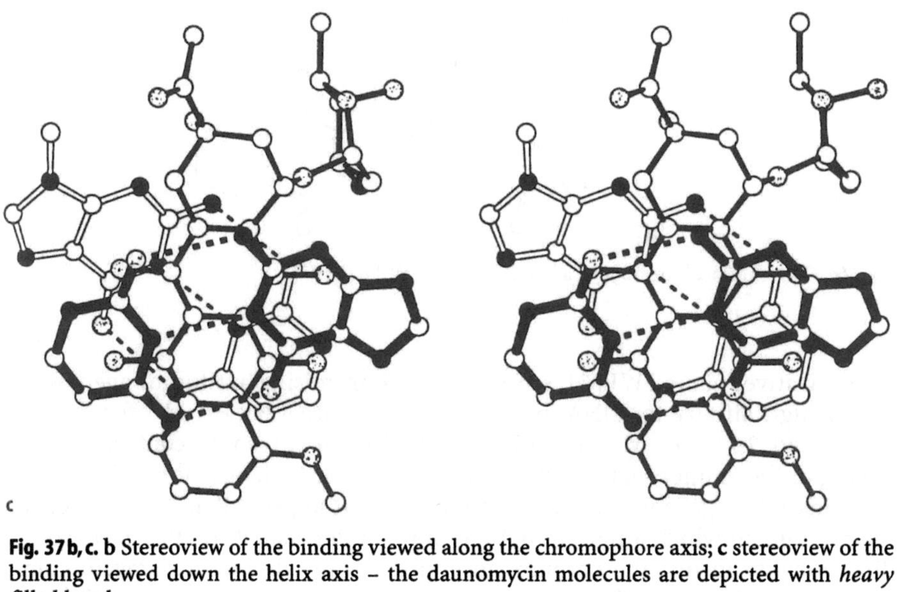

c

Fig. 37 b, c. b Stereoview of the binding viewed along the chromophore axis; **c** stereoview of the binding viewed down the helix axis – the daunomycin molecules are depicted with *heavy filled bonds*

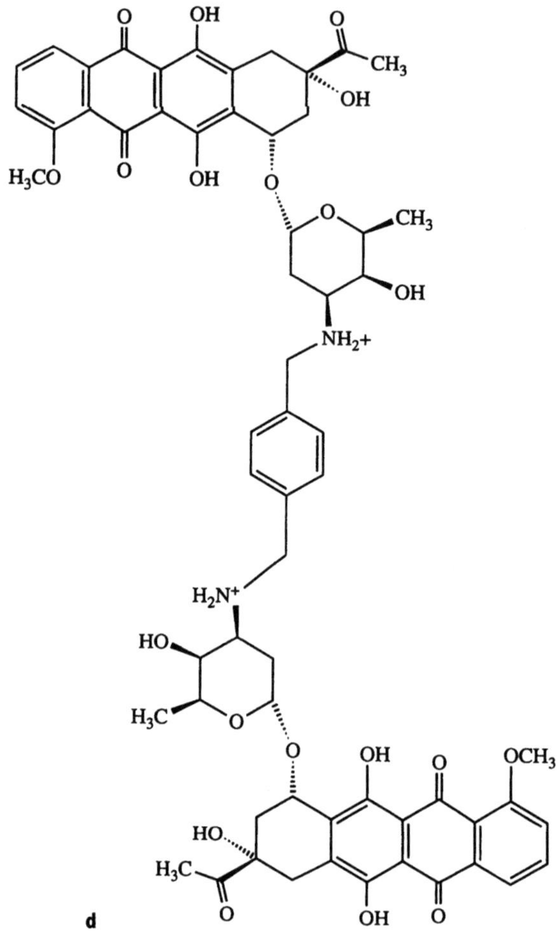

d

Fig. 37 d. The chemical formula of a bisintercalator, WP 631 [97]

cin derivative (named WP631, see Fig. 38) is a promising antitumor agent with a very high affinity for DNA; binding constants are 1.6×10^7 M^{-1} for daunomycin and 2.7×10^{11} M^{-1} for WP631 [98]. Except for some conformational flexing of the aminosugar, the structures are similar with or without the linker. The major change is in the region of the linker. It appears that intermolecular interactions in which a flat aromatic group is intercalated in DNA have some directionality in that the chromophore in daunomycin and adriamycin complexes generally lies in the same position even though some of the peripheral binding interactions are different.

Minor groove binders to DNA are generally A · T specific and fit snugly in the narrow minor groove. Such drugs are replacements for well-ordered water molecules found in the absence of drugs. Netropsin and distamycin A, which are pyrrole-amidine oligopeptide antibiotics, bind very tightly to A · T regions of

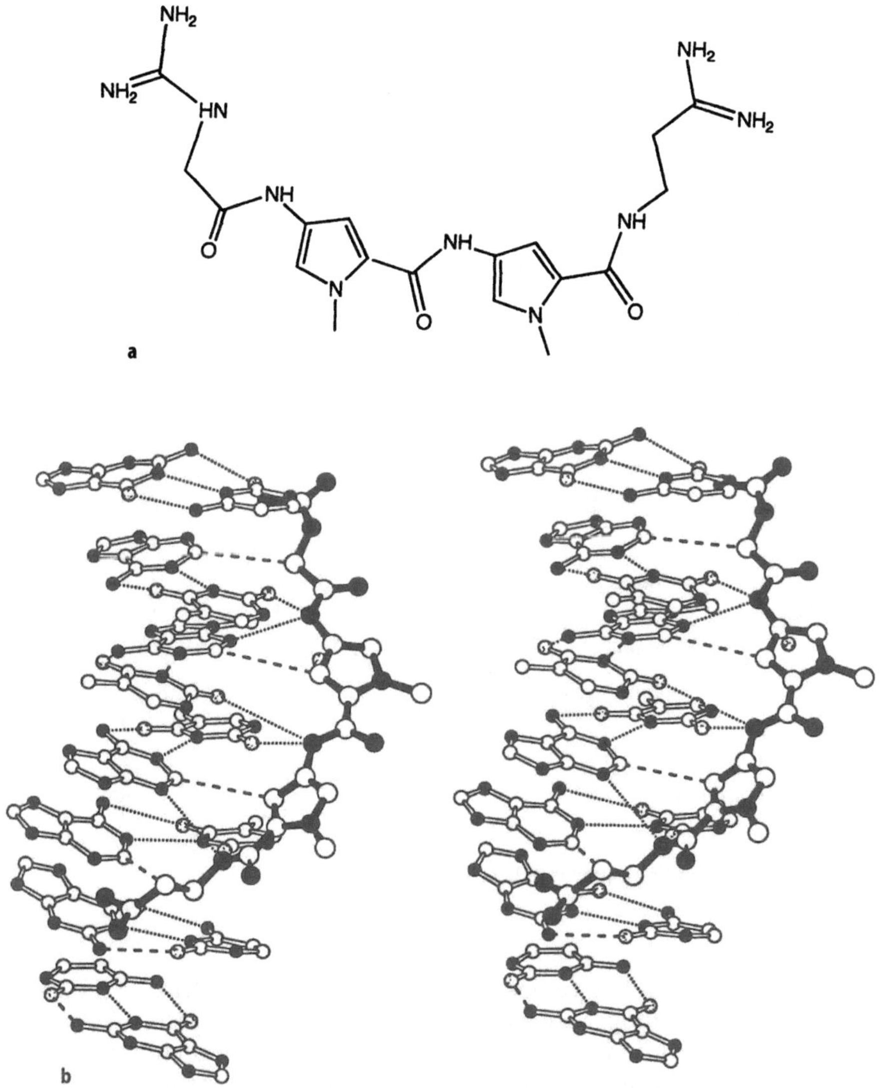

Fig. 38 a, b. Minor groove binding of netropsin in DNA [100]: **a** chemical formula of netropsin; **b** binding of netropsin by way of N-H···O bonds (*dotted lines*) and C-H···C interactions. The latter define the specificity of the drug

duplex DNA, as illustrated in Fig. 38. These compounds are relatively cytotoxic.

In the crystal structure of netropsin bound to a dodecamer with four A·T base pairs in the middle, the four base pairs have high propeller twists [99, 100]. As a result, the amino group of adenine is located half-way between the carbonyl groups of two adjacent thymine residues and some three-center hydrogen bonds

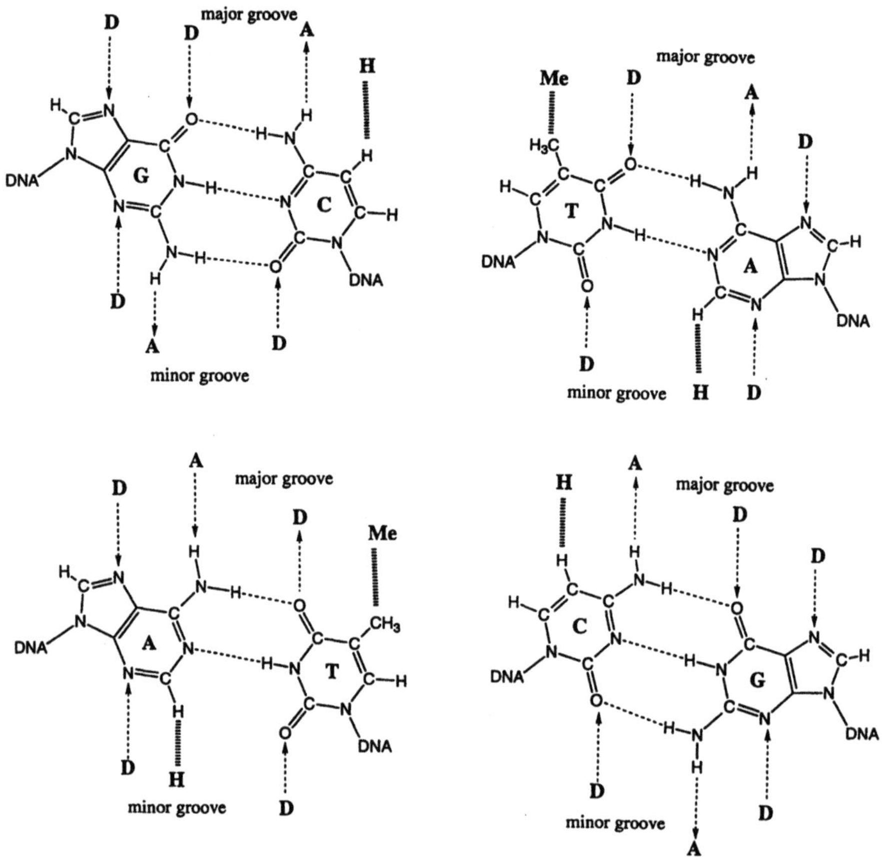

Fig. 39. Hydrogen-bonding patterns of base pairs [103]

are formed. Netropsin NH groups form hydrogen bonds to DNA, as shown in Fig. 38, while ring hydrogen atoms are in contact with the CH groups (on C2) of adenine. If there is an amino group in these positions (giving guanine instead of adenine), netropsin could not bind. The specificity of netropsin for A · T base pairs is determined by a combination of hydrogen bonding and van der Waals interactions. Modification of the chemical formula of netropsin can provide a G · C-specific binding agent [100].

9.3 Protein-Nucleic Acid Interactions

Proteins can interact with nucleic acids by means of a helix-turn-helix motif [101] in the protein structure, by a zinc finger [71, 72], or by a leucine zipper [102] in which leucine side chains that are seven amino-acid residues apart on along each of two α helices interact and force a protein conformation that is suitable for binding to a nucleic acid. In each case these are protuberances from a fairly spherical protein that can then interact with one of the grooves of DNA. Specific interactions of these types allow the protein to regulate gene expression. The recognition of the bases in DNA by hydrogen bonding to the protein involves patterns, shown in Fig. 39. Each combination of base pairs in a nucleic acid – G·C, A·T, C·G, and T·A – can be differentiated in the major groove by a hydrogen-bond pattern [103]. The recognition is less specific in the minor groove. The side chains of proteins available for such recognition are arginine with two hydrogen-bond donor groups in the required orientation, glutamic or aspartic acid with two hydrogen-bond acceptors, and asparagine and glutamine which contain one donor and one acceptor.

Many crystal structures that involve protein-nucleic acid interactions have been published in the scientific literature recently. Some show this recognition scheme as in a zinc finger protein-nucleic acid complex, illustrated in Fig. 40 [71,

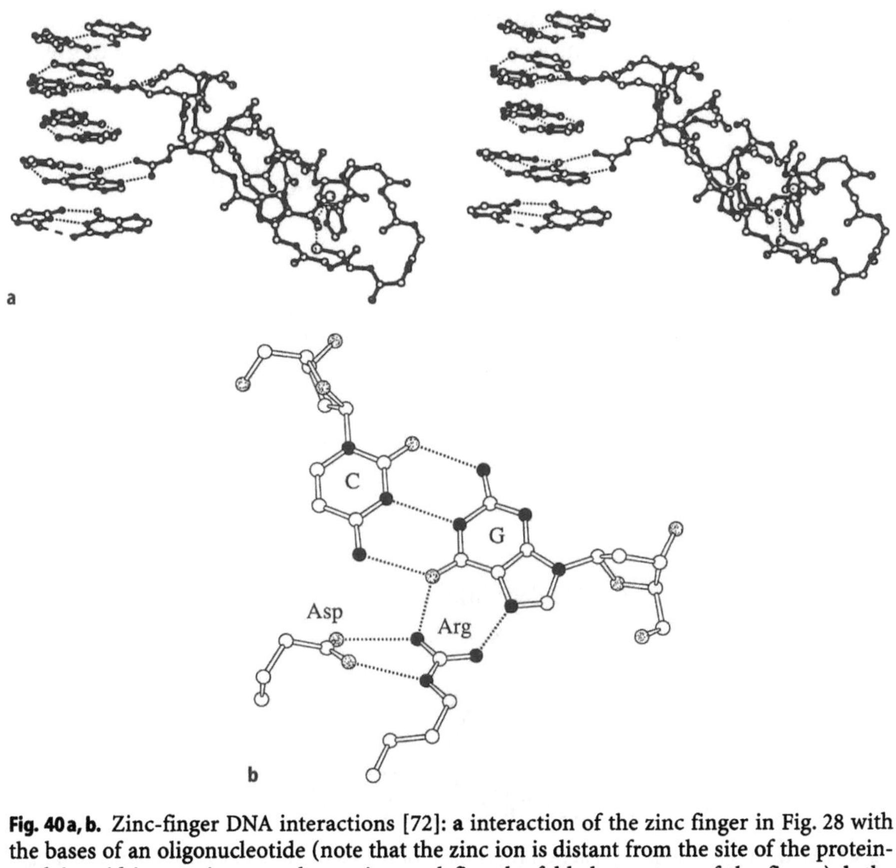

Fig. 40 a, b. Zinc-finger DNA interactions [72]: **a** interaction of the zinc finger in Fig. 28 with the bases of an oligonucleotide (note that the zinc ion is distant from the site of the protein-nucleic acid interaction, merely serving to define the folded structure of the finger), **b** the hydrogen bond pattern involved (see Fig. 22)

72, 104], but others do not. Their interactions are more varied and apparently less specific. This is an exciting area of research and more results are being published each month. There is still a need to establish unequivocally what the recognition rules are for protein-nucleic acid interactions.

10 Incipient Chemical Reaction Pathways

Two interacting molecules may interact with each other. Crystal structure analyses can give information on this in at least two ways. Inspection of coordinates of crystal structures containing interacting groups may reveal the reaction pathway between them [105]. The nucleophilic addition of a nitrogen atom to a carbonyl group showed this (Fig. 41) [106]. The N···C distance varied from 2.91 to 1.49 Å in different crystal structures in such a way that the nitrogen approaches the carbon of the carbonyl group in a direction perpendicular to the plane of the latter. Now that high intensity synchrotron radiation is available, it is possible to follow chemical reactions in proteins, seeing how the molecules interact with each other. This method, which involves taking instantaneous X-ray diffraction photographs as a function of time, has been applied to glyco-

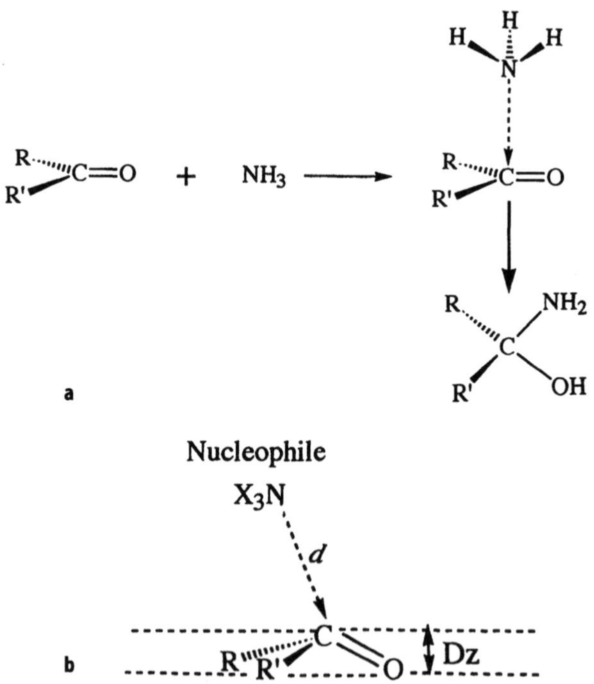

Fig. 41 a, b. Incipient chemical reaction (approach of N to a carbonyl group) [104]: **a** diagram of the interaction involved; **b** geometry of the interaction. As *d* decreases in length, Dz increases in height gen phosphorylase *b* [107] and the blue-light photoreceptor photoactive yellow protein [108].

11
Conclusion

We are now learning much about intermolecular interactions and reactions from the types of studies described here. An understanding of directional preferences of binding of functional groups is an essential starting point for proposals on mechanisms of reactions between two molecules, large or small, and also for the construction of supramolecular assemblies, be they in solution or in the solid state, the latter being in the focus of interest here.

Acknowledgment. This work was supported by grants CA-10925 and CA-06927 from the National Institutes of Health, and by an appropriation from the Commonwealth of Pennsylvania. The contents of this article are solely the responsibility of the authors and do not necessarily represent the official views of the National Cancer Institute.

12
References

1. Legon AC, Millen DJ (1993) Molecular recognition involving small gas-phase molecules. In: Buckingham AD, Legon AC, Roberts SM (eds) Principles of molecular recognition. Blackie, London, p17
2. Buckingham AD (1993) Intermolecular forces. In: Buckingham AD, Legon AC, Roberts SM (eds) Principles of molecular recognition. Blackie, London, p1
3. Jeffrey GA (1997) An introduction to hydrogen bonding. Oxford University Press, New York
4. Allen FH, Bellard S, Brice MD, Cartwright BA, Doubleday A, Higgs H, Hummelink T, Hummelink-Peters BG, Kennard O, Motherwell WDS, Rodgers JR, Watson DG (1979) Acta Cryst B35:2331
5. Bernstein FC, Koetzle TF, Williams GJB, Meyer EF Jr, Brice MD, Rodgers JR, Kennard O, Shimanouchi T, Tasumi M (1977) J Mol Biol 112:535
6. Berman HM, Olson WK, Beveridge DL, Westbrook J, Gelbin A, Demeny T, Hsieh S-H, Srinivasan AR, Schneider B (1992) Biophys J 63:751
7. Bergerhoff G, Hundt R, Sievers R, Brown ID (1983) J Chem Info 23:66
8. Matthews BW (1968) J Mol Biol 33:491
9. Murray-Rust P, Glusker JP (1984) J Am Chem Soc 106:1018
10. Rosenfield RE Jr, Swanson SM, Meyer EF Jr, Carrell HL, Murray-Rust P (1984) J Mol Graphics 2:43
11. Williams DE (1972) Acta Cryst A 28:629
12. Kitaigorodsky AI (1955) Organic chemical crystallography. Consultants Bureau, New York
13. Gavezzotti A (1994) Acc Chem Res 27:309
14. Glusker JP, Carrell HL, Berman HM, Gallen B, Peck RM (1977) J Amer Chem Soc 99:595
15. Vedani A, Dunitz JD (1985) J Am Chem Soc 107:7653
16. Vedani A, Huhta DW (1990) J Am Chem. Soc 112:4759
17. Klebe G (1994) J Mol Biol 237:212
18. Rosenfield RE Jr, Parthasarathy R, Dunitz JD (1977) J Am Chem Soc 99:4860
19. Fukui K, Yonezawa, T, Shingu H (1952) J Chem Phys 20:722
20. Guru Row T, Parthasarathy R (1981) J Am Chem Soc 103:477
21. Chakrabarti P (1989) Biochem 28:6081
22. Ramasubbu N, Parthasarathy R, Murray-Rust P (1983) J Am Chem Soc 105:3206

23. Cody V, Murray-Rust P (1984) J Mol Struct 112:189
24. Lommerse JPM, Taylor R (1997) J Enzyme Inhib 11:223
25. Baker EN, Hubbard RE (1984) Prog Biophys Mol Biol 44:97
26. Pauling L (1940) The nature of the chemical bond, 2nd edn. Cornell University Press, Ithaca, New York
27. Pimentel GC, McClellan AL (1960) The hydrogen bond. Freeman, San Francisco
28. Umeyama H, Morokuma K (1977) J Am Chem Soc 99:1316
29. Taylor R, Kennard O, Versichel W (1983) J Am Chem Soc 105:5761
30. Etter MC (1990) Acc Chem Res 23:120
31. Dunitz JD (1996) Mol Cryst Liq Cryst 279:209
32. Ibers JA (1964) J Chem Phys 40:402
33. Kemnitz E, Werner C, Trojanov S (1996) Acta Cryst C52:2665
34. Roelofsen G, Kanters JA (1972) Cryst Struct Commun 1:23
35. Lommerse JPM, Price SL, Taylor R (1997) J Comput Chem 18:757
36. Chakrabarti P (1990) Biochem 29:651
37. Sutor DJ (1963) J Chem Soc, p 1105
38. Biradha K, Sharma CVK, Panneerselvam K, Shimoni L, Carrell HL, Zacharias DE, Desiraju GR (1993) Chem Commun, p 1473
39. Stallings, WC, Monti CT, Belvedere JF, Preston RK, Glusker JP (1980) Arch Biochem Biophys 203:65
40. Murray-Rust P, Stallings WC, Monti, CT, Preston, RK, Glusker, JP (1983) J Am Chem Soc 105:3201
41. Shimoni L, Glusker JP (1994) Struct Chem 5:383
42. Dunitz JD, Taylor R (1997) Chem Eur J 3:89
43. Brown ID (1978) Chem Soc Res 7:359
44. Ermer O, Eling A (1994) J Chem Soc, Perkin Trans 2:925
45. Allen FH, Hoy VJ, Howard JAK, Thalladi VR, Desiraju GR, Wilson CC, McIntyre GJ (1997) J Am Chem Soc 119:3477
46. Kashino S, Tomita M, Haisa M (1988) Acta Cryst C 44:780
47. Allen FH, Howard JAK, Hoy VJ, Desiraju GR, Reddy DS, Wilson CC (1996) J Am Chem Soc 118:4081
48. Jedrzejas MJ, Singh S, Brouilette WJ, Laver WG, Air GM, Luo M (1995) Biochem 34:3144
49. Ji X, Zhang P, Armstrong RN, Gilliland GL (1992) Biochem 31:10,169
50. Rosenfield RE, Murray-Rust P (1982) J Am Chem Soc 104:5427
51. Harel M, Schalk I, Ehret-Sabattier L, Bouet F, Goeldner M, Hirth C, Axelsar P, Silman I, Sussman J (1993) Proc Natl Acad Sci USA 90:9031
52. Herbstein FH (1971) In: Dunitz JD, Ibers JA (eds) Perspectives in structural chemistry. Wiley, New York, p 166
53. Mayoh B, Prout CK (1972) J Chem Soc, Faraday Trans 1072
54. Einspahr H, Bugg CE (1981) Acta Cryst B37:1044
55. Petersen MR, Csizmadia IG (1979) J Am Chem Soc 101:1076
56. Gandour R (1981) Bioorg Chem 10:169
57. Carrell CJ, Carrell HL, Erlebacher J, Glusker JP (1988) J Am Chem Soc 110:8651
58. Chakrabarti P (1990) Protein Eng 4:49
59. Ramanadham M, Jakkal VS, Chidambaram R (1993) FEBS Lett 323:203
60. Zvelebil MJJM, Steinberg MJE (1988) Protein Eng 2:127
61. Carrell AB, Shimoni L, Carrell CJ, Bock CW, Murray-Rust P, Glusker JP (1993) Receptor 3:57
62. Chakrabarti P (1990) Protein Eng 4:57
63. Kaufman A, Afshar C, Rossi M, Zacharias DE, Glusker JP (1993) Struct Chem 4:191
64. Shimoni L, Glusker JP (1995) Prot Sci 4:65.
65. Glusker JP (1980) Acc Chem Res 13:345
66. Carrell HL, Glusker JP, Percy EA, Stallings WC, Zacharias DE, Davis RL, Astbury C, Kennard CHL (1987) J Am Chem Soc 109:8067
67. Lee J-O, Rieu P, Arnaout MA, Liddington R (1995) Cell 80:621

68. Baur WH (1979) Trans Amer Cryst Assn 6:129
69. Bock CW, Kaufman, A, Glusker JP (1994) Inorg Chem 33:419
70. Bock CW, Katz AK, Glusker JP (1995) J Am Chem Soc 117:3754
71. Párraga G, Horvath SJ, Eisen A, Taylor WE, Hood L, Young ET, Klevit RE (1988) Science 241:1489
72. Pavletich NP, Pabo CO (1991) Science 252:809
73. Katz AK, Glusker JP, Beebe SA, Bock CW (1996) J Am Chem Soc 118:5752
74. Kretsinger RH, Nockolds CE (1973) J Biol Chem 248:3313
75. Glusker JP (1991) Adv Prot Chem 42:1
76. Sekharudu YC, Sundaralingam M (1988) Protein Eng 2:139
77. Duax WL, Hauptman H, Weeks CM, Norton DA (1972) Science 176:911
78. Bright D, Truter MR (1970) J Chem Soc (B) 1544
79. Cram DJ (1992) Nature 356:29
80. Lehn J-M (1988) Angew Chem Int Ed Eng 27:90
81. Carrell HL, Glusker JP, Burger V, Manfre F, Tritsch D, Biellmann J-F (1989) Proc Natl Acad Sci USA 86:4440
82. Yamashita M, Wesson L, Eisenman G, Eisenberg D (1990) Proc Natl Acad Sci USA 87:5648
83. Desiraju GR (1996) Acc Chem Res 29:441
84. Bernstein J, Davis RE, Shimoni L, Chang N-L (1995) Angew Chem Int Ed Engl 34:1555
85. Berkovitch-Yellin Z, van Mil J, Addadi L, Idelson M, Lahav M, Leiserowitz L (1985) J Am Chem Soc 107:3111
86. Weissbuch I, Addadi L, Berkovitch-Yellin Z, Gati E, Weinstein S, Lahav M, Leiserowitz L (1983) J Am Chem Soc 105:6615
87. Gidalevitz D, Feidenhans'l R, Matlis S, Smilgies D-M, Christensen MJ, Leiserowitz L (1997) Angew Chem Int Ed Engl 36:955
88. Bolin JT, Filman DJ, Matthews DA, Hamlin R, Kraut J (1982) J Biol Chem 257:13,650
89. Fontecilla-Camps JC, Bugg CE, Temple C Jr, Rose JD, Montgomery JA, Kisliuk RL (1979) J Am Chem Soc 101:6114
90. Armarego WLF, Waring P, Williams JW (1980) J Chem Soc, Chem Commun p 334
91. Reyes VM, Sawaya MR, Brown KA, Kraut T (1995) Biochem 34:2710
92. Lee H, Reyes VM, Kraut J (1996) Biochem 35:7210
93. Quigley GJ, Wang AH-J, Ughetto G, van der Marel G, van Boom JH, Rich A. (1980) Proc Natl Acad Sci USA 77:7204
94. Wang AH-J, Ughetto G, Quigley GJ, Rich A (1987) Biochem 26:1152
95. Frederick CA, Williams LD, Ughetto G, van der Marel G, van Boom JH, Rich A, Wang, AH-J (1990) Biochem 29:2538
96. Gao Y-G, Robinson H, Wijsman ER, van der Marel GA, van Boom JH, Wang AH-J (1997) J Am Chem Soc 119:1496
97. Hu GG, Shui X, Leng F, Priebe W, Chaires JB, Williams LD (1997) Biochem 36:5940
98. Chaires JB, Leng F, Przewloka T, Fokt I, Ling YH, Perez-Soler R, Priebe W (1997) J Med Chem 40:261
99. Kopka ML, Yoon C, Goodsell D, Pjura P, Dickerson RE (1985) J Mol Biol 185:553
100. Kopka ML, Yoon C, Goodsell D, Pjura P, Dickerson RE (1985) Proc Natl Acad Sci USA 82:1376
101. Anderson WF, Ohlendorf DH, Takeda Y, Matthews BW (1981) Nature 290:754
102. Ellenberger TE, Brandl C-I, Struhl K, Harrison SC (1992) Cell 71:1223
103. Seeman NC, Rosenberg JM, Rich A (1976) Proc Natl Acad Sci USA 73:804
104. Pavletich NP, Pabo CO (1993) Science 261:1701
105. Bürgi H-B, Dunitz JD (1983) Acc Chem Res 16:153
106. Bürgi H-B, Dunitz JD, Shefter E (1973) J Am Chem Soc 95:5065
107. Hajdu J, Machin PA, Campbell JW, Greenhough TJ, Clifton IJ, Zurek S, Gover S, Johnson LN, Elder (1987) Nature 329:178
108. Genick UK, Borgstahl GEO, Ng K, Ren Z, Pradervand C, Burke PM, Srajer V, Teng TY, Schildkamp W, McRee DE, Moffat K, Getzoff ED (1997) Science 275:1471
109. Cox EG, Smith JAS (1954) Nature 173:75

110. Schmidt GMJ (1964) J Chem Soc, p 2014
111. Rosenfield RE Jr, Parthasarathy R (1974) J Amer Chem Soc 96:1925
112. Singh JD, Singh HB, Das K, Sinha UC (1995) J Chem Res, p 312
113. Shimoni L, Carrell HL, Glusker JP, Coombs MM (1994) J Am Chem Soc 116:8162
114. Zacharias DE (1993) Acta Cryst C49:1082
115. Landro JA, Gerlt JA, Kozarich JW, Koo CW, Shah VJ, Kenyon GL, Neidhart DJ, Fujita S, Petsko GA (1994) Biochem 33:635
116. Marsh RE (1958) Acta Cryst 11:654

Supramolecular Synthons and Pattern Recognition

Ashwini Nangia · Gautam R. Desiraju

School of Chemistry, University of Hyderabad, Hyderabad 500 046, India.
E-mail: grdch@uohyd.ernet.in

The aims of crystal engineering are the understanding of intermolecular interactions and their application in the design of crystal structures with specific architectures and properties. In general, all types of crystal structures may be considered but this article is limited to organic molecular solids. Because of the molecular basis of organic chemistry, the obvious question arises as to whether there are simple connections between the structures of molecules and the crystals that they form. Answers to such questions may be found through a better and more comprehensive understanding of the interactions that control crystal packing. These interactions include strong and weak hydrogen bonds. Patterns of interactions, such as would be useful in a predictive sense, can be obtained by manual inspection or more rigorously with the use of crystallographic databases. Such patterns are termed supramolecular synthons and they depict the various ways in which complementary portions of molecules approach one another. The identification of synthons is then a key step in the design and analysis of crystal structures. Such ideas are also important in the understanding of phenomena such as biological recognition and drug-enzyme binding. Pattern identification also leads to the possibility of comparison of crystal structures. The use of the supramolecular synthon concept facilitates such efforts and in this regard it may be mentioned that synthons combine topological characteristics with chemical information, thereby offering a simplification that is optimal to drawing such comparisons.

Keywords: Hydrogen bond, Supramolecular synthesis, Molecular recognition, Structure comparison, Structure prediction.

1 Introduction 58

2 Crystals as Supermolecules 59

3 Supramolecular Synthons and Networks 60

4 From Organic Synthesis to Crystal Engineering 63

4.1 Connection Between Molecular and Crystal Structure 65
4.2 Goals and Challenges 70

5 The Cambridge Structural Database (CSD) 71

6 Multipoint Recognition Patterns 73

7 **Supramolecular Interactions in Biological Systems** 78

7.1 Weak Hydrogen Bonds 78
7.2 Pattern Recognition in Drug-Enzyme Binding 79
7.3 Pharmacophore Model 81

8 **Polymorphism** 83

9 **Comparison of Crystal Structures** 87

10 **"Absence" of Patterns** 90

11 **Conclusions** 92

12 **References** 92

1 Introduction

Supramolecular chemistry is the chemistry of the intermolecular bond based on the underlying theme of mutual recognition. Molecules recognise each other through a complex combination of geometrical and chemical factors and the complementary relationship between interacting molecules is characteristic of the recognition process. Supramolecular chemistry has grown around Lehn's analogy that "supermolecules are to molecules and the intermolecular bond what molecules are to atoms and the covalent bond" [1]. This field has grown into two distinct branches – the study of supermolecules in solution and the study of crystal, that is solid state, structures. The concepts and principles of recognition and the nature of the interactions that mediate supramolecular construction are nearly the same in solution and in the solid state. However, distinctions in ethos in the early stages of development of these two branches has led to studies of solution supermolecules being referred to as molecular recognition and those in the solid state as crystal engineering [2]. Molecular recognition has been studied by physical and synthetic chemists interested in understanding biological processes, biosynthetic mimics and enzyme catalysis. Crystal engineering has been developed by structural chemists and crystallographers to better understand non-covalent interactions for the design of novel materials and solid state reactions. Crystal engineering and molecular recognition are the supramolecular equivalents of traditional organic synthesis, so that instead of building molecules with atoms and covalent bonds, one builds supermolecules with molecules and non-covalent interactions recognising certain repeating patterns of interactions, or supramolecular synthons, that are the key elements of the synthetic exercise. In this article, we attempt to show the importance of pattern recognition in organic crystal chemistry and also the implications of such ideas in related areas.

2
Crystals as Supermolecules

The crystal structure is an ideal paradigm of a supermolecule, a supermolecule par excellence, according to Dunitz [3]. The organic crystal is an example of a nearly perfect periodic self-assemblage of millions of molecules, held together by medium- and long-range non-covalent interactions, to produce matter of macroscopic dimensions. Crystals are ordered supermolecular systems at an amazing level of precision. The high degree of order in a crystal structure is the result of complementary dispositions of shape features and functional groups in the interacting near-neighbour molecules. From the early work of Kitaigorodskii on crystal packing, ideas of shape-induced recognition between molecules became firmly established [4]. Accordingly, even for recognition between identical molecules, as is the case in most crystal structures, it is the dissimilar parts that come into close contact and not the similar surfaces – bumps fit into hollows just as a key fits into a lock. Conversely, identical parts of neighbouring molecules tend to avoid one another, and space groups containing only rotation axes and mirror planes are found much less frequently when compared to those containing inversion centres, screw axes and glide planes. Centrosymmetric close packing is preferred even for those molecules that do not possess an inversion centre, and the four space groups $P\bar{1}$, $P2_1/c$, $C2/c$ and $Pbca$ account for 56% of all organic crystal structures. These close-packing arguments, based on the complementary recognition between molecules, are of an all-pervasive character and packing coefficients in most single-component organic crystal structures lie in the range 0.65–0.77.

The space group preferences of many heteroatom crystals parallel those derived on the basis of the Kitaigorodskii model because the directional requirements of several common heteroatom contacts such as O-H···O, N-H···O, Cl···Cl and S···X (X = halogen) are in accord with the geometrical dictates of the same three symmetry elements that govern close packing: the inversion centre, the screw axis and the glide plane. This is generally not so well-appreciated [5]. For instance, carboxylic acids hydrogen bond across centres of inversion, phenols around 2_1 screw axes and the “L-shaped” geometry of Cl···Cl interactions are optimised when the contacting molecules are related by glide planes or screw axes. These geometrical preferences of the common hetero-atom interactions reinforce the close-packing tendencies with the result that there is a dominance of a very small number of space groups that contain translational symmetry elements and this too, distributed among the low symmetry crystal systems.

The crystal structure of any molecule is the free energy minimum resulting from the overall optimisation of the attractive and repulsive intermolecular interactions which have varying strengths, directional preferences and distance-dependence properties. Therefore understanding the nature, strength and directionalities of intermolecular interactions is of fundamental importance in the design of solid state supermolecules. Intermolecular interactions in organic compounds are of two types: isotropic, medium-range forces that define the shape, size and close packing; and anisotropic long-range forces which are electrostatic and include hydrogen bonds and heteroatom interactions (see also the article by J. P. Glusker in this volume). The observed three-dimensional architecture in the crystal is the result then of the interplay between the demands of isotropic van der Waals forces whose magnitude is proportional to the size of the molecule, and anisotropic hydrogen bond interactions whose strengths are related to donor atom acidities and acceptor group basicities. The crystal structures of hydrocarbon molecules are largely dictated by close packing arguments while the structures of molecules containing heteroatoms and functional groups are dominated by hydrogen bonds and anisotropic interactions [2, 6].

3 Supramolecular Synthons and Networks

The synthesis of complex natural products and aesthetically-pleasing molecules has been practised by organic chemists for decades. It was in 1967 that Corey introduced a formalism in organic synthesis to organise the sequence of steps, and consequently to focus the chemical thought process, from starting material to the target substance [7]. Corey defined *synthons* as "structural units within molecules which can be formed and/or assembled by known or conceivable synthetic operations" and the term has been used since its inception to represent key structural units in target molecules. A synthon is usually smaller and less complex than the target molecule and yet contains most of the vital bond connectivity and stereochemical information required to synthesise the goal substance. The dissection of a complex target molecule into simpler synthons is performed through a series of logical and rational bond disconnections and this exercise is termed *retrosynthetic* analysis [8].

By analogy with synthons, *supramolecular synthons* are structural units within supermolecules which can be formed and/or assembled by known or conceivable intermolecular interactions, and crystal engineering is the solid state supramolecular equivalent of organic synthesis [9]. Just as in traditional organic synthesis, where the retrosynthetic bond disconnections must be carried out in accordance with the reactivity of functional groups and their stereochemical preferences [10], so is the case in *supramolecular retrosynthesis* wherein the complex interplay of close packing, hydrogen bonding and other intermolecular interactions during crystallisation must be analysed and exploited [11]. Supramolecular synthons are spatial arrangements of intermolecular interactions and play the same focussing role in supramolecular synthesis that conventional synthons do in molecular synthesis. Scheme 1 shows a few important supramolecular synthons **1**–**13**.

The advantage of using the synthon approach is that it offers a considerable simplification in the understanding of crystal structures. For example, the same carboxy dimer motif **1** is present in the structures of benzoic acid, terephthalic acid, isophthalic acid, trimesic acid and adamantane-1,3,5,7-tetracarboxylic acid in zero-, one-, two- and three-dimensional arrangements [9] (see also the article by R. E. Meléndez and A. D. Hamilton in this volume).

The emphasis in crystal engineering may be increasingly diverted from the constituent molecules to the topological features and geometrical connectivities

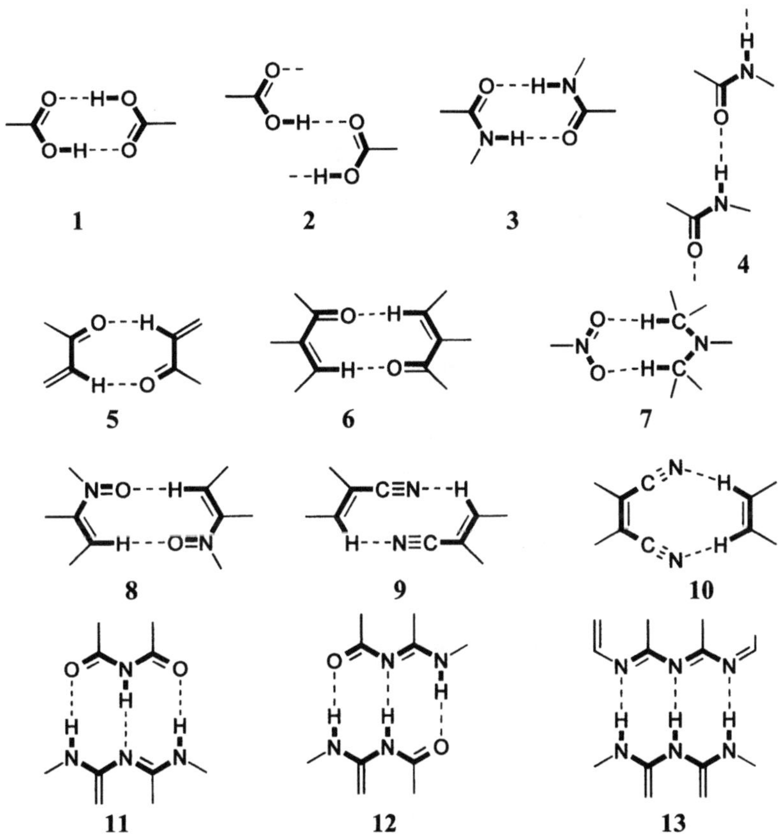

Scheme 1. Some supramolecular synthons selected from the recent literature

of non-bonded interactions between molecules. Networks constituted with node and node connections may be thus defined. It is profitable to describe targets in crystal engineering in terms of such networks [12]. Retrosynthetic analysis may be performed accordingly on the network structure to yield the node structure (molecules) and the node connectivity (supramolecular synthons). The advantages of such an approach in crystal engineering are that: (1) supermolecule ⇒ molecule connections are easily established; (2) comparisons between seemingly different crystal structures are facilitated; (3) the interference between supramolecular synthons can be strategically minimised; and (4) more than one combination of molecular and supramolecular synthons are seen to lead to similar crystal structures.

To illustrate this last point, the structures of *tetrakis*(4-bromophenyl)methane (**15**) and the 1:1 molecular complex 20 of CBr_4 and tetraphenylmethane **14** (Scheme 2) are pertinent [13]. In the crystal structure of **15**, four Br atoms are arranged in a tetrahedral fashion and form the Br_4-synthon **18** (Br···Br 3.91 Å). If the empty centroid in synthon **18** is considered to be a phantom "carbon"

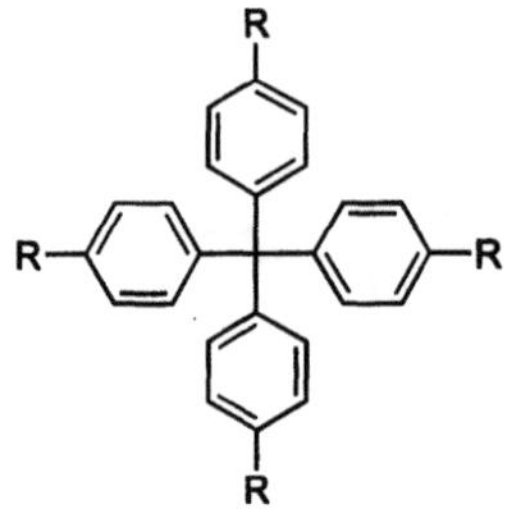

14: R = H
15: R = Br
16: R = I

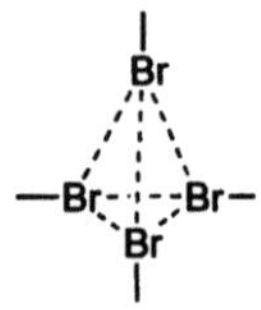

18

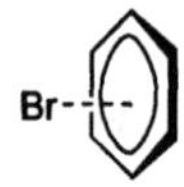

19

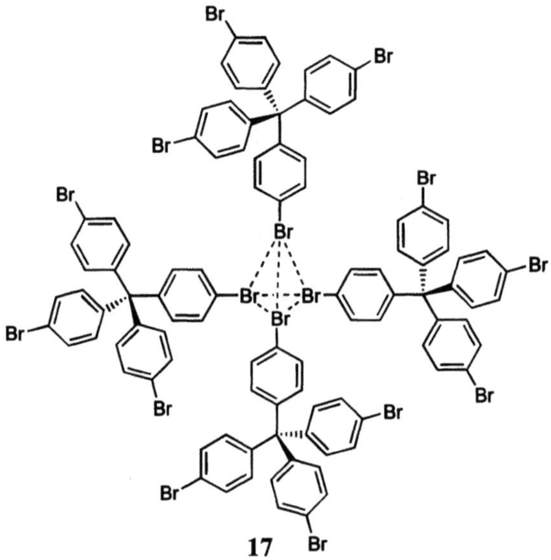

17

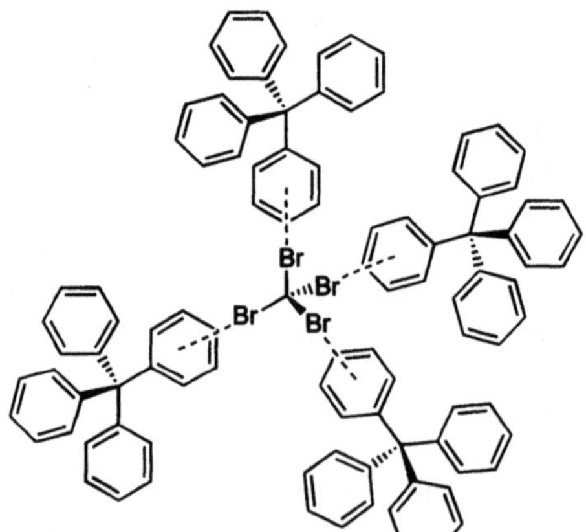

20

Scheme 2. Isostructural supramolecular architectures of single-component **17** and two-component **20** type atom, the cluster approximates to a super-CBr_4 molecule. In retrosynthetic terminology, cluster **17** is easily understood then as the combination of two tetrahedral moities, the tetraphenylmethane molecular synthon **15** and the Br_4 supramolecular synthon **18.** Now, if one interchanges the molecular and supramolecular synthons by replacing the Br_4-cluster **18** with covalently bonded CBr_4 and the C-Br covalent bonds with the Br···Ph interaction **19,** the result is complex **20.** In **20,** four molecules of **14** are linked to a CBr_4 molecule through synthon **19** (Br···Ph centroid 3.67 Å) and the central carbon atoms of CBr_4 and tetraphenylmethane may be treated as spheres and joined to form a distorted diamondoid network. The crystal structures **17** and **20,** derived from single- and two-components respectively, bear close and striking similarities at the supramolecular level and have similar diamondoid networks mediated by Br_4 and Br···Ph supramolecular synthons.

To continue, the crystal structure of *tetrakis*-(4-iodophenyl)methane (**16**) is isomorphous with 17 having an I···I distance of 3.95 and 4.16 Å in the corresponding I_4-synthon. In this family of halo-substituted tetraphenylmethanes with S_4- or pseudo-S_4-symmetry, the molecules exhibit similar columnar packing in the solid state [14]. In summary, if a supramolecular structure is defined as a network and analysed retrosynthetically, general connectivity strategies can be systematically derived.

4
From Organic Synthesis to Crystal Engineering

A smooth and logical progression from molecular synthesis to crystal engineering follows if functional groups are replaced with supramolecular synthons and target molecules with networks. The functional group approach for defining covalent bond connections is suited to the goals of molecular synthesis where carbon-carbon bonds are formed between atoms with unlike polarisations, as between nucleophiles (enolates, carbanions) and electrophiles (ketones, halides) [15, 16]. In traditional organic synthesis, the molecule is assembled in a stepwise and sequential manner such that carbon atoms in two functional groups react to form a covalent bond while the others are deliberately made to stand as innocuous spectators. In cases of potential competition between two or more functional groups, protection of the more reactive group is necessary. The development of mild and selective reagents and reaction conditions has been motivated by the need to minimise the number of steps and improve the stereoselectivity in covalent synthesis.

This approach has served synthetic organic chemistry for close to five decades and has led to the total synthesis of a large number of molecules [17]. Complex natural products such as quinine, cortisone, strychnine, cedrol and reserpine were the landmark syntheses of the 1940s and 1950s. The period from 1960 to 1990 was one of intensive research encouraged by the advent of modern spectroscopic techniques and the ready availability of a large storehouse of starting chemicals. A number of natural [10,16–19] and non-natural [20, 21] target molecules were successfully tackled during this period. The total synthesis of vitamin B_{12} completed by Woodward [22] and Eschenmoser and Winter

[23] in 1973 and 1977 represented the most complex natural product synthesised at that time. The project spanned two continents and occupied an army of researchers during a period of 12 years. The second "chemical Everest" of molecular synthesis was scaled in 1989 with the total synthesis of palytoxin **21** (Scheme 3) by the team of Kishi [24]. The total synthesis of this complex macromolecule took almost 10 years and involved more than 100 researchers. The rapid pace of research in the 1990s [25] was brought to the forefront with the publication in 1994 of two total syntheses of taxol from the laboratories of Holton [26] and Nicolaou [27] within a week of each other.

The essential difference between organic synthesis and crystal engineering is that the stepwise and sequential covalent bond formation in the former is replaced by an organised self-assembly of the supermolecule involving orthogonal functionalities in a single step in the latter [28]. During crystallisation, all functional groups present in the molecule compete for the numerous possible combinations of intermolecular interactions even while it is understood that only some of these recognition events are eventually fruitful. Thus, if a molecule M containing functional groups $F_1, F_2, F_3, \cdots F_n$ approaches another molecule of M, then a matrix of intermolecular interactions, $F_i - F_j$ is theoretically possible. Two or more molecules of M may now come together to form, in principle, several supramolecular synthons $S_1, S_2, S_3, \cdots S_n$, some of which may be very close in energy. However, there is a simplifying feature here. Some combinations of $F_i - F_j$ inherent in S_1, S_2, S_3, $\cdots$ may exclude others with the result that the complex matrix of intermolecular interactions and supramolecular synthons converges rapidly to a free energy minimum (the crystal structure) without

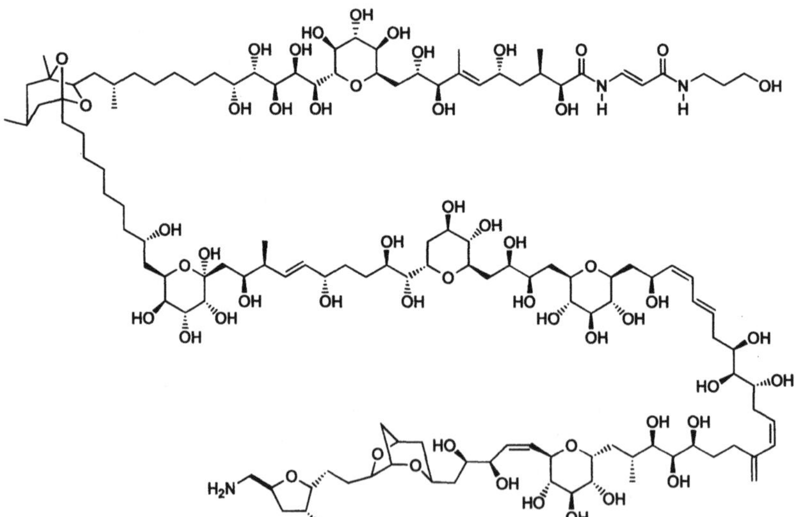

21

Scheme 3. Chemical structure of palytoxin. This giant molecule has 65 dissymmetric carbons and 10^{21} possible isomers, weighs 2680 daltons and has dimensions of the order of 10–100 Å really sampling all the recognition patterns. The absence of rampant polymorphism in molecular crystals suggests that crystallisation is an inherently very efficient process that cascades into stable crystal structures.

Crystals are supermolecules and their structural features are best described in terms of motifs or patterns of interactions, that is supramolecular synthons. A functional group is a molecular descriptor and does not contain any information about its hydrogen bond forming ability that is critical to the design and construction of supermolecules. For example, three of the most common families of hydrogen bonded compounds – carboxylic acids, primary amides and alcohols – each have more than one hydrogen bond recognition motif [29]. The difference between the dimer and catemer motifs **1** and **2** is obvious in acids but in amides, the chain motifs related by translation and screw or glide appear to be similar, and yet lead to distinct packing arrangements. Thus, there is a one to many correspondence between functional groups and synthons. The smallest structural unit that contains crystallisation information is the supramolecular synthon, not the functional group.

It should be emphasised that the identification of the set of interactions that constitutes a synthon is a subjective matter and is at the discretion of the chemist. Within the limit that every possible combination of interactions is defined as a supramolecular synthon, the term will fall into disuse much like its molecular sibling [8, 15]. Interactions or groups of interactions that are needlessly identified as synthons but are unable subsequently to demonstrate and sustain a predictive role in crystal structure design will drop out from practical usage. A convenient working definition of a useful supramolecular synthon is thus a unit consisting of hydrogen bonds and/or non-polar intermolecular interactions that has maximum structural information while maintaining an economy in size.

The terms "maximum" and "economy" are significant. As synthon size is increased up to the entire unit cell, so does its information content. Conversely, as synthon size decreases down to just a single interaction, its information content decreases. However, there is a critical intermediate region wherein all the crucial structural information is present within units of manageable size. It is this region that is of the greatest importance to the structural chemist. Implied in all this of course is synthon robustness, that is the ability of a synthon to reappear in related structures, in effect controlling self assembly within a crystal family. This then is another feature that distinguishes synthons from mere interactions.

In summary, rather than attempt an unambiguous and rigorous definition of the term *supramolecular synthon*, we feel it best that the scope and meaning of the term be allowed to refine during the evolution of crystal engineering into a mature subject.

4.1 Connection Between Molecular and Crystal Structure

In the context of crystal engineering strategies a frequently asked question [30] is "given the molecular structure of an organic substance, what is its crystal structure?" Because there is a one-to-many correspondence between functional groups and supramolecular synthons, this question may be answered in several ways: (1) not possible to predict; (2) difficult to answer; (3) has multiple answers. The first response is perhaps the technically most correct given the state-of-the-art today, but along with its certitude is associated a rather static and pessimistic frame of mind that is not conducive to the improvement of our undertanding of these matters. We choose therefore to explore the second and third options as answers.

The earliest attempt to correlate molecular structures with crystal structures was by Robertson who discussed the packing of planar fused-ring aromatic hydrocarbons in the 1950s [31]. He suggested that aromatic disk-like molecules with an area large in comparison with their thickness (coronene, ovalene) tend to stack in columns while those with a smaller area (naphthalene, anthracene) tend to be steeply inclined. This issue was re-examined by Desiraju and Gavezzotti in 1989 given the enhanced perception of intermolecular forces and a knowledge of the ubiquitous herringbone and stacking motifs in crystal structures of aromatic compounds [32]. These authors divided polynuclear aromatic hydrocarbons into four categories. Small arenes (benzene, naphthalene, anthracene) have herringbone or inclined "T"-geometry structures, while larger arenes pack as sandwich herringbone (pyrene, perylene), γ-structures (Robertson's coronene group renamed) and β-structures (tribenzopyrene).

Based on the above classification along with the carbon to hydrogen atom ratio and the surface area of the molecule, a prediction of the aromatic hydrocarbon crystal structures was shown to be possible. However, one must be aware that such predictions are accurate because the forces in hydrocarbon crystals are only of the isotropic variety. Nonetheless, different packing modes do exist even for aromatic hydrocarbons, and polymorphs of molecules as distinct as toluene, perylene, quaterphenyl and dibenzanthracene have been observed.

Most organic compounds contain heteroatoms and the intermolecular interactions in their crystal are a complex mosaic of forces of varying strengths, directionalities and distance dependence characteristics [2, 5, 6]. In such situations, direct and simple extrapolations from molecular to crystal structure are tricky and difficult in the best cases and impossible in the worst. We present here two examples of molecule $\rightarrow$ crystal extrapolation to convey the mixed feelings associated with such correlations. Both examples are based on compounds containing hydroxy and amino groups. Ermer and Eling [33] and Hanessian et al. [34] pointed out independently that molecular recognition between these groups is robust and predictable because of the complementary 1:2 and 2:1 hydrogen bond donor:acceptor ratios present. The coordination environment about these N- and O-atoms is accordingly expected to be tetrahedral.

Hanessian and co-workers reported the self-assembly of supramolecular hydrogen bonded structures of complexes of C_2 symmetrical chiral 1,2-diols and 1,2-diamines [34]. In the 1:1 complex **25** of (1*S*,2*S*)-1,2-diaminocyclohexane **22** and (1*S*,2*S*)-1,2-cyclohexanediol **23** (Scheme 4), the cyclohexane rings align into four vertical columns and the polar hydrogen bonding groups face inward. The structure is a pleated sheet-like array of eight-membered, square planar, hydrogen bonded units in which the oxygen and nitrogen atoms are tetracoor-

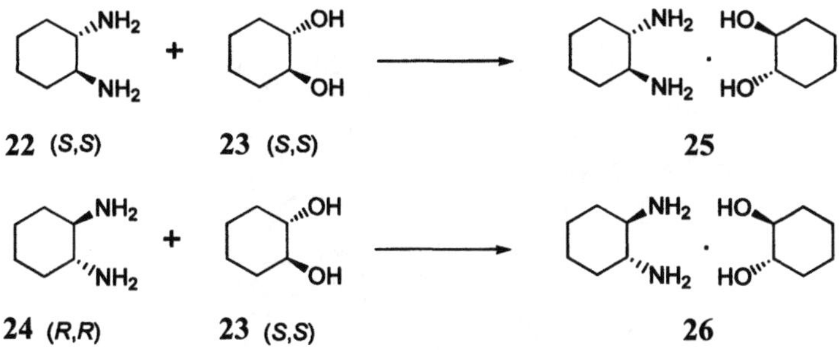

Scheme 4

dinated (Fig. 1a). The remaining hydroxy and amino functional groups are engaged in two symmetrical side rows of tricoordinated zigzag hydrogen bond patterns which flank opposite sides of the central octagonal staircase core. The crystal structure of the 1:1 complex **26** of enantiomerically pure (1*R*,2*R*)-1,2-diaminocyclohexane **24** and (1*S*,2*S*)-diol **23** is pleasingly predictable and similar to that of **25**. It shows virtually identical pleated cores and side rows of hydrogen bonds between alternating oxygen and nitrogen atoms (Fig. 1b). The relative sense of H-donation in the core is opposed to that in the side rows in both **25** and

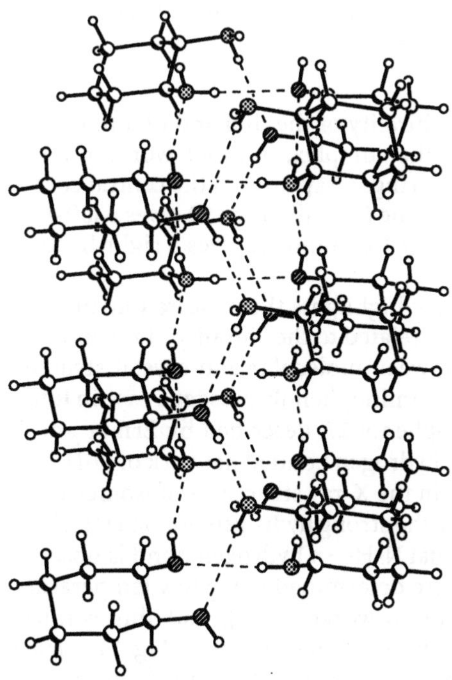

Fig. 1a. View of the hydrogen bonding network in 1,2-diaminocyclohexane·1,2-cyclohexanediol (*SS*,*SS*)-complex **25**

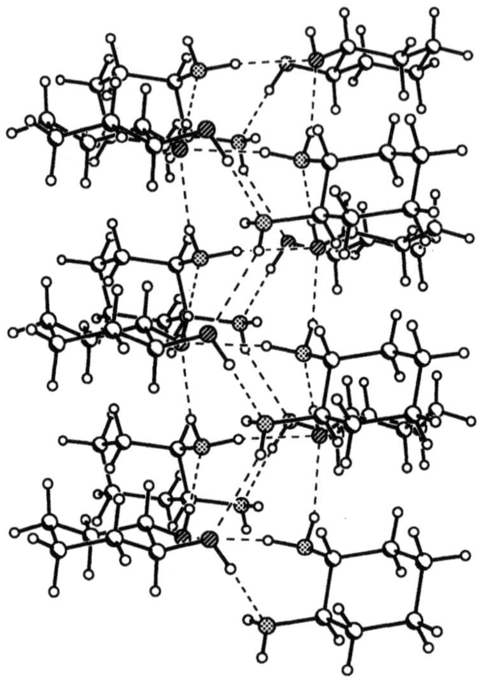

Fig. 1 b. (*RR,SS*)-complex **26**. (Figures adapted from Hanessian et al. [34]). Note the slight differences in the two figures

26. Interestingly, the chirality of the diol and diamine components controls the tertiary structure of the complex, in other words the sense of chirality with which the cyclohexane rings wrap around the central core in the triple-stranded helicate. Thus, whereas the (*S*,*S*)-diamine **22** gives a left-handed helicate **25** with the (*S*,*S*)-diol **23**, the (*R*,*R*)-diamine **24** gives a right-handed helicate **26** with the same diol.

If the above example highlights the delicate way in which the handedness of the crystal structure is related to the chirality of the molecular components, and how relationships between molecular and crystal structure can be traced, the structures of the three aminophenols present a case in total contrast. The crystal structure of 4-aminophenol **27**, described by Ermer and Eling [33], shows the expected tetrahedral hydrogen bonded network of O-H···N and N-H···O hydrogen bonds. However, in the X-ray structures of isomeric 2- and 3-aminophenols **28** and **29** (Scheme 5), the strong hydrogen bonded tetrahedral network is absent and instead, an unusual N-H···π hydrogen bond is observed. The crystal structures of **28** and **29** were determined using low temperature neutron diffraction data by Desiraju and co-workers [35], and this confirmed the presence of N-H···π hydrogen bond in both structures (Fig. 2). Surprisingly, the "normal" hydroxy-amino recognition seen in **27** is not found in **28** and **29**. In 2-aminophenol **28**, each -OH group donates an O-H···N hydrogen bond and accepts N-H···O and C-H···O hydrogen bonds with three different molecules of **28**. Each

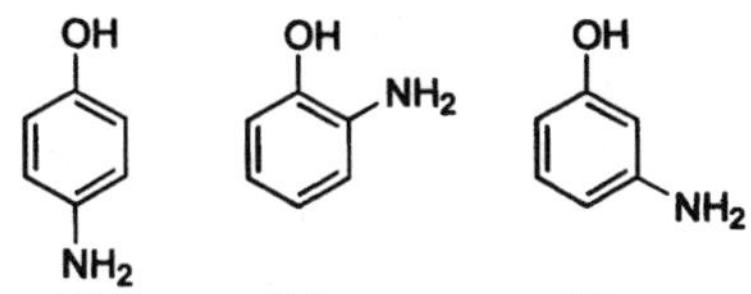

Scheme 5 27 28 29

a

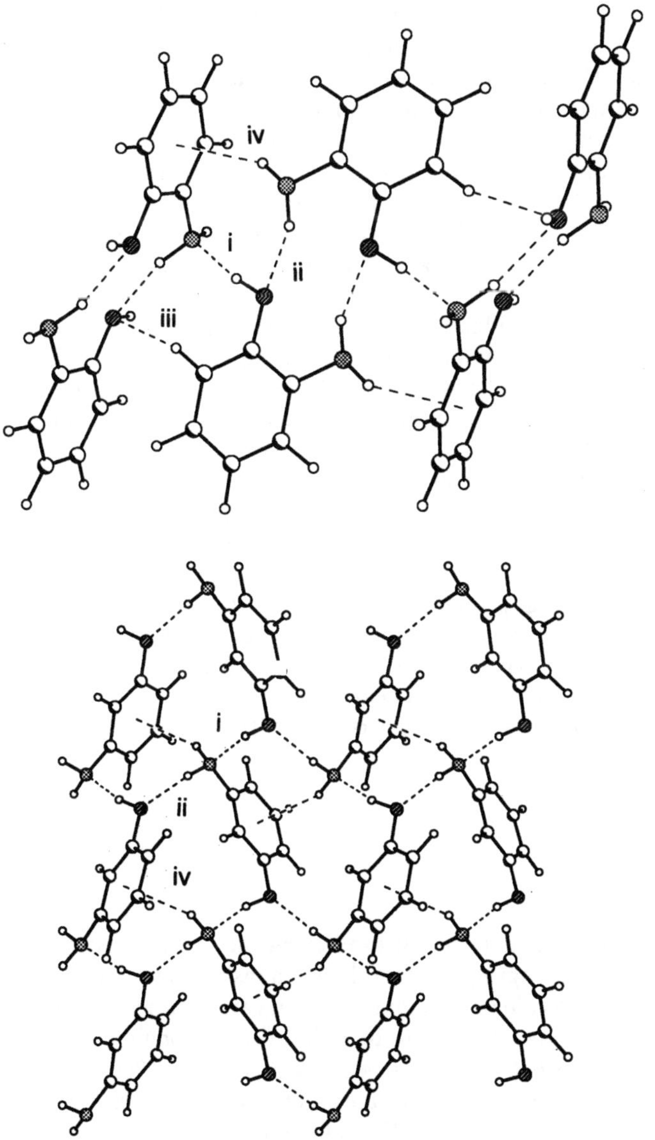

b

Fig. 2a,b. O-H···N (i), N-H···O (ii), C-H···O (iii) and N-H···π (iv) hydrogen bonds in the crystal structures of: **a** 2-aminophenol **28**; **b** 3-aminophenol **29** [35]

-NH_2 group similarly donates and accepts a strong hydrogen bond with two different molecules of **28**. The second H-atom participates in the N-H···π hydrogen bond with a third molecule of **28**. The hydrogen bonding array in 3-aminophenol **29** is similar to **28**. A tetrahedral environment around O- and N-atoms is maintained in both structures, but unlike in **27** it is not exclusively of the strong type.

The reason for the "anomalous" structures of 2- and 3-aminophenols compared with the "normal" structures of 4-aminophenol is understood in terms of the need to attain herringbone or "T"-shaped geometry of the phenyl rings in the two structures. The origin of the weaker N-H···π and C-H···O hydrogen bonds is then the herringbone geometry of the phenyl rings which makes the heteroatoms inaccessible for the formation of conventional N-H···O hydrogen bonds. The preference for two weak instead of one strong hydrogen bond is caused by the optimisation of the herringbone interactions and this is identified as the primary structural effect in these compounds.

In summary, simple molecules can have complex crystal structures and the relationship between molecular and crystal structures can accordingly be difficult to delineate.

The above examples illustrate that connections between molecular and crystal structures are straightforward or difficult depending on whether there is insulation or interference between the different sets of significant intermolecular interactions. Thus, in 4-aminophenol, the crystal structure retains a high degree of fidelity to other hydroxy-amino systems described by Ermer because the competing herringbone and hydrogen bonding interactions are effectively insulated. In 2- and 3-aminophenols, however, the structure determining interactions interfere with one another and this leads to unexpected crystal structures. Hanessian's compounds are somewhat less problematic because they do not contain aromatic groups. Unfortunately, it is often difficult to anticipate from a casual inspection of the molecular structure when such insulation will or will not be present. This then is a major problem in crystal engineering; indeed it is *the* problem, and at its heart lies our inability to predict accurately the supramolecular (crystal) structure of an arbitrary molecule from its molecular structure.

4.2 Goals and Challenges

The "supramolecular Everest" of predicting the crystal structure of any given molecule is as challenging and formidable a task today as the total syntheses of vitamin B_{12}, gingkolide or palytoxin were in earlier times (Table 1). The subject of supramolecular synthesis is in its infancy and probably at the same stage of development as target oriented synthesis was in the 1960s. A set of empirical chemical principles for the solid state assembly of molecules has been formulated but its utility in the construction of complex architectures is yet to be completely demonstrated. This is one of the major goals of current crystal engineering strategies.

Table 1. Comparison of molecular and solid state supramolecular synthesis

Organic synthesis	Crystal engineering
1. Molecular targets	Targets are networks
2. Connectivity is defined in terms of covalent bonds	Hydrogen bonds, chemical and van der Waals interactions define supramolecular connectivity
3. Structure descriptors are functional groups	Structure descriptors are supramolecular synthons
4. Reactant → Transition state → Product	Molecule → Crystal nucleus → Crystal
5. Build molecule in separate and discrete steps	Crystallisation occurs in a single step, intermediates are not isolable
6. Isomers (geometrical, stereo-) are common in organic molecules	Polymorphism is uncommon but still a nuisance in particular systems
7. Goals: structure elucidation, new reactions and reagents, stereo- and enantiocontrol, total synthesis of complex molecules	Goals: Understanding of intermolecular interactions and crystal packing, connections between molecular and crystal structure
8. Motivations: aesthetics, utilitarian (pharmaceuticals, agrochemicals, fine chemicals)	Motivations: aesthetics, utilitarian (nanostructures, materials and functional solids, NLO, molecular electronics)
9. Chemical Everest: Vitamin B_{12}, Gingkolide, Palytoxin, Taxol	Supramolecular Everest: Anticipate the crystal structure(s) of a molecule

5 The Cambridge Structural Database (CSD)

Supramolecular synthons or patterns in crystals are the key structural units that help in focussing supramolecular synthetic effort and also in facilitating comparisons between crystal structures. It therefore becomes important to review methods that are available for their identification and analysis. Synthons are built from hydrogen bonds and non-covalent interactions and contain information about the directional preferences of intermolecular interactions and the topology of attached groups. For an effective and predictable utilisation of synthons in crystal structure design, a better understanding of the constituent hydrogen bonds and intermolecular interactions is essential. X-ray crystallography provides precise and unambiguous information on these interactions and this information is contained in the Cambridge Structural Database (CSD) (see also the article by J. P. Glusker in this volume).

The CSD (April 1997, version 6.2, 167,797 entries) contains accurate crystal structure data for organic, inorganic and organometallic compounds [36]. The retrieval of geometrical data from the CSD and its subsequent analysis by statistical procedures allows for extremely reliable conclusions pertaining to the nature of intermolecular interactions. Only conclusions based on examination of a statistically significant number of crystal structures are chemically meaningful. This is because: (1) the crystal structure of any molecule must have a few intermolecular contacts in the repulsive van der Waals region – a few such forced contacts can be comfortably accommodated in any crystal structure provided they are compensated for by the free energy gained from the other numerous dispersive forces, hydrogen bonds and heteroatom interactions; (2) the weak, polarisable hydrogen bonds (C-H···O, C-H···N, C-H···X, O-H···π) and other heteroatom interactions (X···X, N···X, S···X) have approach distances and angle characteristics scattered over a wider range – they are also more susceptible to the mutually interfering effects in a crystal, effects that can deform considerably the approach vector from the ideal geometry.

In the statistical approach, such as is possible with the CSD, a large number of observed interaction geometries cluster in a narrow region. A few outliers are deformed and may be repulsive in nature. The larger the number of data points, the greater is the confidence in the intermolecular contact or structural hypothesis under study. As for the outliers, they could furnish an additional bonus in that their occurrence is often indicative of an unusual or different chemical effect.

The utility of the CSD in identifying supramolecular synthons is revealed in a study of N-H···O hydrogen bonding [37]. The patterns of interest are **34** and **35** and the expected precursor urea and imide structures, **30** and **31** (Scheme 6), were retrieved initially from the April 1997 update. Examination of the two populations revealed a significant number of common hits, all of which contain fragment **32**. These were excluded from subsequent analysis. Of the 230 remaining acyclic ureas, there were 44 occurrences of synthon **34** but only one occurrence of synthon **35** in the 32 acyclic imides. It is clear then that **34** occurs more frequently in ureas than does **35** in imides. This raises the question as to whether **34** is inherently more stable than **35**. Both synthons have three-centre hydrogen bonds and while **34** contains a bifurcated acceptor, **35** contains a bifurcated donor. The stabilisation from the urea and imide hydrogen bonded patterns was calculated (AM1) to be 6.9 and 6.1 kcal mol^{-1}. It is unlikely that the more

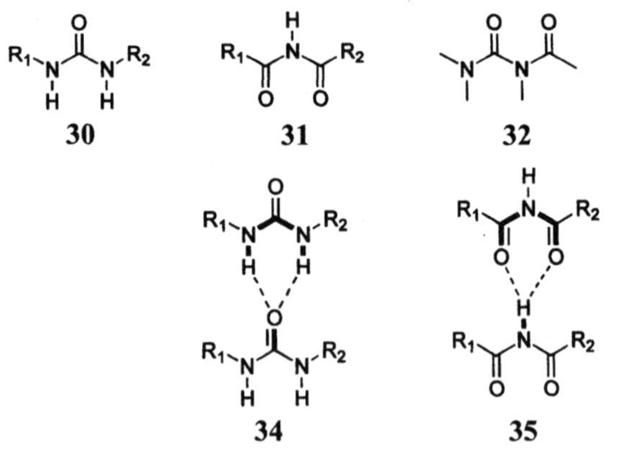

Scheme 6. Hydrogen bonding motifs in ureas 34 and imides 35

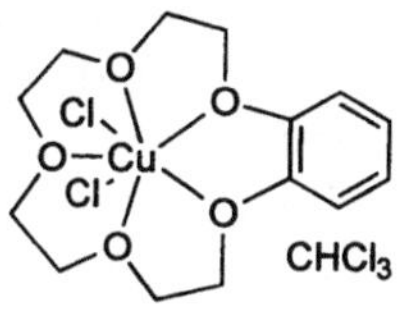

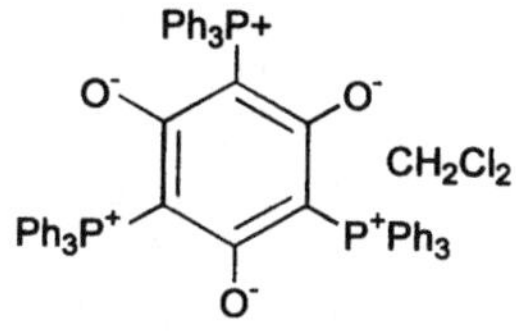

Scheme 7. Outlier refcodes in the CSD study of C-H···O hydrogen bonding in $CHCl_3$ and CH_2Cl_2 solvates

frequent occurrence of the urea motif is a result of this marginally higher stability. A more likely reason is that the synthon **35** is not common in imides such as **31** because it can occur only when both the carbonyl groups are *syn* to each other. Examination of the crystal structures of the retrieved imides shows that a large number of these compounds (such as, say, **33**) exist in the conformation wherein the two carbonyl groups are *anti* to one another. Thus, in ureas the stablest molecular conformation is in consonance with the requirements for the target tape motif, whereas in imides the conformation of the molecule is in conflict with that necessary for the extended, linear hydrogen bonded tape. Another possible reason for the formation of the infinite tape **34** in the ureas and the non-formation of the infinite tape **35** for the imides is the match of donor (2 H atoms) and acceptor (1 C=O group with 2 acceptor sites) groups in the former and mismatch (1 donor, 4 acceptor sites) in the latter.

The examination of outlier structures in a CSD analysis is often instructive. It is well-known that C-H···O bonds formed by $CHCl_3$ are in general shorter than those formed by CH_2Cl_2 because of carbon acidity effects [38]. However, histograms of C-H···O distances of these two populations show some degree of overlap so that some of the shorter contacts in the CH_2Cl_2 sample are shorter than the longer contacts in the $CHCl_3$ sample. This corresponds to the outlier region. For example, a long and bent contact is formed by $CHCl_3$ in the crystal structure of BOLVAZ (Scheme 7). Here, the acceptor oxygen is electron-deficient because it is bonded to a Cu(II) atom and this reduces its propensity to form a strong C-H···O hydrogen bond ($d=2.88$ Å, $\theta=125.3$). Conversely, a short and linear contact is formed by CH_2Cl_2 in the structure of PIDDOV. In this phosphonium salt, the acceptor phenoxide O^--atom is unusually basic and forms a strong hydrogen bond ($d=1.96$ Å, $\theta=162.4$). In both instances, the outlier status of the points is confirmed by examination of the individual structures and this provides additional insight about the chemical nature of the intermolecular contact.

6
Multipoint Recognition Patterns

The presence of two or more complementary hydrogen bond donor and acceptor sites in a recognition event, that is multipoint recognition, confers stability and robustness to the resulting supramolecular synthon. Multipoint recognition between the DNA base pairs with Watson-Crick hydrogen bonding schemes may be represented as DA:AD for adenine-thymine and DDA:AAD for guanine-cytosine (D=hydrogen bond donor, A=hydrogen bond acceptor). The distinct arrays DDA:AAD, DAD:ADA and DDD:AAA in three-point recognition motifs have been investigated separately by Zimmerman and Murray [39] and Mingos et al. [40]. The association constant (K_{assoc}) for the complex **36** of the DAD:ADA type is weak and in the 10^2 M^{-1} range (Scheme 8). The DDA:AAD type complex **37** is more stable with K_{assoc} in the 10^4 M^{-1} range. The DDD:AAA complex **38** was the tightest examined and has $K_{assoc}>10^5$ M^{-1}.

Studies on triply hydrogen bonded complexes are of additional interest. Jorgensen et al. have interpreted the trends in the stabilities of these bonded arrays as arising from the different arrangement of D and A sites [41]. Since the primary hydrogen bonds are similar in the three systems, differences in association energies are held to arise from differences in secondary electrostatic interactions. In the DDA:AAD complex there are two attractive and two repulsive secondary interactions, in the DAD:ADA complex all four secondary interactions are repulsive, while in DDD:AAA all four secondary interactions are attractive. Thus, for similar hydrogen bonds between D and A, the DDD:AAA complex is expected to be the strongest, the DDA:AAD complex of intermediate stability, and the DAD:ADA complex the weakest. However, cooperativity of π-effects predicts the opposite trend [39, 40]. The delocalised π-system is able to enhance the hydrogen bonding interaction in DDA:AAD and in DAD:ADA but not in the DDD:AAA system.

In addition to the function of ordered hydrogen bonded arrays in the biological cell [42], the design of new materials is closely connected with the organised self-assembly of supramolecular structures [43]. The physical and chemical properties of molecular aggregates depend on the nature of the constituent molecules as well as on the manner in which the molecules assemble in the solid state. This is because the properties and architectural features of the supramole-

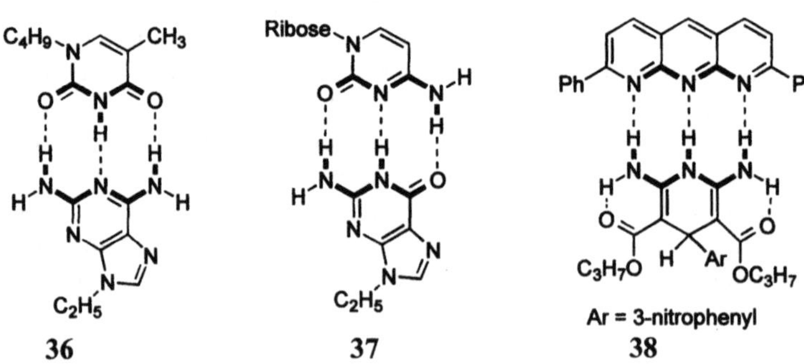

36 $K_a = 140\ M^{-1}$

37 $K_a = 1X10^4\ M^{-1}$

38 $K_a > 10^5\ M^{-1}$

Scheme 8. Donor-acceptor complexes with multi-point recognition motifs DAD:ADA **36**, DDA:AAD **37** and DDD:AAA **38** (from Zimmerman and Murray [39]) cular structures result from the information stored in the components and the interactions between them. The supramolecular aggregates generated by self-assembly principles are good examples of nanostructures. The goal of nanochemistry is to evolve design strategies to express a desired property or function in a material of a given structure [43, 44].

In this context, the zigzag array of DAD:ADA hydrogen bonds in the solid state was examined by Lehn and co-workers in the 1:1 complex **39** of 3-butyl-2,4,6-triaminopyrimidine and 5,5-diethylbarbituric acid [45]. Each component forms six hydrogen bonds with its two complementary neighbours as illustrated in Scheme 9. The consequence of such a pattern of hydrogen bonds is that all pyrimidines are located on one side of the strand and all the barbituric acids on the other. Such an ordered and oriented arrangement of molecules in the crystal offers the opportunity for inducing specific NLO, electronic, ionic and magnetic properties (For more examples of H-bonded tapes and similar motifs, see the article by R. E. Meléndez and A. D. Hamilton in this volume).

The triply hydrogen bonded motifs discussed in the context of biomacromolecules and materials have been constructed purely with N-H···O and N-H···N hydrogen bonds. Notwithstanding the pivotal role of strong hydrogen bonds in biological systems and supramolecular structures, the contribution of weak hydrogen bonds and interactions should not be underestimated [46–48]. This is because weak interactions have more gradual distance fall-off characteristics and are operative at much longer intermolecular distances. Early recognition events between molecules could be influenced by the soft and polarisable weak interactions at long distances. Although much weaker (<3 kcal mol^{-1}) than the conventional hydrogen bonds (5–10 kcal mol^{-1}), these interactions are able to steer the approach orientation of molecules and the formation of supramolecular synthons, and thereby exert an influence on observed crystal structures. Therefore, a consideration of the weak hydrogen bonds with the strong ones gives a more faithful description of the actual crystallisation events. We therefore examined the triply hydrogen bonded DAD:ADA array constructed solely with weak hydrogen bonds in the complexes of dibenzylidene ketones and trinitrobenzenes with the following specific objectives: (1) to extend the strong hydrogen bonded motif **11** to the weak hydrogen bond category; (2) to assess the robustness of the weakly hydrogen bonded array **49**; (3) to compare hydrogen

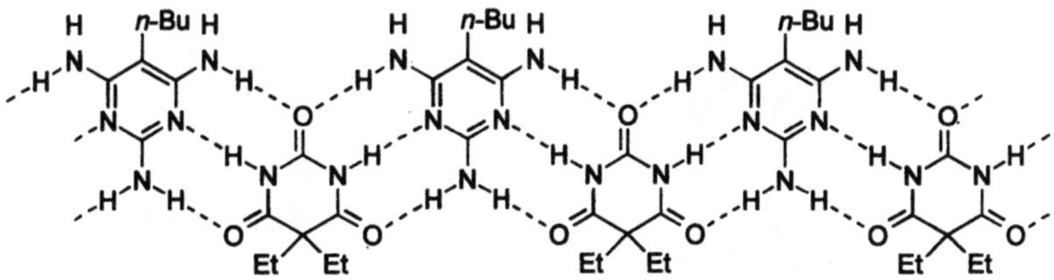

39

Scheme 9. 1:1 Complex of the barbituric acid-triaminopyrimidine type. Note that like residues, Et and *n*-Bu are located on either side of the hydrogen bonded tape bonds in the three-point array with isolated hydrogen bonds; and lastly (4) to demonstrate the construction of supramolecular architectures using the synthon concept.

The crystal structure of the 1:2 complex **46** of dibenzylidene ketone **40** and trinitrobenzene **43** showed the expected DAD:ADA motif **49** constituted with C-H···O hydrogen bonds (Scheme 10) [49]. The overall structure is fortified by additional C-H···O hydrogen bonds from the acidic vinyl and phenyl hydrogen atoms to the nitro oxygen atom on the other side of the carbonyl group. The 1:2 complex **47** between dibenzylidene cyclopentanone **41** and **43** contains the triply hydrogen bonded array **49** along with C-H···O hydrogen bonds between the allylic H-atoms of **41** and the nitro O-atoms of **43**. The additional fortification from bifurcated C-H···O hydrogen bonds on the hydrocarbon side of the molecule is present in complexes **46** and **47** but not in **48**. This may be ascribed to acidity effects. A comparison of the three crystal structures **46–48** reveals also that the C-H···O hydrogen bonds that constitute synthon **49** are shorter and more linear compared to the isolated C-H···O hydrogen bonds elsewhere in the structure. This strengthens the idea that C-H···O hydrogen bonds which are part of a multipoint array are more significant (see also the article by J.P. Glusker in this volume).

The idea of robustness of a synthon has been alluded to earlier. The ability of synthon **49** in controlling self-assembly within this family of crystal structures was evidenced in the related complexes with picryl chloride **44** where it was found that the chloro group does not interfere with the formation of the DAD:ADA array. However, when the -Cl was replaced with an -OH group, the multipoint recognition motif is absent. This suggests that the strongly hydrogen

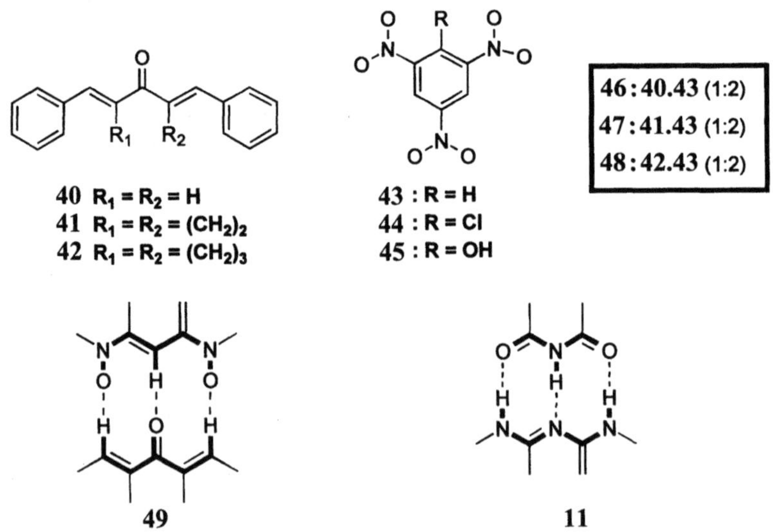

Scheme 10. DAD:ADA complexes **46–48** of dibenzylidene ketones and trinitrobenzene (see [49])

bonding -OH group in picric acid **45** is capable of disrupting the recognition motif **49** of dibenzylidene ketones with trinitrobenzenes. There is thus a limit to synthon robustness and an appreciation of these limits is advantageous in the design of newer crystal structures within the same family.

Two recent examples illustrate the consequences of N-H···O and C-H···O hydrogen bonds co-existing in a crystal structure. MacGillivray and Atwood observed that in doubly protonated **50**, the *in-in* conformation is more stable than the *out-out* form by 21 kcal mol^{-1} stabilised, in part, by the two intraionic N^+-H···O hydrogen bonds [50]. In the 1:2 complex **52** of [2.2.2]cryptand **50** and 2,6-dihydroxybenzoic acid (**51**) (Scheme 11a), there are two interionic N^+-H···O hydrogen bonds (N···O 2.68, 2.68 Å; N-H···O 177, 166°) on the outer surface between the cryptand and the acid and four intraionic C-H···O hydrogen bonds (C···O 3.01, 2.91, 3.04, 2.96 Å; C-H···O 111, 110, 117, 117°) inside the cavity. Since the *in-in*, *in-out* and *out-out* conformations of the cryptand are in rapid equilibrium in solution, the stabilisation of the *out-out* conformation is attributed to the four weak hydrogen bonds. This example underscores the idea that if the strength of the strong hydrogen bonds in two distinct orientations are comparable, then the weak interactions play a dominant structure-determining role in the crystal.

In an example that illustrates "supramolecular weaving", Stoddart and coworkers found that in the X-ray structure of the 1:2 complex **55** between BPP34C10 **53** and the ammonium salt **54**, the two cations are threaded co-directionally through the centre of the macrocycle (Scheme 11b) [51]. This arrangement is stabilised by a total of six N^+-H···O hydrogen bonds between the two NH_2^+ centres and three oxygen atoms from each polyether macrocycle (N···O

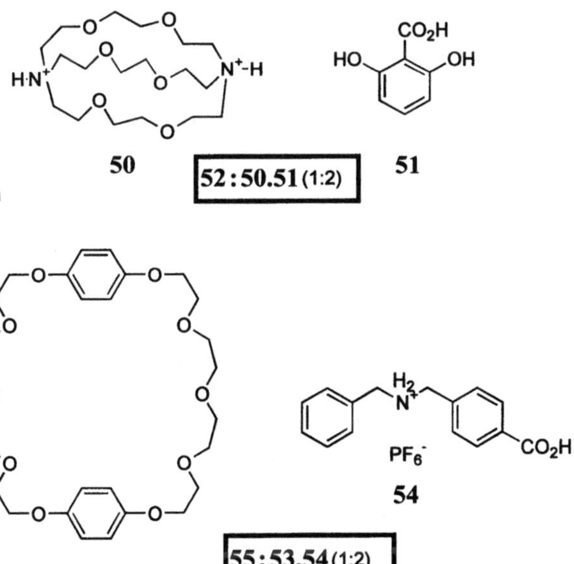

a **50** **52 : 50.51** (1:2) **51**

b **53** **54** **55 : 53.54** (1:2)

Scheme 11

2.86–3.18 Å). The more significant observation is that secondary stabilisation is conferred upon the superstructure by a "T"-shaped C-H···π contact between one of the methylene H-atoms of the 4-carboxybenzyl group and one of the hydroquinone rings of the BPC34C10 polyether (H···π 2.88 Å, C-H···π 143°). The PF_6^- counterion assists in maintaining the co-directionality of the two carboxyl groups of the [3]pseudorotaxane within the macrocyclic polyether, possibly through C-H···F hydrogen bonds. These three-component complexes aggregate to generate doubly encircled, six-component organic supermolecules that are linked by the robust carboxyl dimer supramolecular synthon through pairs of strong O-H···O hydrogen bonds (O···O 2.61, 2.68 Å). All this goes to show that in large and complex supramolecular architectures, numerous strong and weak hydrogen bonds act in concert to stabilise the crystal structure.

7 Supramolecular Interactions in Biological Systems

7.1 Weak Hydrogen Bonds

The first report on the existence of C-H···O hydrogen bonds in organic crystal structures was published by Sutor in 1962 [52]. The potential of these weak interactions in nucleotides was emphasised by Rubin et al. in 1972 to explain conformational preferences in nucleotide bases [53]. They suggested that specific C-H···O interactions between the purine C(8)-H or pyrimidine C(6)-H and the O(5′) oxygen are responsible for the *anti* conformation of these bases. The existence of weak C-H···O hydrogen bonds in proteins has been described by Derewenda et al. in 1995 [54]. The crystal structures of carbohydrates also exhibit a large number of short intermolecular C-H···O contacts [55]. The traditional view has been that only hydrogens bonded to oxygen and nitrogen atoms should be considered in the analysis of crystal structures of biological macromolecules [42]. There is now convincing evidence that there are a significant number of weak C-H···O and C-H···N hydrogen bonds in such structures which co-exist with the strong O-H···O, O-H···N and N-H···N hydrogen bonds. A statistical analysis of protein crystal structures shows that the C-H···O interactions exhibit characteristics associated with weak hydrogen bonds: the acidic C-H groups are more frequent; the acceptor O-atom is of the carbonyl or carboxyl type, and the median d and θ values are about 2.4 Å and 140°. The current view, expressed by Wahl and Sundaralingam [56], is that weak hydrogen bonds play a significant role in determining the structure as well as the function of biological macromolecules like nucleic acids, proteins and carbohydrates.

Watson-Crick (WC)and Hoogsteen (HG) base pairs between adenine and thymine or uracil have been traditionally constructed with strong hydrogen bonds. The participation of weak interactions in the recognition between adenine and thymine or uracil bases leads to the triply hydrogen bonded motif DAD:ADA consisting of two N-H···O and one C-H···O hydrogen bonds [57]. The conventional WC and HG hydrogen bonding schemes are strengthened by additional C(2)-H···O(2) and C(8)-H···O(2) interactions forced by the geometry of the A.T/U base pair (Scheme 12). Moreover, the repulsive secondary interaction between the O(2) atom of T/U and N(1) or N(7) atom of A is now balanced by a stabilising hydrogen bond. In recognition motifs where the C-H···O hydrogen bond is a viable contributor, the hydrogen bonding abilities of thymine and uracil must be considered in a different light. The sp^2 C-H of uracil has a greatly enhanced hydrogen bonding capability compared to the CH_3 hydrogens of thymine because of its higher acidity. It has been shown that the geometries and energies of DNA base triplets (CG):C, (TA):T, (AT):A and (GC):G and their matching in the homologous recombination can be explained by the auxiliary C-H···O and C-H···N hydrogen bonds [58]. It is clear that C-H···O and C-H···N hydrogen bonds are abundant in biological macromolecules and despite their small enthalpic contribution (0.5–2.0 kcal mol^{-1}) [59], they seem to exert pronounced structural and functional effects [60].

7.2 Pattern Recognition in Drug-Enzyme Binding

The idea of recognition between enzyme and substrate goes back more than one hundred years to Fischer's "lock-and-key" analogy of enzyme catalysis [61]. Drug-receptor recognition and binding is, however, not as simple as the lock-and-key analogy and may be more likened to a hand fitting into a glove. Both the drug (or ligand) as well as the receptor (or active site) are flexible and yet have very specific requirements for mutual recognition [62]. It is important to note that the docking of a drug molecule into the active site of a receptor protein is governed by the same combination of forces that control crystallisation, namely electrostatic forces, hydrogen bonding interactions, van der Waals forces and hydrophobic effects, the last factor being responsible for the exclusion of water

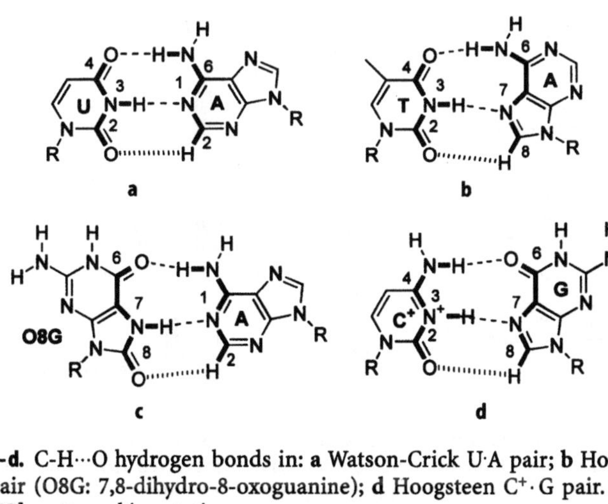

Scheme 12a–d. C-H···O hydrogen bonds in: a Watson-Crick U·A pair; b Hoogsteen T·A pair; c O8G·A pair (O8G: 7,8-dihydro-8-oxoguanine); d Hoogsteen C^+·G pair. For patterns a, b and c, see [57]; pattern d is putative

from nonpolar regions of the macromolecule. It is well known that the difference between a molecule that binds to the active site with minimal side effects, and hence a potential drug candidate, and the numerous ligands that do not is usually a very small structural difference in atom type, position or stereochemistry. The objective of rational drug design is therefore to predict new molecules or analogues of known drugs which will have favourable interactions with a protein of known structure [63, 64]. The three crucial inputs for structure-based ligand design are: (1) the molecular structure of the ligand; (2) the shape of the macromolecular receptor protein; and (3) the typical intermolecular interaction patterns between ligand and receptor. The last of these is within the scope of this article and is discussed here.

The number of accurate crystallographically resolved ligand-protein structures is still not large and hence a systematic, statistical analysis of their interaction geometry is not possible. However, the enormous wealth of information on small molecule crystallography in the CSD can be retrieved to extract orientational preferences for interaction around different functional groups. Klebe has presented composite crystal field environments for the approach of different donor hydrogens to various acceptor atoms, in effect compiling a hydrogen bond geometry data file relevant to drug-protein binding [65]. The histograms and scattergrams for four selected functional groups from the CSD and Protein Data Bank (PDB) are given in Figs. 3 and 4. A comparison of these figures suggests that patterns and motifs observed in the low molecular weight ligand structures are a faithful representation of the spatial orientation of interaction geometry between ligands and proteins. The application of the CSD also extends to ligand conformation. It has been shown that ligand conformations in small-molecule

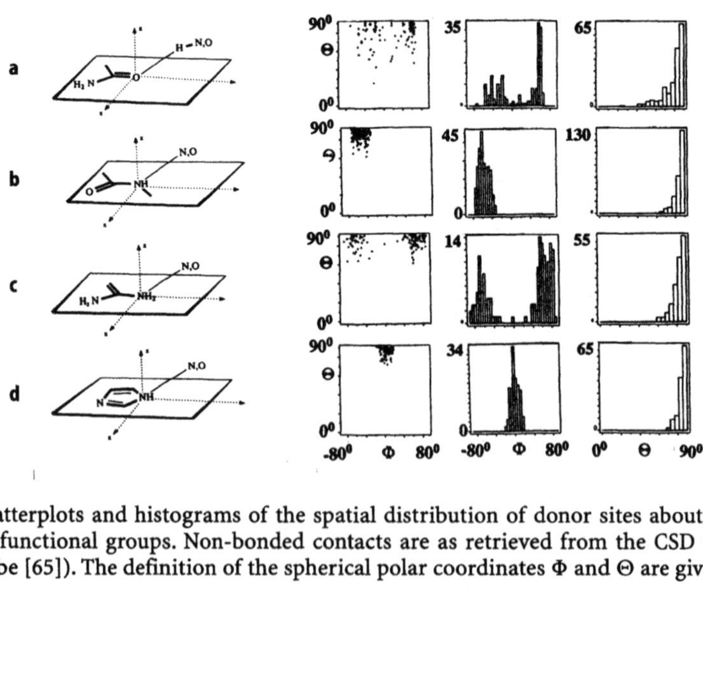

Fig. 3. Scatterplots and histograms of the spatial distribution of donor sites about different acceptor functional groups. Non-bonded contacts are as retrieved from the CSD (Adapted from Klebe [65]). The definition of the spherical polar coordinates Φ and Θ are given in this paper

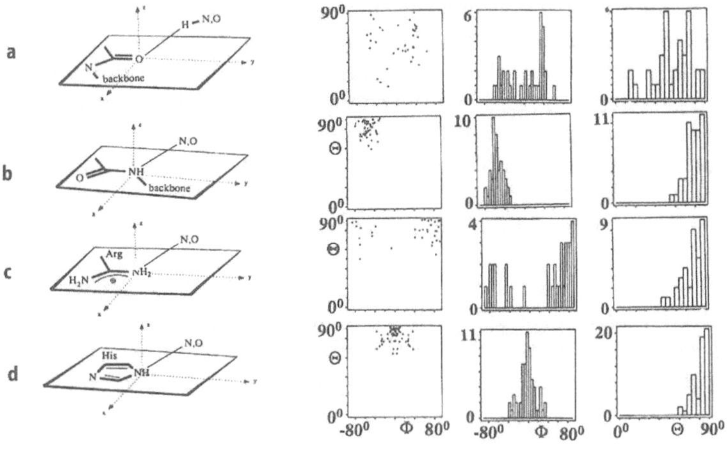

Fig. 4. Scatterplots and histograms of the spatial distribution of acceptor or donor sites in ligand molecules about different functional groups of active site residues in ligand-protein crystal structures (PDB) (Adapted from [65]). Note that the distributions are more smeared when compared with those in Fig. 3

crystal structures offer a good representation of their conformations in ligand-enzyme complexes [64]. While adopting this approach, it is reassuring to note that the average densities found for crystal structures of small organic compounds and for the inner regions of macromolecular proteins are in the same range. In summary, therefore, CSD derived geometries offer a considerable simplification in the analysis and understanding of drug-receptor interaction patterns (see also the article by J. P. Glusker in this volume).

7.3 Pharmacophore Model

The repetitive nature of enzyme-ligand patterns leads to the concept of the pharmacophore model which postulates that there is a minimum and essential set of functional groups that a molecule must possess in a specific geometrical arrangement to be recognised by the receptor [66]. The pharmacophore model can be generated by identifying the chemically important functional groups that are common to the molecules that bind to a given receptor.

We have accordingly examined the β-lactam class of antibiotics [67]. Geometrical details from a total of 114 β-lactams containing the fragment **56** were retrieved from the CSD (Version 5.05) and categorised into biologically active (80) and inactive (30) compounds. The superposition plot of 12 randomly selec-

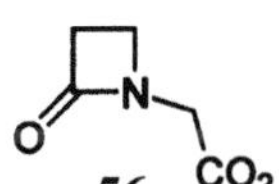

56

ted active compounds containing the β-lactam-C-CO_2 fragment shows that the lactam carbonyl and the carboxylate groups are bunched in their specific, narrow regions and do not show much variation in the positioning of the four-membered β-lactam ring and and the connecting carbon linkage (Fig. 5a). This depiction of functional groups is therefore a good pharmacophore model for the binding of the β-lactam with the receptor protein. The corresponding superposition plot for the 12 randomly selected inactive β-lactams shows considerable variation in the positioning of the lactam ring and the connecting -C- linkage (Fig. 5b).

Thus, the 3D structures of a class of drug molecules, together with their biological activity, can be profitably utilised to develop a pharmacophore model for the drug. Furthermore, the observation that the active molecules are tightly bunched and adopt a nearly similar conformation in the pharmocophore

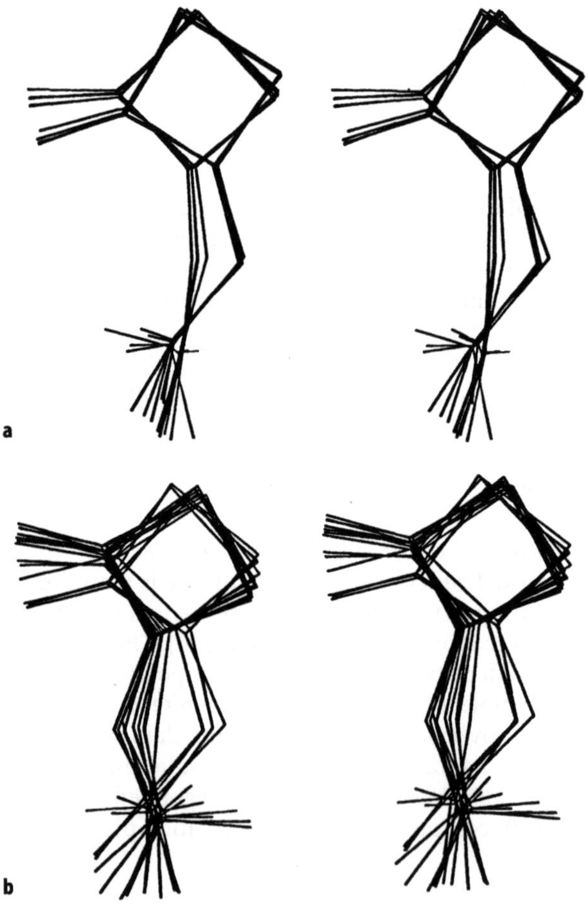

a

b

Fig. 5 a, b. Superposition plot of 12 randomly chosen β-lactams containing fragment **56** from: **a** the active diagonal; **b** the inactive diagonal of the Woodward (h) – Cohen (c) scatterplot (Taken from Nangia et al. [67])

portion of the antibiotic indicates that the receptor cavity has a well-defined geometry and that shape recognition occurs without much ligand-induced fit.

8
Polymorphism

Polymorphism is defined as the phenomenon wherein the same chemical substance exists in different crystalline forms [68]. In the context of the present article, polymorphism corresponds to the existence of different crystalline patterns for the same molecule. With the current levels of intense activity in crystal engineering, there is much interest in this well-known though little-understood phenomenon [69]. There is as yet no clear consensus on the exact definition of polymorphism. McCrone defined it as the existence of at least two crystalline arrangements in the solid state of the same chemical substance [70] while Buerger and Bloom defined it on the basis that the two polymorphs must have distinctive properties [71]. Some of the criteria employed for assessing the existence of polymorphs are differences in unit cell parameters, crystal packing arrangements and physical properties. Among the techniques used for identifying polymorphs, single crystal and powder X-ray diffraction are the most powerful, while others like hot-stage microscopy, IR spectroscopy and differential scanning calorimetry provide supporting evidence.

Dunitz has expressed a more contemporary view of polymorphism [3]. Since a crystal is a supermolecule, polymorphic forms are superisomers and polymorphism is a kind of superisomerism. Similar ideas are implicit in the molecule → supermolecule analogy and crystalline polymorphs are the supramolecular equivalents of molecular structural isomers [9]. Polymorphism in crystals is a very complex issue, and this complexity relates not only to causes for its occurrence but also to criteria that can judge as to whether it is present at all. Conformational polymorphism [72] and configurational polymorphism are two different types of polymorphism that arise because molecules adopt different conformations or crystallise as geometrical isomers, diastereomers, enantiomers or tautomers in the solid state. The suggestion of Byrn that polymorphism is very common in pharmaceutical compounds [73] is illustrated in the recent high-profile legal battle concerning the polymorphs of one of the best selling drugs, ranitidine [74]. Pseudopolymorphism, the inclusion of different solvents in the crystal structures of the same chemical substance, is not, strictly speaking, polymorphism at all because it involves different chemical species. This notwithstanding, the phenomenon is of immediate concern to the pharmaceutical industry while filing patent applications because different solvates of the same drug can exhibit widely differing solubilities and bioavailability properties. The unpredictable and capricious nature of polymorphism has been discussed in a recent article, humorously entitled "Disappearing Polymorphs" [75].

The existence of polymorphism in crystals can, at one stroke, demolish a well-planned and logically devised strategy to obtain a target supramolecular structure. Instead of treating this phenomenon as a complication or nuisance in crystal engineering strategies, a more positive approach is to view it as an excellent opportunity to study the same chemical substance in different crystalline environments. The existence of polymorphic forms may also render the molecule more susceptible to the formation of molecular complexes and this is a helpful feature. By calculating the lattice energies of polymorphs, the energies of molecules in different conformations can be estimated. Different supramolecular synthons in polymorphs provide good comparisons of distinct intermolecular interactions that can arise from the same functional group(s) and their effects on crystal packing.

A number of factors have been suggested as possible causes for polymorphism. Polymorphism can occur when the free energy differences between alternative crystal structures are very slight. This may happen: (1) because the intermolecular forces are feeble; (2) when the entropic contribution to the free energy is high; and (3) because kinetic rather than thermodynamic factors control the events leading to crystallisation. The degree of difference between polymorphs may itself vary. While there are polymorphs that contain completely different supramolecular synthons, there are others which contain the same synthon occurring in slightly different ways. Three distinct situations are possible: (1) the same synthons are formed by the same functional groups but the differences in overall packing are caused by variations in the rest of the crystal; (2) the same synthons are formed by the same functional groups but there are multiple occurrences of these groups in distinctive molecular locations leading to different packing arrangements; (3) and different synthons are formed leading to radically different packings. The three situations are discussed with representative examples [69].

In many cases, the same synthons are formed by the same functional groups in the various polymorphs. Both forms of resorcinol, RESORC, contain synthon 57. However, the difference between the two forms is that while linear chains are formed in form B, tetrameric helices are present in form A (Fig. 6). 4-Hydroxya-

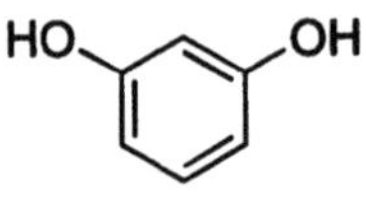

RESORC

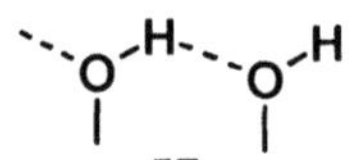

57

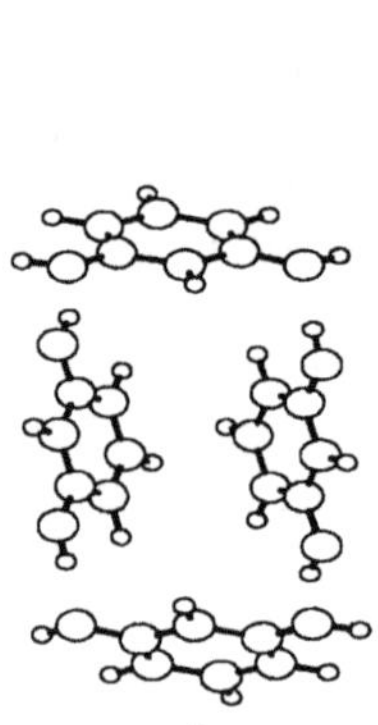

A

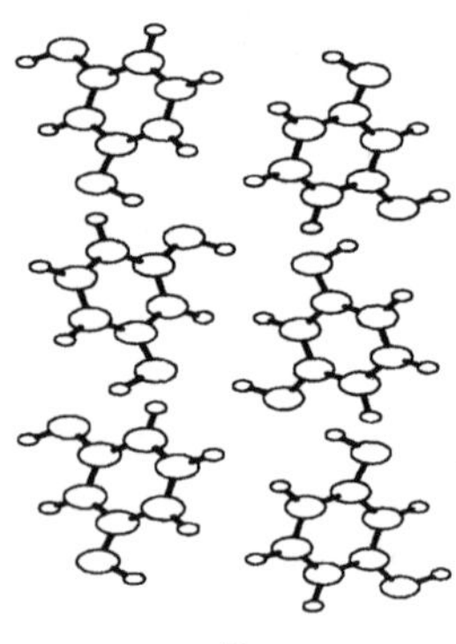

B

Fig. 6. Polymorphs A and B of resorcinol (RESORC)

cetanalide, HXACAN, is also dimorphic with synthon **58** being present in both the forms. As in resorcinol, linear and nearly tetramer geometries are obtained in forms B and A respectively. Alternating molecules that are utilised in the generation of synthon **58** are related by translation in form B, while such a relation exists between successive molecules in form A (Fig. 7).

In some cases, there are distinctive occurrences of the same functionality within a molecular skeleton and so the same synthons can be formed in different ways and locations, leading to polymorphism. Synthons **3** and **59** are typical for primary carboxamides and synthons **60** and **61** are common in aza-heterocycles and imines (Fig. 8). Pyrazine carboxamide, PYRZIN, an anti-tuberculosis drug, is a rare example of a compound with four or so different polymorphic forms (A–D; it is possible that there also exists a fifth form, E, very similar to form A). In each of these forms, a combination of N-H···O synthons **3** and **59** and N-H···N and C-H···N synthons **60** and **61** are optimised by the carboxamide and pyrazine functional groups. Form C has a planar sheet structure whereas the molecules are arranged to form linear ribbons in the other polymorphs. Forms A, B and C contain the common dimer synthon **3**, but it should be noted that one or more of the other synthons **59**–**61** are also optimised and this is what leads to the differences between these polymorphs. Form D has an interesting non-centrosymmetric packing that is characterised by synthon **61** and the unusual synthon **62**.

Of course, there are some cases where the polymorphs have completely different crystal structures, that is totally different synthons are seen in the various forms. Kinetic factors become significant here. 3-Hydroxybenzoic acid, 2,5-dihydroxybenzoic acid and 2,6-dihydroxybenzoic acid (LEZJAB) are all

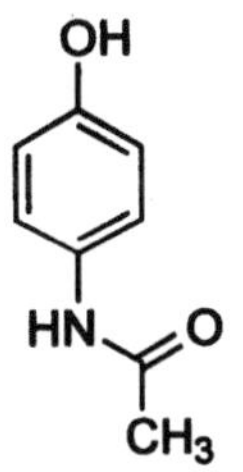

HEXCAN

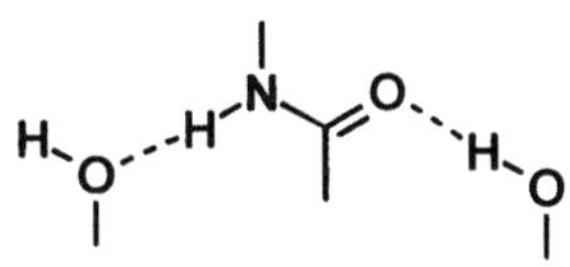

58

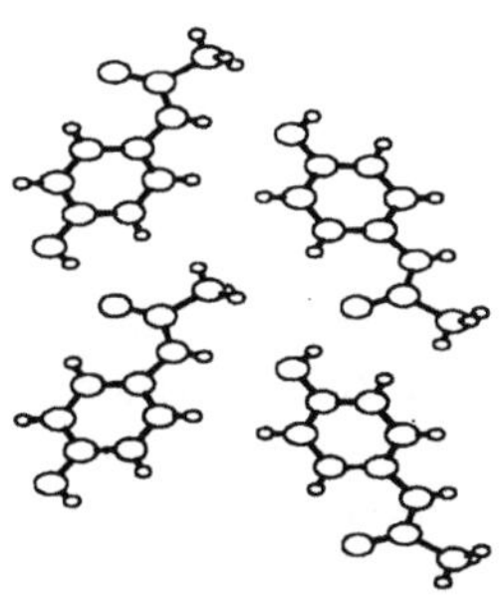

A

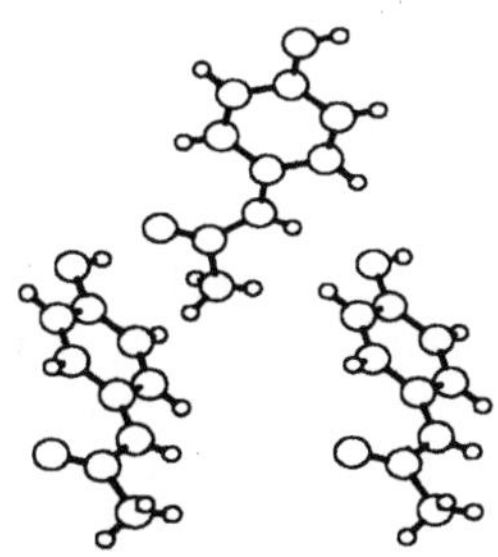

B

Fig. 7. Polymorphs A and B of 4-hydroxyacetanilide (HEXCAN)

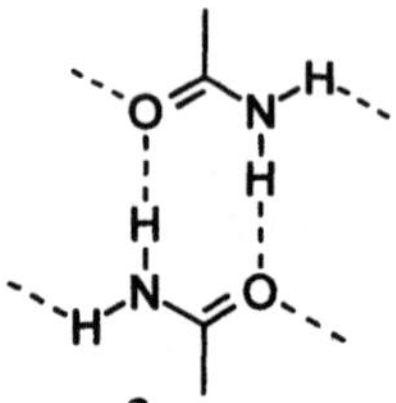

3

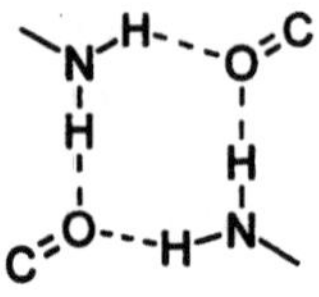

59

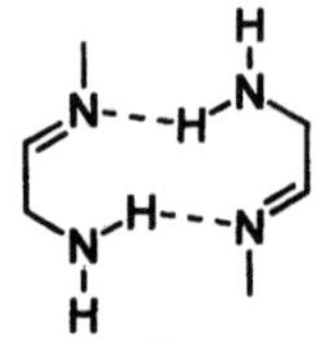

60

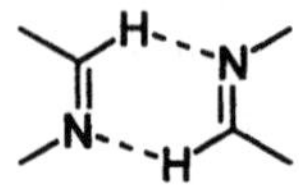

61

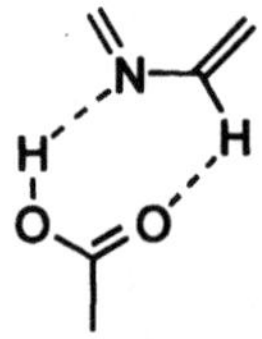

62

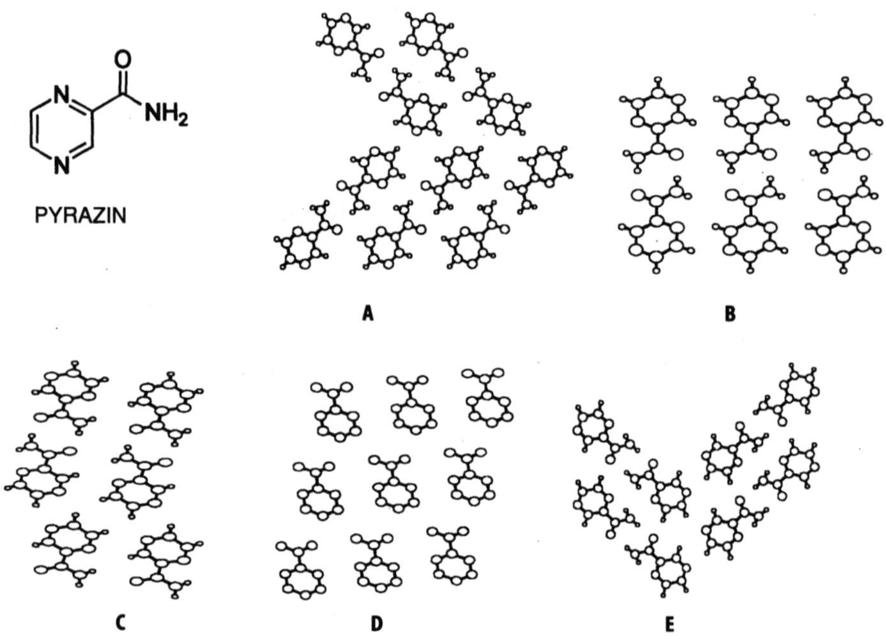

Fig. 8. Polymorphs A–D of pyrazine carboxamide (PYRAZIN). A possible fifth polymorph E is very similar to A

dimorphic and polymorphism is easy because there are several types of competing hydrogen bond donor (phenolic and carboxylic H-atoms) and acceptor (phenol, carboxyl and carbonyl O-atoms) groups and also because the carboxyl groups can have dimer 1 and catemer 2 arrangements. Inspection of Fig. 9 shows that the two forms of LEZJAB are distinctive. The relationships between the polymorphs of the other hydroxybenzoic acids are similar and involve different combinations of intra- and intermolecular hydrogen bonding. Essentially, the

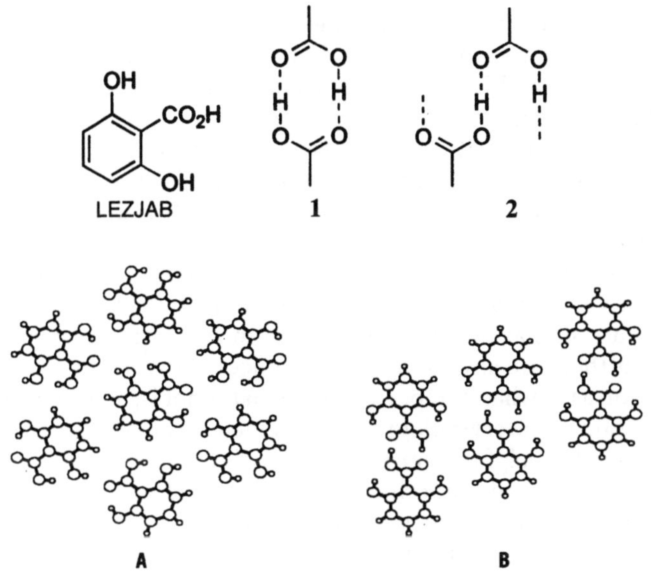

Fig. 9. Polymorphs A and B of 2,6-dihydroxybenzoic acid (LEZJAB)

presence of both phenolic and carboxylic OH groups provides alternative hydrogen bonding motifs of comparable energy and with it a route towards polymorphism. (A more detailed discussion of this topic is given in the article by M. R. Caira in this volume).

9 Comparison of Crystal Structures

Crystals are composed of molecules but crystal structures cannot be derived or anticipated in simple ways from molecular structures. Similar molecules can have very dissimilar crystal structures, and dissimilar molecules can have very similar crystal structures. This is because the core constituents of a crystal are the patterns and topologies of intermolecular interactions that result from complementary approaches of molecular functional groups and because the exact patterns formed depend not just on the functional groups present in the molecules but also on their relative juxtapositioning. The complexity of this issue will be further appreciated when it is realised that *all* portions of a molecule are supramolecular functionalities. Therefore, a detailed understanding of crystal packing and crystal design depends very substantially on viewing the molecule as an organic whole. Indeed, the supramolecular paradigm is particularly appropriate in the crystalline world because the essential structural attributes of a crystal are supramolecular rather than molecular in nature.

Given such realities, an immediate need in crystal engineering is to be able to compare crystal structures. Many will appreciate that the structure of, say, naphthalene resembles that of anthracene more than it resembles benzene. Is it possible to quantify such comparisons? If so, such quantification would amount to pattern matching and becomes important because crystals that are structurally similar are also likely to have similar properties. Ideally, one would like to arrive at an index of similarity between two crystal structures. In order that two or more structures are deemed to be similar or dissimilar, two steps are involved: (1) identification of the core structural features; and (2) evaluation of the extent of their likeness.

Such an exercise can be carried out at varying levels of scrutiny. The traditional approach is to analyse manually several crystal structures and decide whether they are similar or not. The problem in such a complex and detailed analysis is that there are always minor differences between any two structures and the decision as to what is important and what is not is, in the end, quite subjective. Inspection of the crystallographic parameters can obscure the focus and need not always be helpful. Conversely, crystals with different crystal symmetries, space groups and unit cell parameters may be structurally quite similar. For these and related reasons, manual comparison of complete crystal structures is not practical. Some simplification is necessary.

The graph set notation for comparing crystal structures was suggested originally by Etter et al. [76, 77]. Several clarifications of earlier ambiguities appeared subsequently in a review of Bernstein et al. [78] (see also the article by M.R. Caira in this volume). This method recognises that crystal structures need to be simplified before they can be compared. Accordingly, the essential hydrogen bond connectivity information is retained while the covalent framework on which the functional groups are mounted is neglected (see, however, [79]). The graph set representation offers an exact network depiction of hydrogen bonded patterns but the rules for its definition are difficult to implement. It has been noted that, while the rigour in the graph set definition provides a precise topological description, the same rigour can also obscure general similarities in hydrogen bonding patterns that would need to be revealed during a comparison of crystal structures [80]. Among other problems with the graph set notation are the definition of acceptors as single atoms [81] and the inapplicability of the method to the many interactions that cannot be considered as being of the donor-acceptor type.

The concept of a supramolecular synthon also recognises the need to be able to simplify a three-dimensional crystal structure into modular units prior to structural comparison. Again, the emphasis is on the hydrogen bonds and intermolecular interactions between functional groups, neglecting the molecular skeleton that is deemed to be passive [11, 82]. However, the definition of a supramolecular synthon is deliberately left unconstrained and non-quantitative. Synthons range from a single interaction to multipoint recognition patterns that contain hydrogen bonds and non-directional interactions, and the term encompasses both chemical and geometrical recognition. Such flexibility is advantageous and allows the chemist to select crystal patterns *not only on the basis of topological attributes but also through chemical intuition.* This in-built subjectivity in defining the term "synthon" and a certain flexibility in its usage are necessary because the entities being described, that is the repeating patterns in crystals, are not rigorously quantifiable. In this respect, the usage of the synthon concept in crystal engineering and supramolecular chemistry follows very closely its usage in classical organic synthesis. In both these usages, simplification is combined with chemical focus. Given that crystal structures need to be simplified before they can be compared and analysed, the graph set notation doubtless offers an accurate topological description of hydrogen bonded patterns. However, the simplification is drastic and is achieved at the cost of obscuring the chemical nature of the recognition events that are the primary causes of crystallisation. On balance, synthons appear to offer a middle ground wherein simplification is obtained without compromising the chemical information contained in the supramolecular system [83–90].

A common deficiency of the graph set and synthon approaches to crystal structure description is that they represent interactions in a crystal without any indication as to whether they are strong or weak and as to their importance in controlling crystal packing. Such information is available in a new pictorial method called NIPMAT (*N*onbonded *I*nteraction *P*attern *MAT*rix) that is used to display intermolecular interactions in crystals [91]. A two-dimensional quantitative representation gives a single-view visualisation of the relative strengths of the hydrogen bonding and van der Waals forces operating in the crystal. For example, the crystal structures of 2-, 3- and 4-aminophenols detailed earlier in this article may now be compared through their NIPMAT plots [92]. Examination of Fig. 10 shows that hydrogen bonding between the amino and hydroxy groups is the common, dominant intermolecular interaction as evidenced from the dark grey squares for H6-O1 and H7-N1 in the upper right hand corner of the plot. The contribution from the N-H$\cdots\pi$ interaction in 2- and 3-aminophenols is revealed by the grey squares for H5-C3/4/5 and H5-C4/5/6 respectively (Figs. 10a, b), whereas this interaction is absent in 4-aminophenol (Fig. 10c).

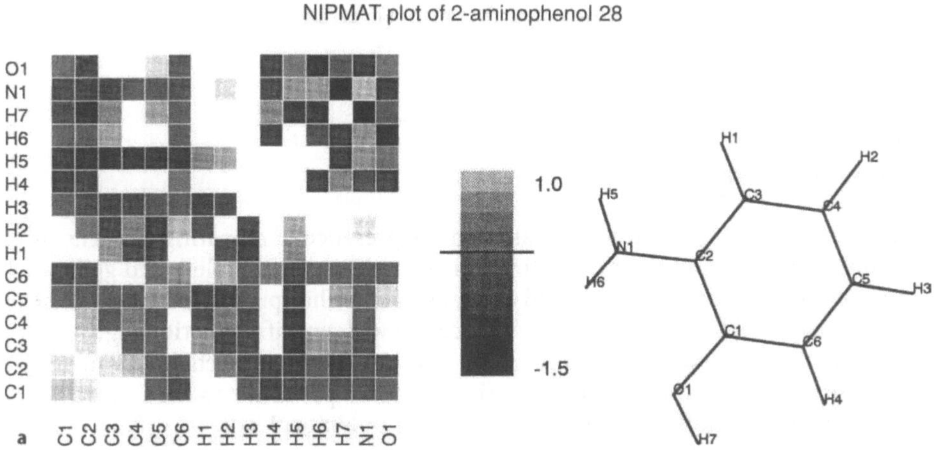

Fig. 10a–c. Comparison of crystal structures with NIPMAT plots: **a** 2-aminophenol **28**

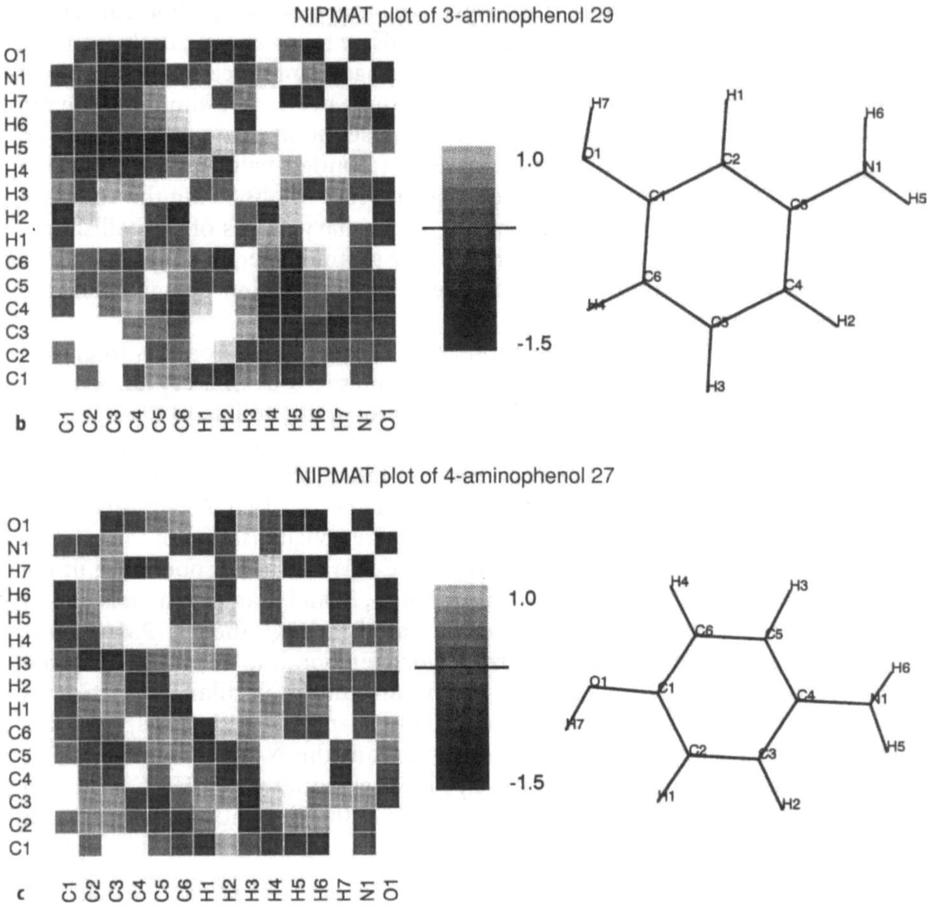

Fig. 10b,c. 3-aminophenol **29**; **c** 4-aminophenol **27**. *The darker squares* correspond to the shorter intermolecular contacts. *The heavy black line* on the bar grey scale corresponds to a contact at van der Waals separation

10
"Absence" of Patterns

The theme of the article thus far has been the presence of repeating patterns in crystal structures and their identification. Even though a complete and general connection between molecular and crystal structure has proved to be elusive, it is seen that certain patterns may be associated with specific functional groups. Carboxyl groups tend to form dimers, phenols form catemeric chains, aromatic rings form herringbone patterns and so on. Known patterns can then associate in distinct ways, leading to the phenomenon of polymorphism.

More intriguing are cases where the pattern identification itself is difficult. There are instances when, because of competing isoenergetic interactions in a certain functional group category, no dominant motif is adopted by the molecules. Such appears to be the case for acetylenic alcohols containing the fragment **63** [93]. Retrieval of data for sub-structure **63** from the CSD produced a list of 76 compounds. The hydrogen bonding networks of these molecules did not produce a single recurring pattern that can be associated with this sub-structure [94]. Rather, four different motifs could be identified (Scheme 13), with none being predominant. This is in contrast to the behaviour of simple monoalcohols that crystallise mainly in two type of O-H···O aggregates – chains and rings [95]. There is significant structural interference in these acetylenic alcohols. Instead of the C≡C-H and OH groups associating in their isolated networks, complex patterns are formed wherein C≡C-H···OH **65** and OH···C≡C-H **66** hydrogen bonds are found in addition to the expected C≡C-H···C≡C-H **67** and O-H···O-H **64** hydrogen bonds. This may be due to the fact that the four different hydrogen bonding patterns between the two functional groups, O-H···O **64**, C-H···O **65**, O-H···π **66**, and C-H···π **67** have comparable energies, steric requirements and topologies.

At a certain level, the absence of patterns in groups of crystal structures is a manifestation of the fact that the molecular and crystal structures need not be related in obvious ways. After all, the geminal positioning of hydroxy and ethynyl groups in **63** is only a molecular structural feature. There is no need that

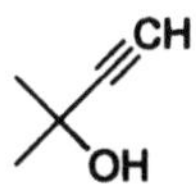

63

hits = 76

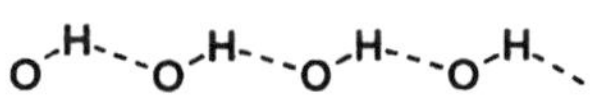

64

hits = 8
fragments = 22

C≡C–H- - -O

65

hits = 25
fragments = 27

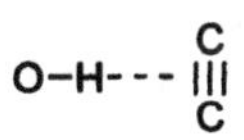

66

hits = 4
fragments = 5

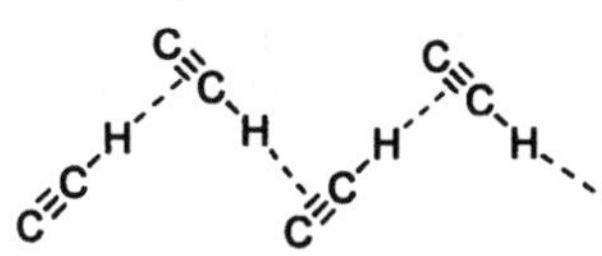

67

hits = 12
fragments = 15

Q–H- - -Q

68

Q–H- - -Q–H- - -Q–H- - -Q—- -

69

Scheme 13. Four common supramolecular synthons **64**–**67** in 76 occurrences of the acetylenic alcohol fragment **63**

all acetylenic alcohols should have related crystal structures. At yet another level, though, it is possible that the "absence" of patterns is only indicative of the fact that the crystal structures are being compared at an inappropriate frame of reference. If, for instance, the hydroxy and ethynyl groups were considered as being equivalent to Q-H, then the four patterns **64**–**67** reduce to Q-H···Q interaction **68** and its repeating array **69**.

11 Conclusions

Patterns in crystals reflect recognition features of molecules and the recurring nature of these patterns reveals their robustness, which is a very desirable attribute in crystal engineering and design. Patterns of interactions are important not only in molecular crystals but also in the related biological areas of drug design and receptor-ligand binding.

The concept of a supramolecular synthon places the geometrical notion of a pattern in a chemical context. More significantly, this concept lends itself to retrosynthetic analysis and to the identification of similar crystal structures that can be derived from dissimilar molecules. Supramolecular synthons are critical sub-structural units that contain the maximum of structural information encapsulated in patterns of the most economical size. A full three-dimensional structure contains all possible structural information for a given crystal but is too detailed to permit realistic comparisons with other crystal structures. Structural simplification is therefore necessary.

The graph set notation of a crystal structure provides precise topological information but inherent in its definition is a drastic reduction in the chemical information underlying molecular recognition. The supramolecular synthon combines chemical and geometrical features and occupies a middle ground between the full crystal structure and its graph set notation.

With these and similar ideas, it is hoped that the task of designing and building a crystal from smaller structural units will progress to a more reliable and manageable exercise.

Acknowledgements. Financial assistance from the Department of Science and Technology (SP/S1/G19/94) and the Council of Scientific and Industrial Research (1/1431/96/EMR-II), Government of India is acknowledged. We thank Prof. G. Klebe for the originals of Figs. 3 and 4 and Dr. R. S. Rowland for making available a copy of the NIPMAT program.

12 References

1. Lehn J-M (1994) Perspectives in supramolecular chemistry: from the lock-and-key image to the information paradigm. In: Behr JP (ed) Perspectives in supramolecular chemistry, vol 1. The lock-and-key principle. The state of the art – 100 years on. Wiley, Chichester, p 307
2. Desiraju GR, Sharma CVK (1995) Crystal engineering and molecular recognition. Twin facets of supramolecular chemistry. In: Desiraju GR (ed) Perspectives in supramolecular chemistry, vol 2. The crystal as a supramolecular entity. Wiley, Chichester, p 31

3. Dunitz JD (1995) Thoughts on crystals as supermolecules. In: Desiraju GR (ed) Perspectives in supramolecular chemistry, vol 2. The crystal as a supramolecular entity. Wiley, Chichester, p 1
4. Kitaigorodskii AI (1973) Molecular crystals and molecules. Academic Press, New York
5. Desiraju GR (1989) Crystal engineering. The design of organic solids. Elsevier, Amsterdam
6. Desiraju GR (1996) Review of general principles. In: MacNicol DD, Toda F, Bishop R (eds) Comprehensive supramolecular chemistry, vol 6. Solid-state supramolecular chemistry: crystal engineering. Pergamon, Oxford, p 1
7. Corey EJ (1967) Pure Appl Chem 14:19
8. Corey EJ (1988) Chem Soc Rev 17:111
9. Desiraju GR (1995) Angew Chem Int Ed Engl 34:2311
10. Corey EJ, Cheng X-M (1989) The logic of chemical synthesis. Wiley, New York
11. Desiraju GR (1998) Connections between molecular and supramolecular structure. Implications for self-assembly. In: Wuest JD (ed) Self-assembly in synthetic chemistry. NATO ARW Series, in press
12. Desiraju GR (1997) Chem Comm 1475
13. Reddy DS, Craig DC, Desiraju GR (1996) J Am Chem Soc 118:4090
14. Thaimattam R, Reddy DS, Xue F, Mak TCW, Nangia A, Desiraju GR (1998) New J Chem 143
15. Seebach D (1990) Angew Chem Int Ed Engl 29:1320
16. Fuhrhop J, Penzlin G (1994) Organic synthesis. Concepts, methods, starting materials, 2nd edn. VCH, Weinheim
17. Hanessian S, Franco J, Larouche B (1990) Pure Appl Chem 62:1887
18. Ho T-L (1988) Carbocycle construction in terpene synthesis. VCH, Weinheim
19. Nicolaou KC, Sorensen EJ (1995) Classics in total synthesis. VCH, Weinheim
20. Prinzbach H, Weber K (1994) Angew Chem Int Ed Engl 33:2239
21. Osawa E, Yonemitsu O (ed) (1992) Carbocyclic cage compounds. Chemistry and applications. VCH, New York
22. Woodward RB (1973) Pure Appl Chem 33:145
23. Eschenmoser A, Winter CE (1977) Science 196:1410
24. Armstrong RW, Beau J-M, Cheon SH, Christ WJ, Fujioka H, Ham W-H, Hawkins LD, Jin H, Kang SH, Kishi Y, Martinelli MJ, McWhorter WW, Mizuno M, Nakata M, Stutz AE, Talamas FX, Taniguchi M, Tino JA, Ueda K, Uenishi J, White JB, Yonaga M (1989) J Am Chem Soc 111:7530
25. Trost BM (1991) Science 254:1471
26. Holton RA, Somoza C, Kim H-B, Liang F, Biediger RJ, Boatman PD, Shindo M, Smith CC, Kim S, Nadizadeh H, Suzuki Y, Tao C, Vu P, Tang S, Zhang P, Murthi KK, Gentile LN, Liu JH (1994) J Am Chem Soc 116:1597
27. Nicolaou KC, Yang Z, Liu JJ, Ueno H, Natermet PG, Guy RK, Claiborne CF, Renaud J, Couladouros EA, Paulvannan K, Sorensen EJ (1994) Nature 367:630
28. Philp D, Stoddart JF (1996) Angew Chem Int Ed Engl 35:1154
29. Gavezzotti A, Filippini G (1994) J Phy Chem 98:4831
30. Gavezzotti A (1994) Acc Chem Res 27:309
31. Robertson JM (1951) Proc Roy Soc London Ser A 207:101
32. Desiraju GR, Gavezzotti A (1989) Acta Crystallogr B45:473
33. Ermer O, Eling A (1994) J Chem Soc Perkin Trans 2 925
34. Hanessian S, Simard M, Roelens S (1995) J Am Chem Soc 117:7630
35. Allen FA, Hoy VJ, Howard JAK, Thalladi VR, Desiraju GR, Wilson CC, McIntyre GJ (1997) J Am Chem Soc 119:3477
36. Allen FH, Davies JE, Galloy, JJ, Johnson O, Kennard O, Macrae CF, Mitchell EM, Mitchell GF, Smith JM, Watson DG (1991) J Chem Inf Comput Sci 31:187
37. Katz AK, Glusker JP, Nangia A, Desiraju GR (1997) unpublished results
38. Desiraju GR (1989) J Chem Soc Chem Commun 179
39. Zimmerman SC, Murray TJ (1993) Phil Trans R Soc Lond A 345:49
40. Burrows AD, Chan C-W, Chowdhry MM, McGrady JE, Mingos DMP (1995) Chem Soc Rev 329
41. Pranata J, Wierschke SG, Jorgensen WL (1991) J Am Chem Soc 113:2810

42. Jeffrey GA, Saenger W (1991) Hydrogen bonding in biological systems. Springer, Berlin Heidelberg NewYork
43. Lehn J-M (1990) Angew Chem Int Ed Engl 29:1304
44. Bryce MR (ed) (1997) J Mater Chem 7:1069 (Special issue on molecular assemblies and nanochemistry)
45. Lehn J-M, Mascal M, DeCian A, Fischer J (1990) J Chem Soc Chem Commun 479
46. Steiner T (1996) Cryst Rev 6:1
47. Desiraju GR (1996) Acc Chem Res 29:441
48. Steiner T (1997) Chem Commun 727
49. Biradha K, Nangia A, Desiraju GR, Carrell CJ, Carrell HL (1997) J Mater Chem 1111
50. MacGillivray LR, Atwood JL (1997) Chem Commun 477
51. Ashton PR, Collins AN, Fyfe MCT, Menzer S, Stoddart JF, Williams DJ (1997) Angew Chem Int Ed Engl 36:735
52. Sutor DJ (1962) Nature 195:68
53. Rubin J, Brennan T, Sundaralingam M (1972) Biochemistry 11:3112
54. Derewenda ZS, Lee L, Derewenda U (1995) J Mol Biol 252:248
55. Steiner T, Saenger W (1992) J Am Chem Soc 114:10,146
56. Wahl MC, Sundaralingam M (1997) Trends Biochem Sci 22:97
57. Leonard GA, McAuley-Hecht K, Brown T, Hunter WN (1995) Acta Crystallogr D51:136
58. Zhurkin VB, Raghunathan G, Ulyanov NB, Camerini-Otero RD, Jernigan RL (1994) J Mol Biol 239:181
59. Starikov EB, Steiner T (1997) Acta Crystallogr D53:345
60. Auffinger P, Louise-May S, Westhof E (1996) Faraday Discuss 103:151
61. Behr J-P (ed) (1994) Perspectives in supramolecular chemistry, vol 1. The lock-and-key principle. The state of the art 100 years on. Wiley, Chichester
62. Jorgensen WL (1991) Science 254:954
63. Boyd DB (1995) Computer-aided molecular design. In: Kent A, Williams JG (eds) Encyclopedia of computer science and technology, vol 33. Marcel Dekker, New York, p 41
64. Böhm H-J, Klebe G (1996) Angew Chem Int Ed Engl 35:2588
65. Klebe G (1994) J Mol Biol 237:212
66. Hahn M (1995) J Med Chem 38:2080
67. Nangia A, Biradha K, Desiraju GR (1996) J Chem Soc Perkin Trans 2 943
68. Threlfall TL (1995) Analyst 120:2435
69. Sarma JARP, Desiraju GR (1998) Polymorphism and pseudopolymorphism in organic crystals. A Cambridge Structural Database study. In: Seddon KR, Zaworotko MJ (eds), Crystal engineering. The design and applications of functional solids. NATO ASI Series, in press
70. McCrone WC (1965) Polymorphism. In: Fox D, Labes MM, Weissberger A (eds) Physics and chemistry of the organic solid state, vol 2. Interscience, New York, p 725
71. Buerger MJ, Bloom MC (1937) Z Kristallogr A96:182
72. Davey RJ, Blagden N, Potts GD, Docherty R (1997) J Am Chem Soc 119:1767
73. Byrn SR (1982) Solid-state chemistry of drugs. Academic Press, New York, p 79
74. DeCamp WH (1996) Regulatory considerations in crystallisation processes for bulk pharmaceutical industry. A reviewer's perspective. In: Myerson AS, Green DA, Meenan P (eds), Proceedings of 3rd International Workshop on Crystal Growth of Organic Materials. ACS Series, Washington DC, p 66
75. Dunitz JD, Bernstein J (1995) Acc Chem Res 28:193
76. Etter MC (1990) Acc Chem Res 23:120
77. Etter MC, MacDonald JC, Bernstein J (1990) Acta Crystallogr B46:256
78. Bernstein J, Davis RE, Shimoni L, Chang N-L (1995) Angew Chem Int Ed Engl 34:1555
79. Coupar PI, Ferguson G, Glidewell C (1996) Acta Crystallogr C52:2524
80. Kubicki M, Kindopp TW, Capparelli MV, Codding PW (1996) Acta Crystallogr B52:487
81. Subramanian K, Lakshmi S, Rajagopalan K, Koellner G, Steiner T (1996) J Mol Struct 384:121
82. Gavezzotti A (1996) Curr Opin Solid State Mater Sci 1:501

83. Hulliger J, Rogin P, Quintel A, Rechsteiner P, König O, Wübbenhorst M (1997) Adv Mater 9:662
84. Corey EJ, Barnes-Seeman D, Lee TW (1997) Tetrahedron Lett 38:1699
85. Pan F, Wong MS, Gramlich V, Brosshard C, Günter P (1996) Chem Comm 1557
86. Hoss R, König O, Kramer-Hoss V, Berger U, Rogin P, Hulliger J (1996) Angew Chem Int Ed Engl 35:2204
87. Navon O, Bernstein J, Khodorkovsky V (1997) Angew Chem Int Ed Engl 36:601
88. Carlucci L, Ciani G, Gudenberg DWv, Proserpio DM, Sironi A (1997) Chem Commun 631
89. Michaelides A, Skoulika S, Kiritsis V, Raptopoulou C, Terzis A (1997) J Chem Res (S) 204
90. Kräutler B, Müller T, Maynollo J, Gruber K, Kratky C, Ochsenbein P, Schwarzenbach D, Bürgi H-B (1996) Angew Chem Int Ed Engl 35:1204
91. Rowland RS (1995) Am Cryst Assoc Abstr 23:63
92. Thalladi VR, Nangia A, Desiraju GR (1997) unpublished results
93. Steiner T, Starikov EB, Amado AM, Teixeira-Dias JJC (1995) J Chem Soc Perkin Trans 2 1321
94. Madhavi NNL, Nangia A, Desiraju GR (1997) unpublished results
95. Brock CP, Duncan LL (1994) Chem Mater 6:1307

Hydrogen-Bonded Ribbons, Tapes and Sheets as Motifs for Crystal Engineering

Rosa E. Meléndez · Andrew D. Hamilton

Department of Chemistry, Yale University, New Haven CT 06520-8107, USA.
E-mail: ahamilton@ursula.yale.chem.edu

The design of new architectures for the purpose of crystal engineering has generated great interest in recent years. In particular, organic compounds have been the focus of many studies due to the presence of functional groups that can form strong and stable intermolecular interactions. Therefore, recognizing geometry and functionality at the molecular level has relevant implications in the design of supramolecular patterns. These patterns in turn can be translated to physical or chemical properties in a solid. This article is a presentation of compounds that have been designed for the purpose of molecular recognition and crystal engineering. In particular our focus will be directed towards organic structures based on the tape, ribbon, and sheet motifs. Several approaches in the design of functional solids will be presented, emphasizing the use of certain complementary intermolecular interactions for this purpose. It is not our attempt to define all concepts used for crystal engineering, but to present recent advances in this field.

Keywords: Crystal engineering, Molecular recognition, Hydrogen bonding, Tapes, Sheets.

1	**Introduction**	97
2	**Crystal Engineering and Molecular Recognition**	98
2.1	Strategies for Crystal Engineering	99
2.2	Patterns in Crystal Engineering: Tapes, Ribbons, and Sheets	100
3	**Hydrogen Bonding**	102
3.1	Carboxylic Acids	105
3.2	Amides	112
3.3	Other Strong Hydrogen Bonds	120
3.4	Weak Interactions	122
4	**Conclusion**	127
5	**References**	127

1 Introduction

Traditional synthetic chemistry is based upon the controlled formation and cleavage of covalent bonds. However, complicated nanoscale systems would be extremely difficult to synthesize using traditional covalent methods. As a result, the development of a field known as *synthetic supramolecular chemistry* has recently begun to emerge. Investigations on the assembly of small organic molecules in solution and the solid state will be important in expanding these new notions for the assembly of well-defined supramolecular architectures. The purpose of this article is to review recent developments in the field of synthetic supramolecular chemistry for the purpose of crystal engineering. We will place particular focus on the use of intermolecular interactions to form tape, ribbon, and sheet structures in the solid state.

2 Crystal Engineering and Molecular Recognition

The term crystal engineering was first coined by Schmidt in connection with his work on the topochemical reactions of crystalline cinnamic acids in 1971 [1]. However, recently this field has developed rapidly due to its important implications in materials science. Desiraju defines crystal engineering as "the understanding of intermolecular interactions in the context of crystal packing and the utilization of such understanding in the design of new solids with desired physical and chemical properties" [2]. In this way crystal engineering, first designed for solid state reactions, has been applied to the creation of solids that exhibit properties such as nonlinear optical activity, ferroelectricity, piezoelectricity, triboluminescence, and porosity.

Crystal engineering is based on concepts that have been broadly used in supramolecular chemistry [3]. Crystals are not just collections of molecules and their structural properties are different from those of their molecular constituents. Crystals are a repetitive arrangement of molecules in three dimensions with an impressive level of precision and have been regarded by Dunitz as "supermolecules *par excellence*" [4].

A large amount of effort has been invested in the study of crystal growth [5, 6] and crystal packing [7]. The potential energy of a crystal has been factored into component parts and has been attributed to various kinds of interactions including, electrostatic, hydrogen bonding, donor-acceptor, steric repulsions, and van der Waals attractions. While intermolecular interactions have been classified in different ways, the most meaningful criteria are their distance dependence and their directionality. Desiraju has classified intermolecular interactions in organic solids into two types: medium-range isotropic forces (close-packing) and long-range anisotropic forces (electrostatic interactions) [8]. Isotropic forces include C···C, C···H and H···H interactions and anisotropic forces include ionic interactions, strong hydrogen bonds (O-H···O, N-H···O), weak hydrogen bonds (C-H···O, C-H···N, O-H···π) and other forces such as halogen···halogen interactions (see also the article by J.P. Glusker in this volume).

Non-covalent interactions have been extensively used in crystal engineering, since they are the fundamental cause of the formation of crystals [4]. Hydrogen bonding, π-π interactions, and π-hydrogen interactions have been used in particular for this purpose. It is not our intention to describe every feature of the different intermolecular interactions, but we will briefly mention aspects of their strength and directionality that are important in their use in crystal engineering. Intermolecular interactions have been regarded as synthetic vectors or "glue" in designing new solids [9, 10]. Therefore if geometry, strength, and directionality can be recognized in the functional groups of a molecule, it is possible to rationalize its supramolecular solid state structure. Such a view would have been difficult to defend as recently as 1988 when Maddox made the following statement: "One of the continuing scandals in the physical sciences is that it remains in general impossible to predict the structure of even the simplest crystalline solids from a knowledge of their chemical composition" [11].

2.1 Strategies for Crystal Engineering

There are several approaches to the design of functional solids. Different strategies have been applied by several groups in order to obtain the desired intermolecular recognition. In 1954, Wald proposed that each component molecule could spontaneously assemble into an intact cell and suggested that each should contain all the necessary information to recognize and interact with other appropriate molecules [12].

Lawrence has suggested that certain factors that promote formation of small assembling complexes can also be applied to crystal engineering. The use of the appropriate solvent and conditions can enhance the directionality and geometry of non covalent interactions which are responsible for maintaining the structural integrity of a complex. He also suggests ways in which a supramolecular complex can be selectively isolated from solution in order to prevent dissociation of the aggregate to its component parts [13].

Most recently, Desiraju has drawn a parallel between organic synthesis and crystal engineering in an attempt to identify molecular recognition patterns in solids. Stoddart has also discussed approaches to synthetic supramolecular chemistry of nanosystems [14]. Traditionally structural chemists examined a large number of known crystal structures and attempted to eliminate certain possibilities on the basis of previous knowledge of crystal packing. However, it has been repeatedly observed that certain building blocks or "supramolecular synthons" display a clear pattern preference. Molecules that contain these building blocks tend to crystallize in specific energetically favorable arrangements that can coexist with efficient close packing. In this way, by identifying these synthons, and with the aid of the Cambridge Structural Database (CSD), it is possible to work backwards and "retrosynthetically" formulate empirical rules about the recognition patterns of various geometrical and functional groups [15]. Some examples of the synthons identified by Desiraju are depicted in Fig. 1.

The dimer and the catemer motifs are patterns commonly observed for carboxylic acids [16] as well as for amides (Fig. 1a) [17]. A dimer is formed whenever hetero-functional group recognition between a carboxylic acid and an amide is available (Fig. 1b). This finding has been well documented and used in connection with carboxylic acid receptors [18]. Desiraju has also identified synthons based on the same geometry but with different hydrogen bonding

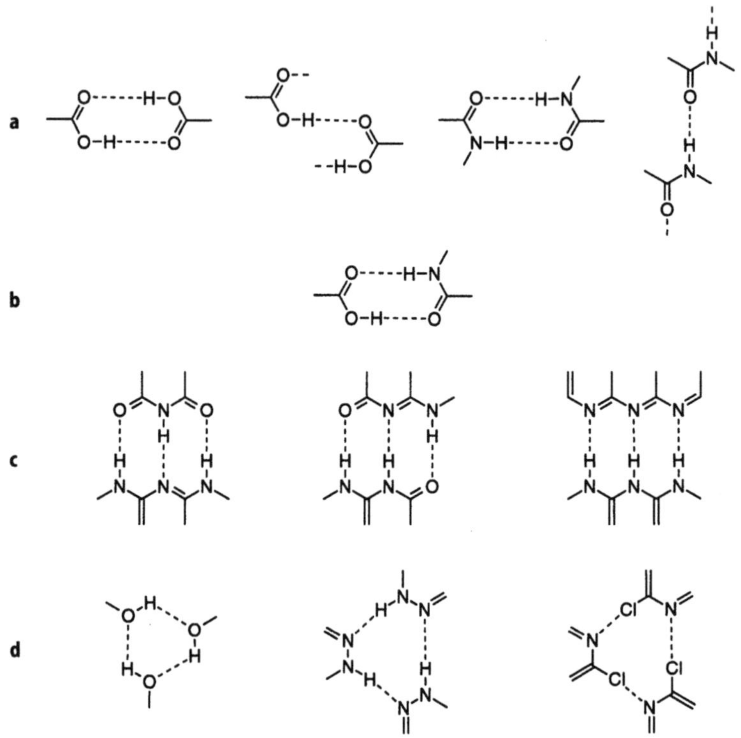

Fig. 1. Representative supramolecular synthons

directionality. For example, the top components in Fig. 1c have, in left to right order, acceptor-donor-acceptor, acceptor-acceptor-donor, and acceptor-acceptor-acceptor hydrogen bonding orientation. Their bottom counterparts have complementary functionalities. Figure 1d shows the formation of cyclic trimers based on strong hydrogen bonds or polarization interactions. The synthons described by Desiraju have been observed repeatedly in solid state structures and have been used in the design of crystals, as well as in other areas of supramolecular chemistry (see also the article by A. Nangia and G.R. Desiraju in this volume).

2.2 Patterns in Crystal Engineering: Tapes, Ribbons, and Sheets

In discussing tapes, ribbons, and sheets we should not ignore the competing formation of cyclic aggregates in many of the designed systems. There has been extensive discussion of the enthalpic and entropic influences on cyclic vs linear aggregate formation [19]. As a consequence, different approaches have been used to promote formation of cyclic aggregates rather than linear aggregates. These approaches will be presented throughout the sections of this article.

The terms tape and ribbon have not yet been defined in a way that allows them to be easily distinguished. In general, ribbons or tapes have been used to describe particular types of packing. For example, MacDonald and Whitesides have distinguished tapes from ribbons formed by amides in the following manner: "a tape motif is generated when each molecule is hydrogen bonded to two neighboring molecules, and when the hydrogen bonds between any two molecules form an eight membered ring; a ribbon motif is generated when each molecule is hydrogen bonded to three or more molecules, regardless of the connectivity of hydrogen bonds between the molecules" [20].

Since this article is a compilation of structures that possess different functionalities and geometries, we will distinguish between tapes and ribbons in the following manner. A tape will form when the functional groups involved in intermolecular interactions on a molecule are antiparallel or at 120° to each other – thus the molecule can interact with only two adjacent molecules in a linear or zigzag fashion. In a ribbon pattern, the functional groups may be oriented in different directions, giving some flexibility in the recognition to adjacent molecules. A sheet will involve the formation of two-dimensional arrays stabilized by intermolecular interactions. Most of the examples presented here will involve compounds with functional groups built around a cyclic template. A schematic representation of tapes, ribbons, cyclic and sheet aggregates is shown in Fig. 2. The triangle represents a template which may be the same or a different

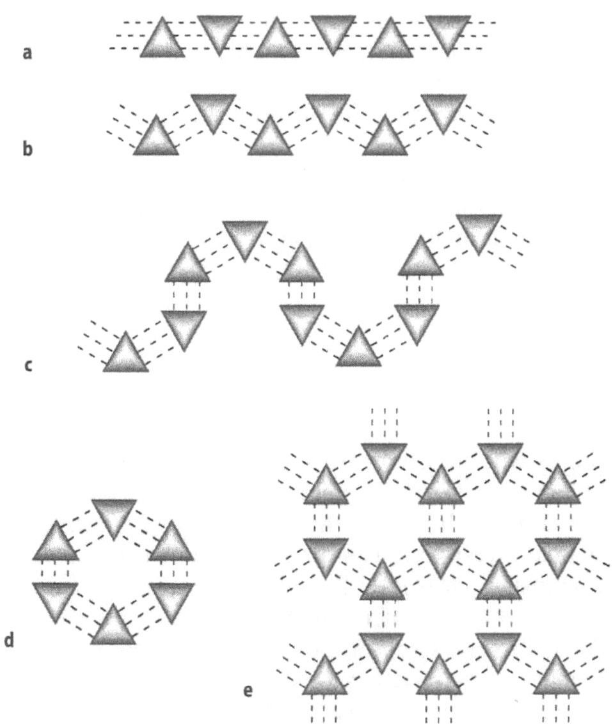

Fig. 2a–e. Schematic diagrams of: **a, b** tapes; **c** ribbons; **d** cyclic aggregates; **e** sheet aggregates

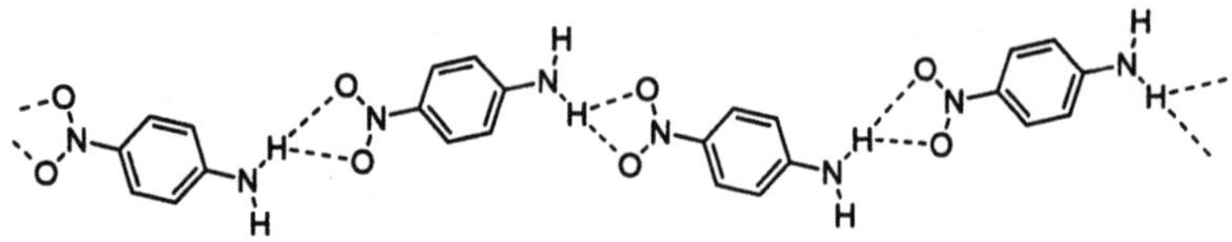

Fig. 3. Tape structure formed by *p*-nitroaniline (**1**)

component and the dashed lines represent intermolecular interactions. In terms of crystal engineering, the ribbon motif is less appealing since at this stage it is the more complex to understand and control.

Tapes and ribbons have been of interest in crystal engineering due to various applications in material sciences. Stabilization of the translation of molecules through intermolecular forces in a solid can generate polarity, which is a necessary condition for a number of physical properties. For example, small-molecule nitroaniline compounds show preference for a motif that involves one amino proton associating with both oxygens of a nitro group, leading to the formation of a tape structure (Fig. 3). In particular, *p*-nitroaniline (**1**) has been studied for its non-linear optical properties [21].

The sheet archetype has received much interest because of its potential to generate structural patterns that contain voids, which may in turn have applications in the area of zeolitic materials. Trimesic acid (**2**) (Fig. 9) has been extensively studied for this purpose [22]. However, Etter and Frankenbach have been able to form polar sheets when 2,5-dinitrobenzoic acid (**3**) and 4-aminobenzoic acid (**4**) are cocrystallized by using the nitro-amine and carboxylic acid recognition motif (Fig. 4) [23].

We will present examples of tapes, ribbons, and sheets based on hydrogen bonding and other intermolecular interaction patterns. The use of both strong hydrogen bonding groups such as carboxylic acids and amides, as well as, weak interactions such as C-H···O hydrogen bonds and iodo···nitro polarization interactions will be covered.

3 Hydrogen Bonding

Hydrogen bonding is without doubt the most studied intermolecular interaction, presumably due to its frequent presence in organic solids and biological molecules. Various attempts have been made to define the hydrogen bond [24, 25] and many detailed investigations of its properties have been carried out [9, 26]. Pioneering scientists such as Etter [27, 28], Leiserowitz and Schmidt [16, 17], Jeffrey and Saenger [29, 30], and Taylor and Kennard [31] have conducted extensive studies of the solid-state structures of organic compounds in a search for patterns, directionality, and strength in the occurrence of hydrogen bonding.

The selectivity and directional nature of hydrogen bonding has led to its extensive use in the construction and stabilization of large non-covalently bonded molecular and supramolecular structures. Hydrogen bonds are formed when

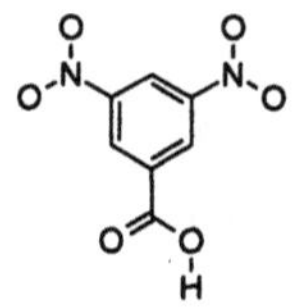

3

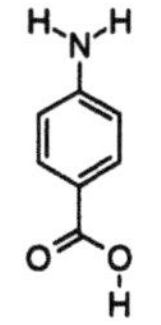

4

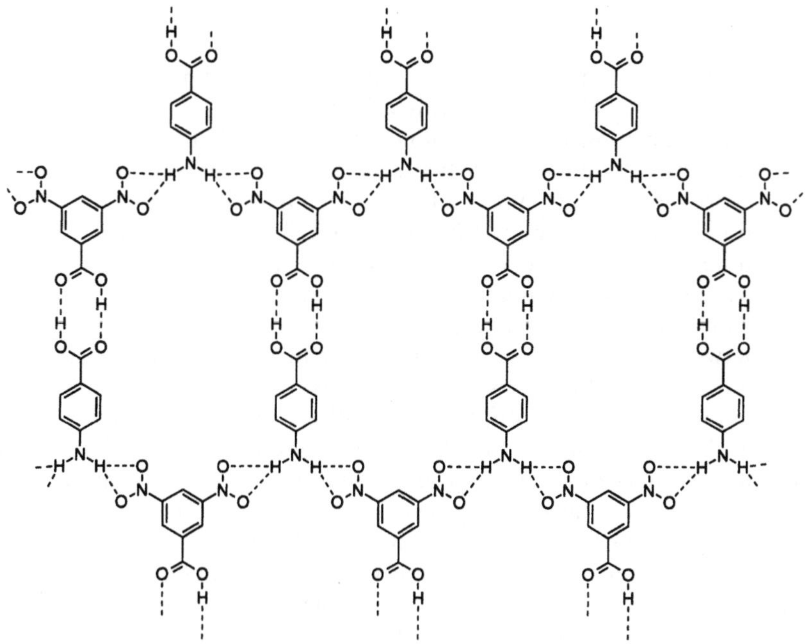

Fig. 4. Polar sheets formed by 3 and 4

a donor (D) with an available acidic hydrogen atom is brought into intimate contact with an acceptor (A) as shown in Fig. 5 [32]. The strength of different hydrogen bonds can vary from 1 kcal/mol for C-H···O [33] interactions to 40 kcal/mol for $[F\text{-}H\text{-}F]^-$ [34, 35]. In terms of geometry, the hydrogen bond has been studied by statistical investigations [31] and by X-ray diffraction analysis [36] (see also the article by J.P. Glusker in this volume).

One of the first detailed studies of hydrogen bonding patterns in the solid state was conducted by Etter. Her studies showed that functional groups display a preference for certain recognition patterns, leading to the conclusion that hydrogen bonding is not random, but controlled by the directional strength of the intermolecular interactions [27]. Thereby illustrating that hydrogen bonding can be used as a synthetic tool to affect profoundly the spatial arrangement of molecules and ions in the solid-state. Another important contribution by Etter was the development of graph sets in order to define the morphology of hydrogen bonded arrays. Recently, Bernstein et al. have expanded the approach

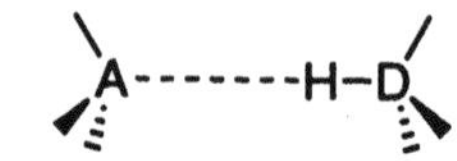

A = F, Cl, O, S, N, π-system
D = F, Cl, O, S, N, C

Fig. 5. Hydrogen bonds formed between an acidic hydrogen atom (D-H) and an acceptor (A)

for those who wish to apply this method to their hydrogen bonding analysis [37] (see also the article by M.R. Caira in this volume).

Perhaps the best known hydrogen bonding pattern is that of carboxylic acids. Carboxylic acids are divided into two sets based on the symmetry of their O-H···O hydrogen bonding patterns. They can form hydrogen bonding patterns that contain a center of inversion (the dimer motif) and also aggregate in acentric one-dimensional chains (catemers) resulting from the formation of hydrogen bonds to two or more neighboring acids (Fig. 6).

A set of 139 benzoic acids was analyzed by Frankenbach and Etter, and a relationship between the presence of inversion in the hydrogen bonding pattern and inversion symmetry in their crystal structure was observed. Of the 118 crystals that contained the dimer motif, 98% crystallized in centrosymmetric space groups. On the other hand, of the 21 crystals that contained the chain motifs, 52% crystallized in acentric space groups. This study suggested that hydrogen-bonded aggregates bias the crystal growth process [38]. Dicarboxylic acids with an antiparallel relationship between the hydrogen bonding groups can be expected to form one-dimensional (1-D) strands [16]. For example, terephthalic acid (1,4-benzene dicarboxylic acid) (5) self-assembles via the dimer motif to form 1-D tapes (Fig. 7). This packing is predictable because of the geometry and functionality of terephthalic acid. However, the arrangement of the chains with respect to their adjacent neighbors is less predictable, because there are no strong directional forces between them. Thus, terephthalic acid has at least two polymorphs [39].

a 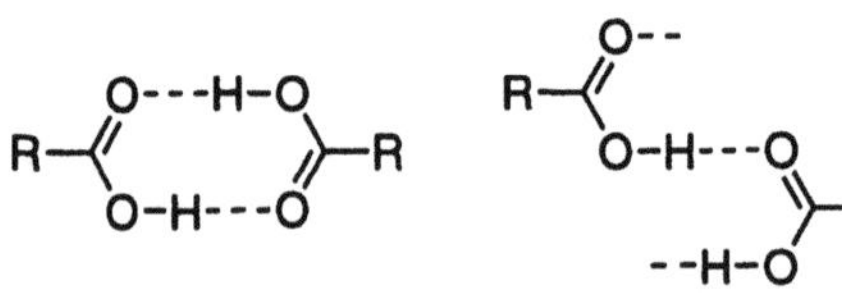 b

Fig. 6 a, b. Hydrogen bonding patterns of carboxylic acids: **a** dimer; **b** catemer

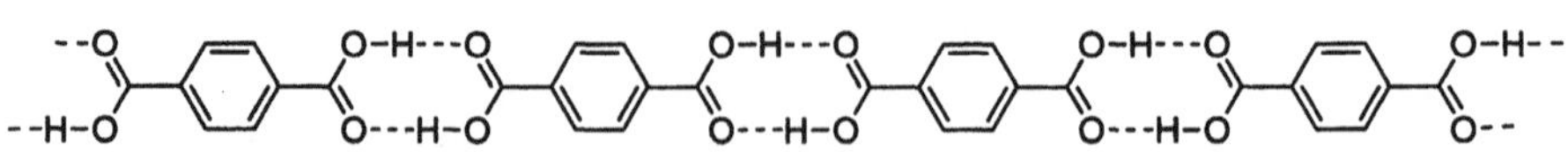

Fig. 7. Hydrogen bonded tape formed by terephthalic acid (5)

3.1
Carboxylic Acids

Carboxylic acids are commonly used as pattern controlling functional groups for the purpose of crystal engineering. The most prevalent hydrogen bonding patterns formed by carboxylic acids are dimers and catemers. The preference of pattern formation is based on the size of the R group in RCOOH. Acids containing small substituent groups (formic acid, acetic acid) form the catemer synthon, while most others (especially aromatic carboxylic acids) form dimers, although not exclusively [16]. Figure 8 shows the hydrogen bonding pattern of benzoic acid (**6**) and 3-(4-chlorophenyl)prop-2-ynoic acid (**7**) [15].

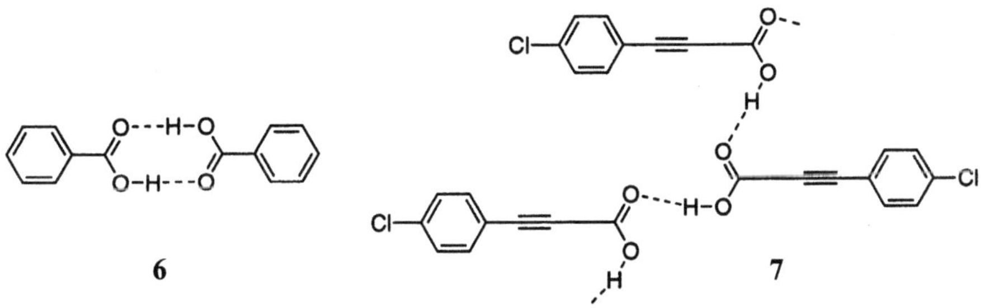

6 7

Fig. 8. Carboxylic acid dimer formed by **6** and polymeric hydrogen bonds formed by **7**

The dimer synthon formed by the carboxyl group has been used to assemble a variety of supermolecules due to its bidentate character which increases the strength of the interaction. For example, terephthalic acid (**8**) [39] and isophthalic acid (**9**) [40] form one dimensional tapes. Trimesic acid (**2**) with its threefold molecular symmetry forms a two dimensional hydrogen bonded sheet [41] and adamantane-1,3,5,7-tetracarboxylic acid (**10**) with its tetrahedrally disposed carboxy functionality forms a diamondoid network (Fig. 9) [42].

Our group has induced the formation of a hexameric cyclic array by appropriately modifying isophthalic acid. The reasoning behind this approach is that a bulky substituent at C-5 on isophthalic acid might disrupt the tape packing motif in the solid state. The structure of 5-decyloxyisophthalic acid (**11**) was successfully characterized by X-ray crystallography yielding the anticipated hydrogen bonding pattern (Fig. 10) [43]. Vapor pressure osmometry experiments indicated the presence of the cyclic hexamer in solution as well as in the solid state. The crystal structure of 5-octyloxyisophthalic acid (**12**) has also been determined and is observed to form the hydrogen bonding pattern in the same space group ($R\bar{3}$) [44].

The crystal structures of several 5-alkoxyisophthalic acids form tapes and ribbons when cocrystallized with pyrazine, pyrimidine and ethanol [45]. The carboxylic acid units of **13** engage in hydrogen bonding to other acids as well as to the bases and alcohol (Fig. 11 a,b). In Fig. 11 c the pyrimidine acts as a spacer between **14**, extending the tape formation. However, when acid **14** is recrystallized in its pure form, the formation of the carboxylic acid dimer motif occurs

8

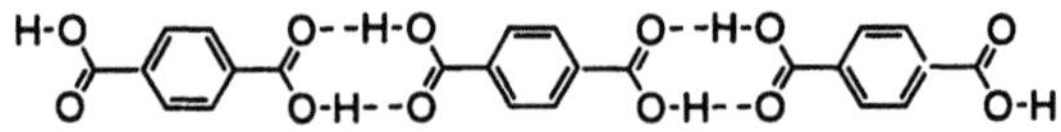

9

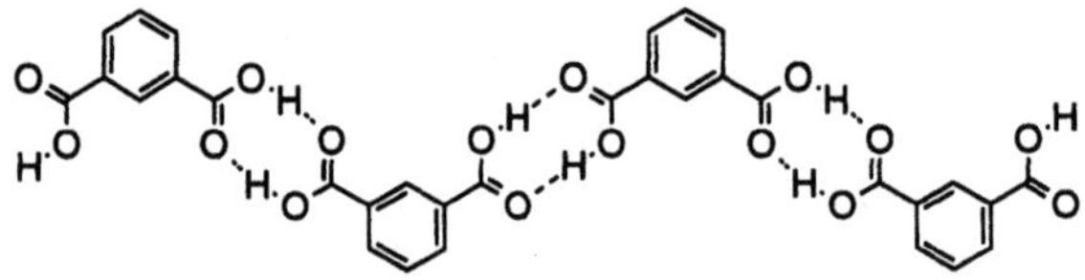

2

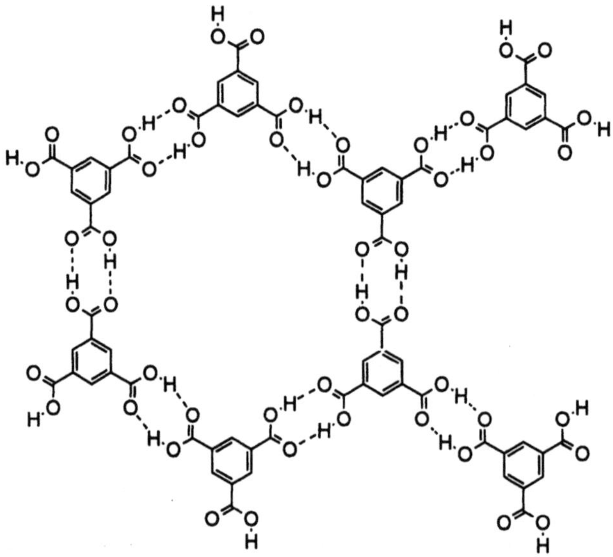

10

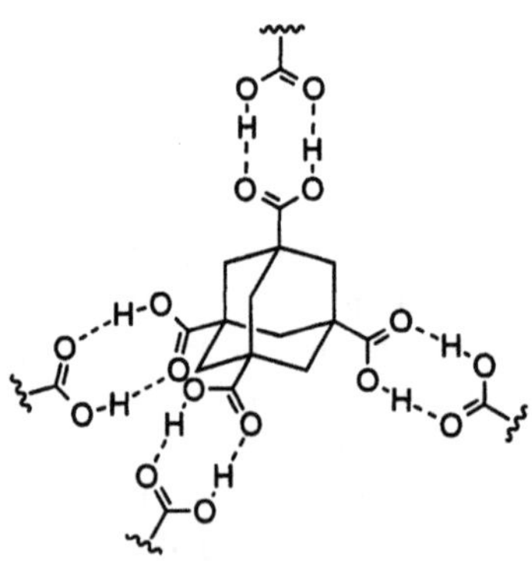

Fig. 9. Examples of one-dimensional tapes (**8**) and (**9**), two-dimensional sheets (**2**) and three-dimensional structures (**10**) held together by the carboxylic acid dimer

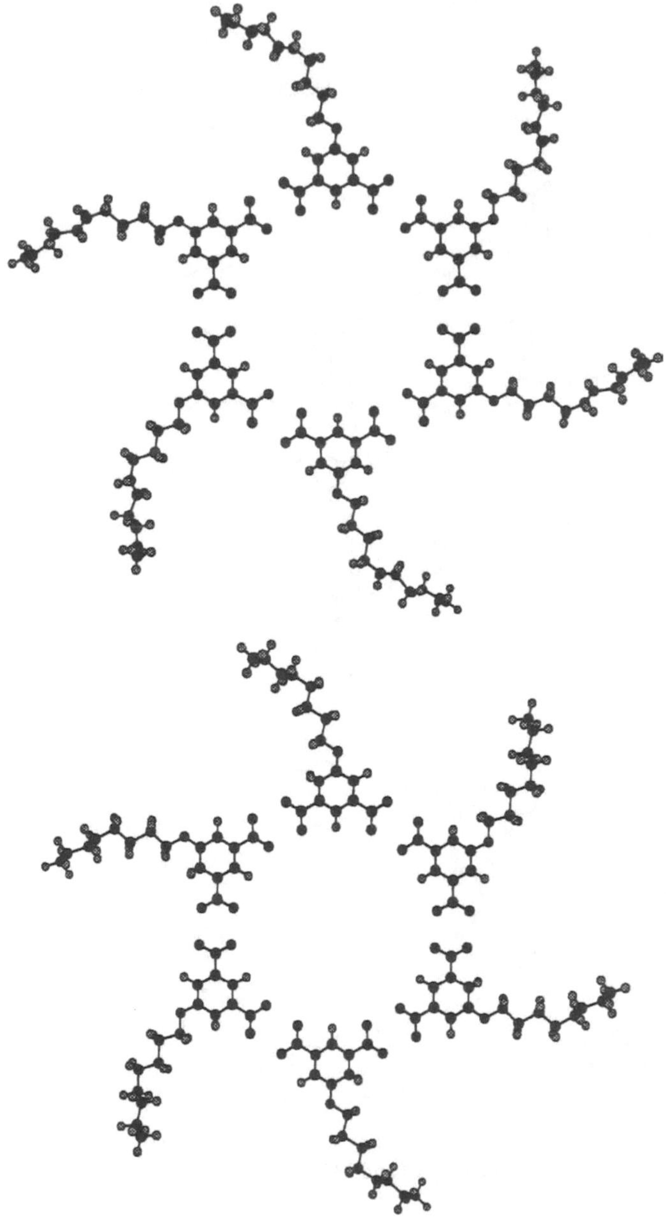

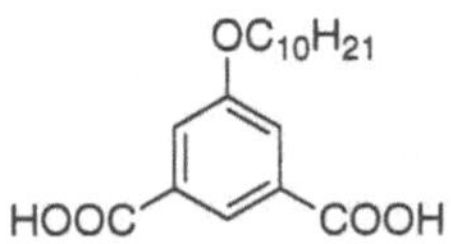

11

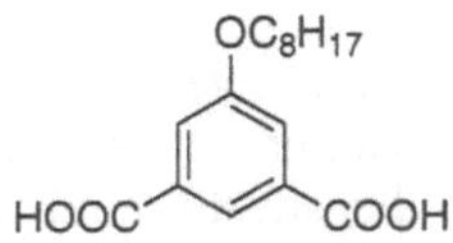

12

Fig. 10. Cyclic hexameric aggregates formed by 5-alkoxyterephthalic acids **11** and **12**

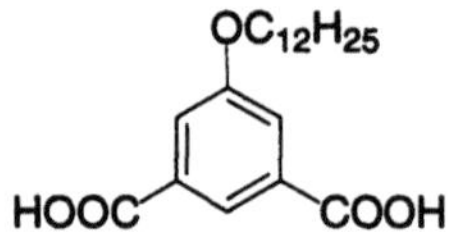

13

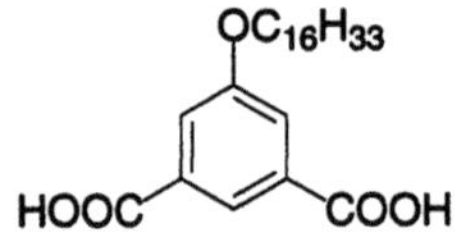

14

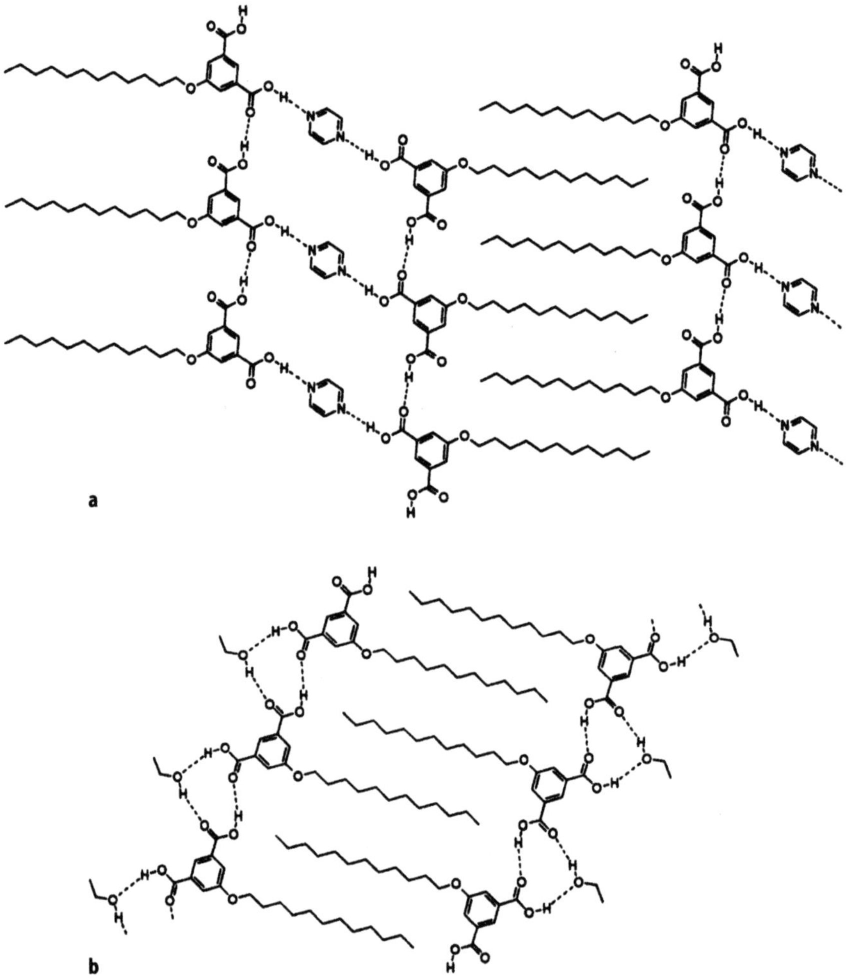

Fig. 11. Tapes and ribbons formed by **13** and **14**

between adjacent molecules. The aggregation of these molecules does not occur in a cyclic hexamer, but as tapes (Fig. 11 d) [46]. It is likely that, despite the presence of a bulky substituent in the 5-position, formation of tapes optimizes van der Waals interactions between the longer hydrophobic chains.

The primary characteristic in the crystal packing of trimesic acid (**2**) (Fig. 9) that has caught the interest of chemists is the formation of cavities that are 14 Å

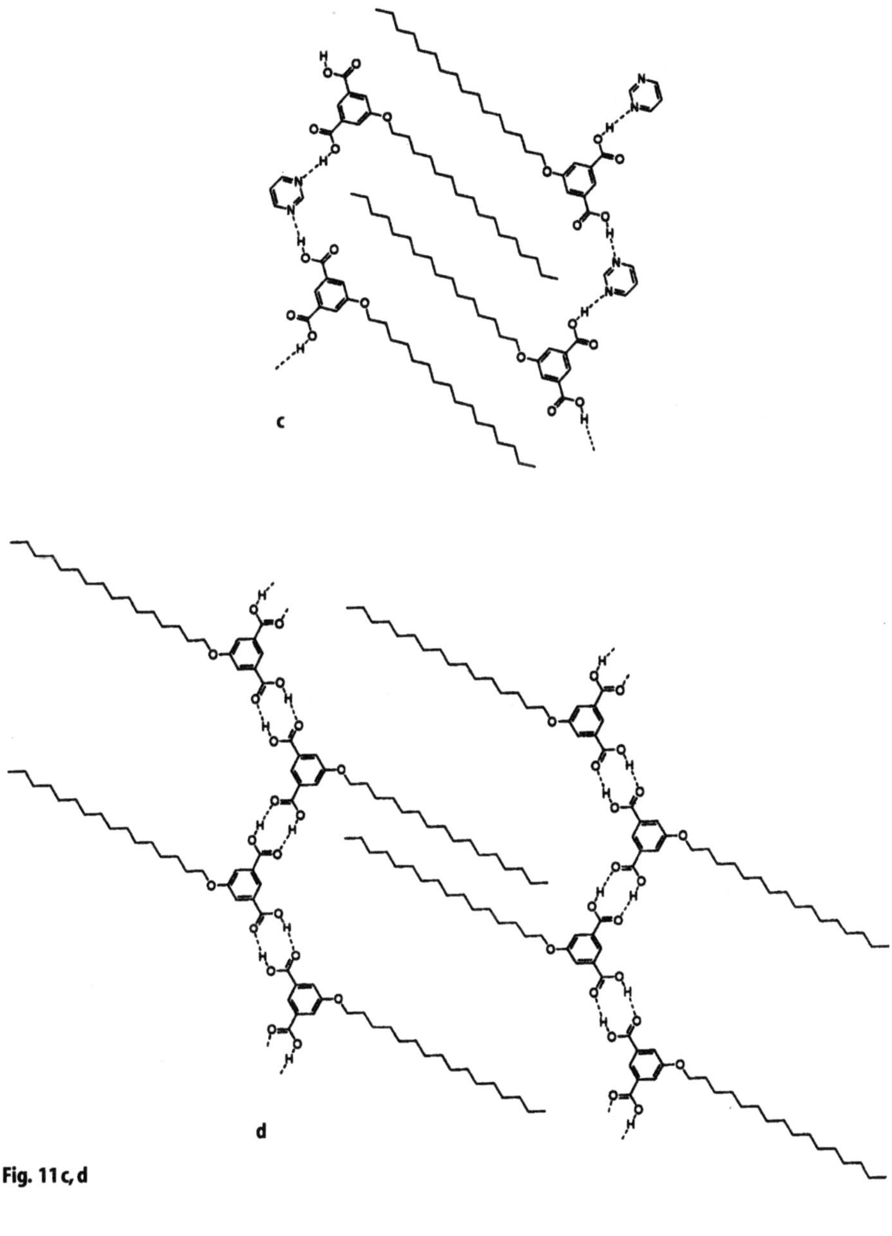

Fig. 11c, d

in diameter. These cavities have potential for the formation of inclusion compounds. However, in the room temperature crystal structure of **2**, the 14 Å holes are filled by perpendicular sheets of equivalent trimesic acid molecules in a triple catenation arrangement [41]. Herbstein has included a variety of guests in these pores – unbranched and branched paraffins (*n*-tetradecane, isooctane), long-chain alcohols (heptanol, octanol, decanol, and oleyl alcohol), alkenes

(1-octene, squalene), alicyclic (camphor) and other guests like methoxyethylether, propellane, and epichlorohydrin. Although in some cases catenation is prevented from occurring, at present only five hexagonal networks stacked on top of each other have been observed [22]. Sections of the crystal structures of **2** with *n*-tetradecane and oleyl alcohol included in the cavities are shown in Fig. 12 [47].

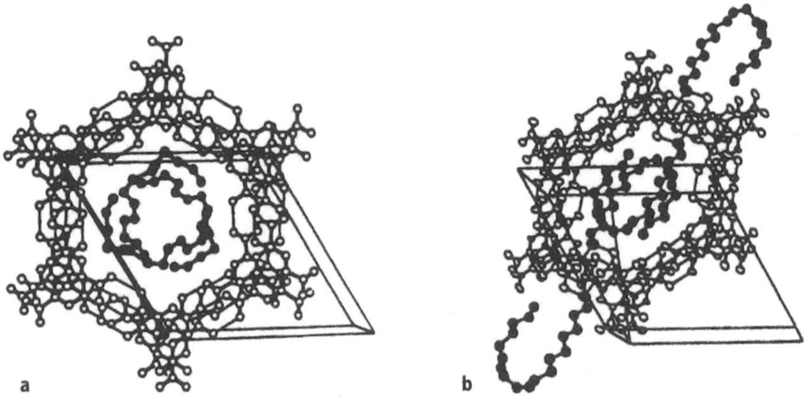

Fig. 12 a, b. Structures of inclusion compounds of **2** with: **a** *n*-tetradecane; **b** oleyl alcohol. (Reproduced with permission of Pergamon (1996) from Comprehensive supramolecular chemistry, vol 6, chap 3)

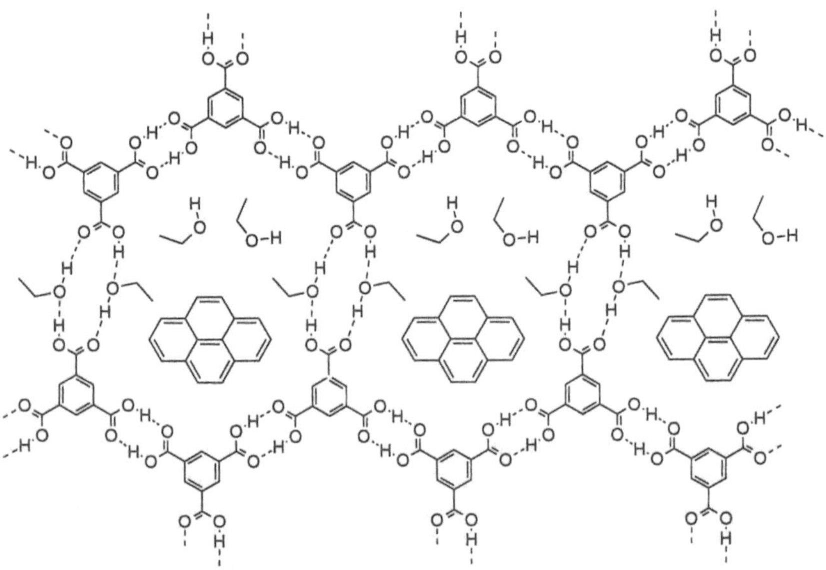

Fig. 13. Sheets formed in the crystal structure of [2 **2** · pyrene · 2EtOH]

Based on Herbstein's results using long chain guests, Zimmerman proposed to prevent interpenetration of the trimesic acid networks by including a large and relatively non-flexible guest molecule such as pyrene in the cavities. Indeed the pyrene is incorporated in the cavity as well as some ethanol solvent molecules (Fig. 13). The hydrogen bonded dimer in **2** is actually expanded by the presence of ethanol molecules, although the formation of the hexamer prevails in the sheet formation [48].

In the context of generating porosity, Zaworotko's group has used a modular approach in order to propagate the symmetry of trimesic acid. This strategy involves the use of hydrogen bonding spacers between units of **2** to expand the size of the cavities. In this case secondary amines were reacted with **2** to form salts **15** and **16** (Fig. 14). The resulting solid lattices were observed in the crystal

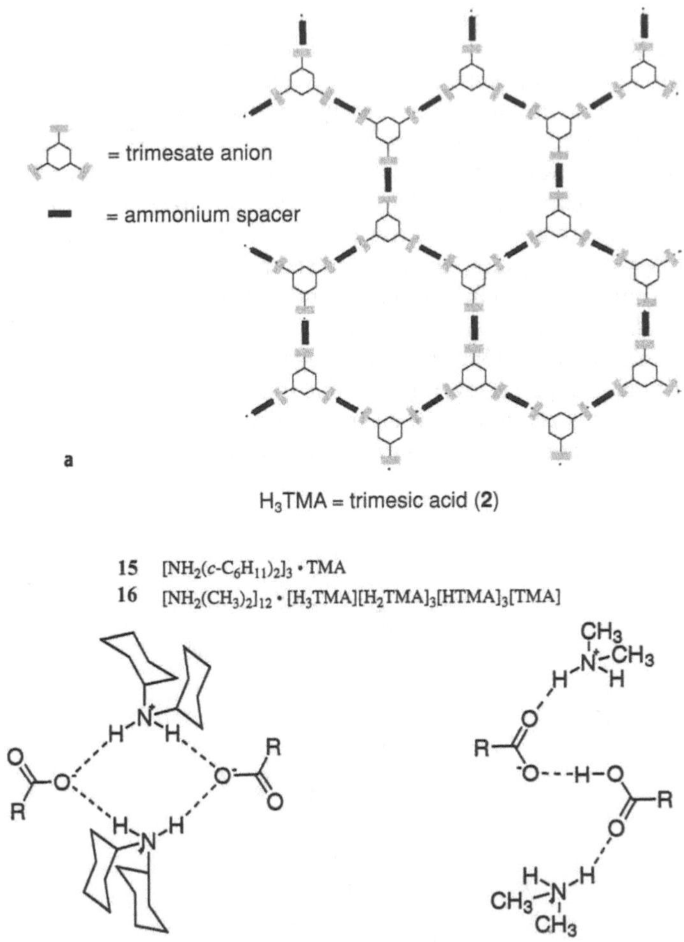

Fig. 14 a, b. Schematic representation of the modular strategy for increasing cavity size in **2**. **b** Hydrogen bonding pattern in salts **15** and **16**

structure of **15** to have expanded their cavities from 11 Å (taking into account van der Waals radius) to 12.7 Å. In **16** the hexamer cavities are distorted and therefore the diameter is reduced to 10.4 Å (see Fig. 14) [49].

3.2 Amides

A functional group that can parallel the hydrogen bonding directionality of the carboxylic acid synthon is the amide group. Primary and secondary amides [17], pyridones [50], and 2-aminopyrimidines [51] form eight-membered ring hydrogen bonded dimers via the bidentate interaction of the N-H and C=O groups. However, amides can also form linear chains through hydrogen bonding to adjacent molecules. For example, in the solid state benzamide (**17**) forms dimers and catemers (Fig. 15) [52].

Lauher et al. have successfully predicted the formation of a bidentate chain pattern formed by ureylenedicarboxylic acids. The analysis of six different structures has led to the conclusion that this interaction persists even in the presence of other strong hydrogen bonding groups (carboxylic acid) [53]. Compound **18** shows the formation of catenated bidentate urea interactions. However, in ureylenedicarboxylic acids of type **19**, both urea-urea and carboxylic acid dimer motifs are present creating a stable sheet arrangement. Although the bidentate interaction is polar, the adjacent strands in **19** take up an antiparallel orientation, therefore cancelling polarity in the sheets (Fig. 16).

Wuest et al. have synthesized dipyridones to take advantage of the amide hydrogen bonding dimer [50, 54]. In an elegant study they separated the pyridone functional groups by several different spacers. When the dipyridones are asymmetric, such as **20**, they form discrete hydrogen bonded dimers. However, the symmetric isomer **21** cannot satisfy its hydrogen bonding potential in the

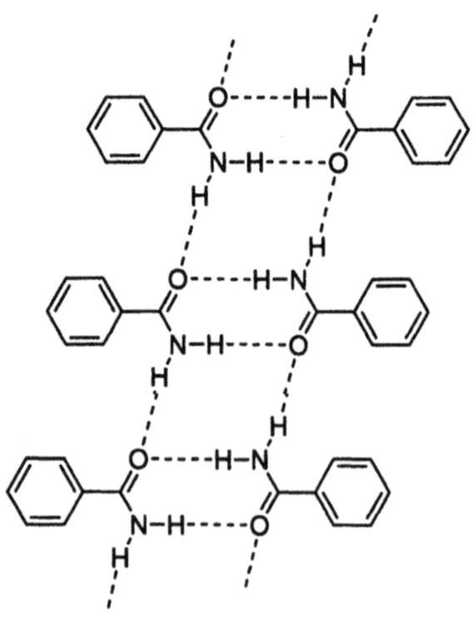

Fig. 15. Formation of dimers and catemers through hydrogen bonding in 17

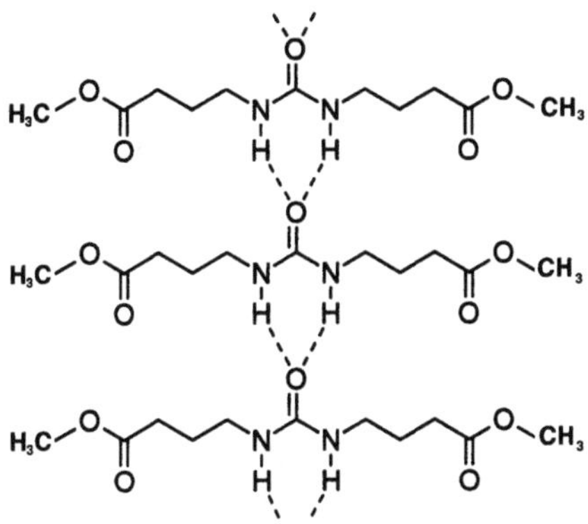

18

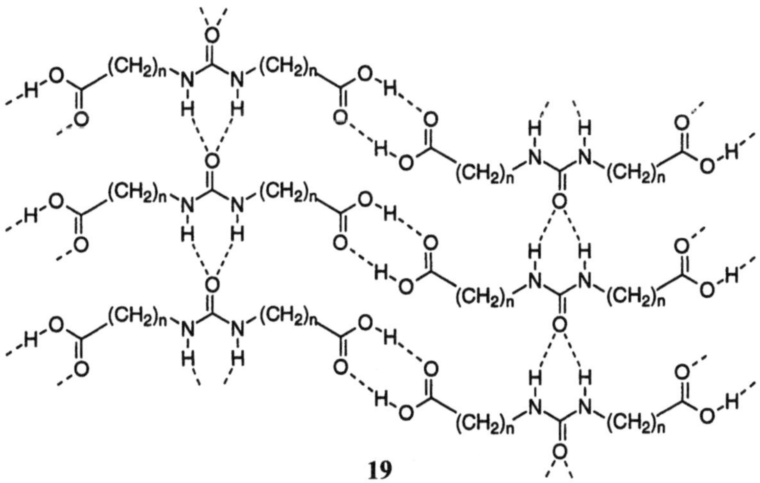

19

n = 1, 2, 3

Fig. 16. Formation of bidentate hydrogen bonded chains in **18** prevails in the presence of other hydrogen bonding groups (**19**)

dimer and is forced to form a tape (Fig. 17). Both of these patterns have been confirmed by vapor pressure osmometry as well as by X-ray crystallography.

Lehn et al. have used knowledge of the formation of the amide dimer to design cyclic hexamers. This group synthesized and separated the (+) and (–) forms of lactam **22**, [55], which were anticipated to form the cyclic hexamer depicted in Fig. 18a. However, the solid state structure of (–) **22** produced a tetramer. The racemate in contrast is found to assemble into a ribbon of alternating (+) and (–) subunits held together via hydrogen bonds (Fig. 18b).

The crystal structure of melamine (**23**) and cyanuric acid (**24**) shows, as expected, the formation of a two-dimensional sheet (Fig. 19) [56]. Whitesides et al. exploited the complementarity between substituted melamines and barbituric/cyanuric acids, which leads to efficient hydrogen bonding between such

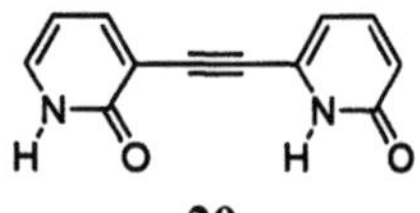

20

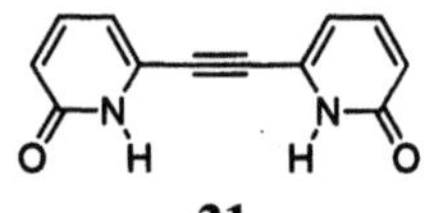

21

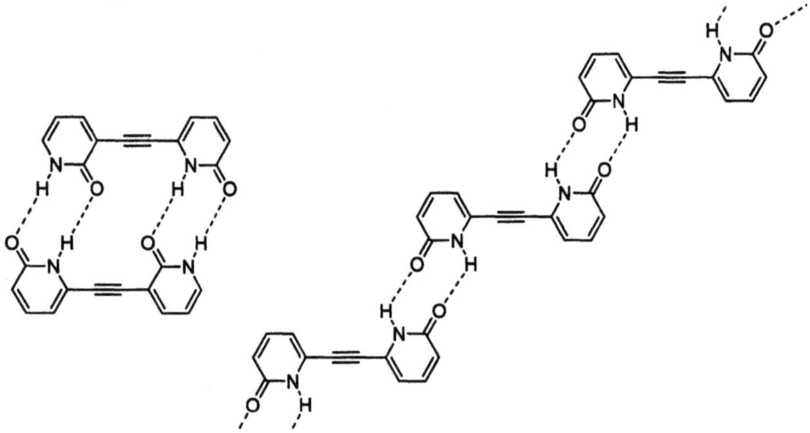

Fig. 17. Dimers and ribbons formed by asymmetric **20** and symmetric dipyridones **21**

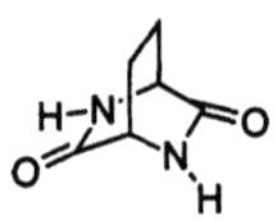

(±) **22**

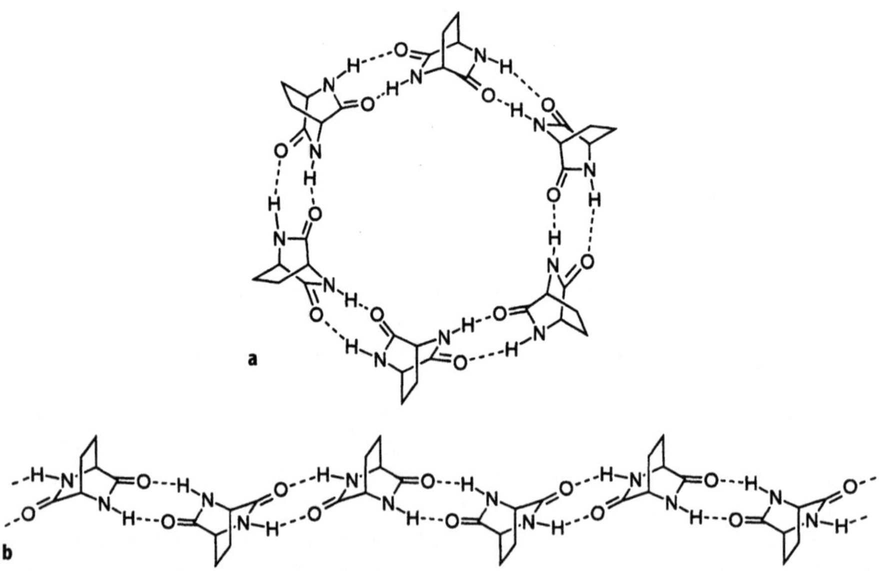

Fig. 18 a, b. Lactam **22** has potential to form: **a** cyclic hexamer in pure enantiomeric form; **b** ribbon when (+) and (–) units alternate

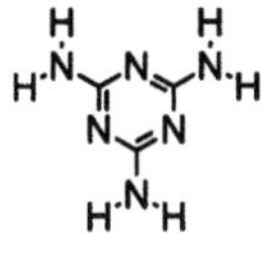

23

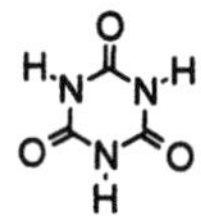

24

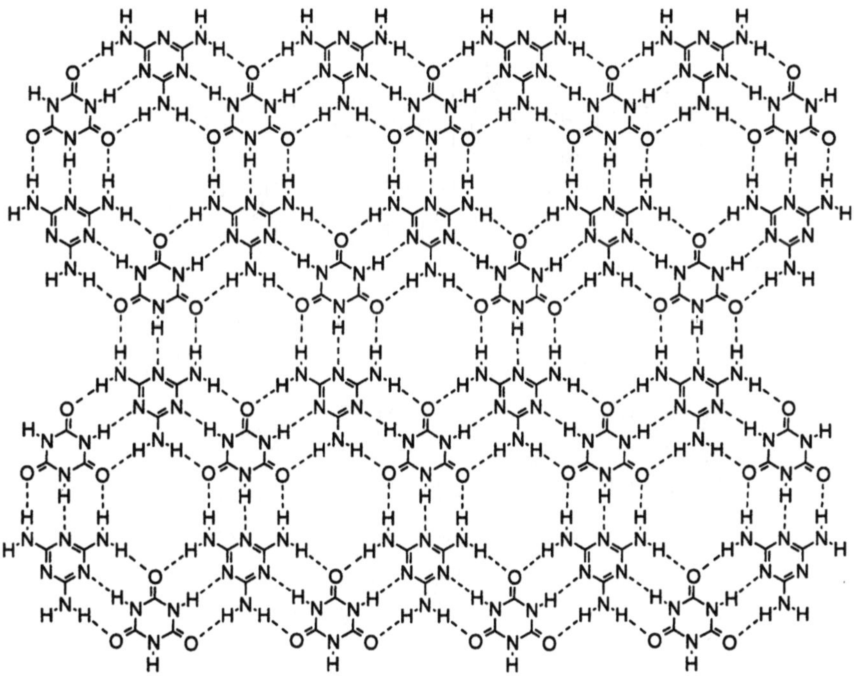

Fig. 19. Sheets formed by 23 and 24

compounds [57, 58]. They have found that the solid state structure formed by compounds of **25** and **26** is dependent upon steric interactions. Variations in the structure of each of the monomers were shown to have a significant effect on the architecture of the assembled system. For example, when the substituents on the phenyl rings in the melamine subunit are small (CH_3, F, Cl, Br, I) (Fig. 20a) tapes are formed. In the case of methyl and ethyl esters the components aggregate as ribbons (Fig. 20b). However, when the substituents are *t*-butyl groups a cyclic hexamer is formed (Fig. 20c) [59].

Lehn and coworkers designed a supramolecular cyclic aggregate based on subunits **25** and **27** [60]. The butyl substituent on the triaminopyrimidine was not bulky enough to induce cyclic aggregate formation and instead formed the tape shown in Fig. 21 (see also the article by A. Nangia and G.R. Desiraju in this volume).

Based on the previous result, Lehn et al. developed an elegant strategy for promoting the formation of cyclic aggregates based on hydrogen bonding function-

25 26

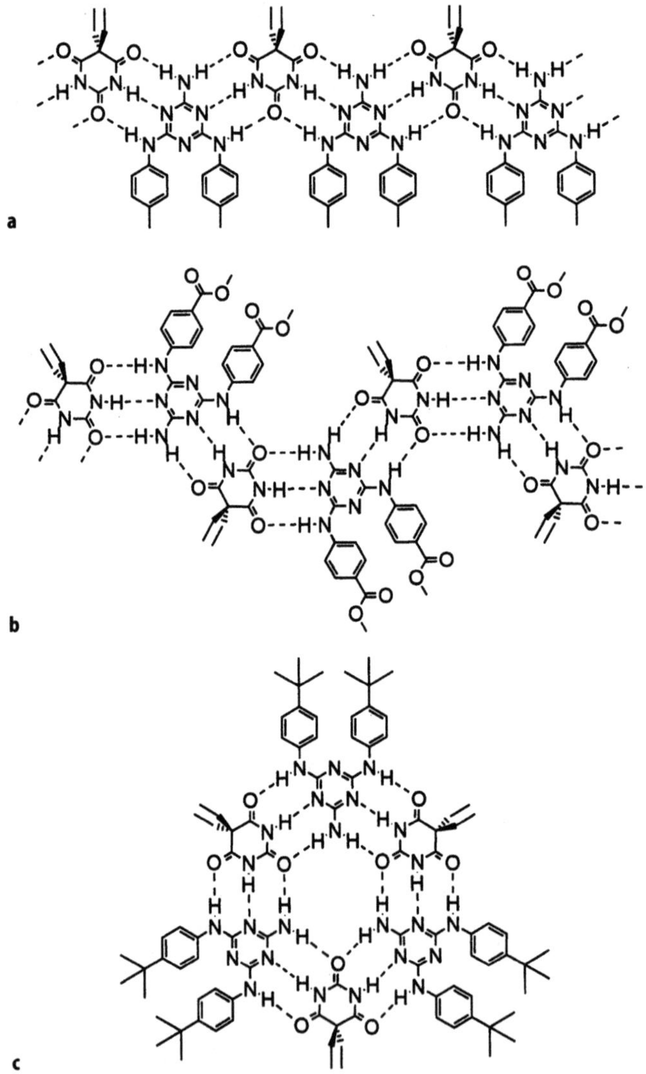

Fig. 20. Formed by **25** and **26**: **a** tapes; **b** ribbons; **c** cyclic aggregates

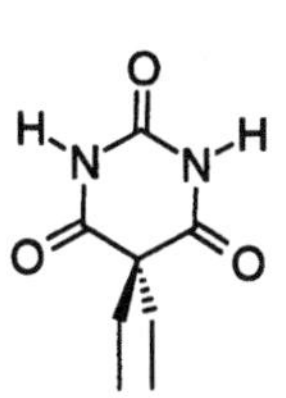

25

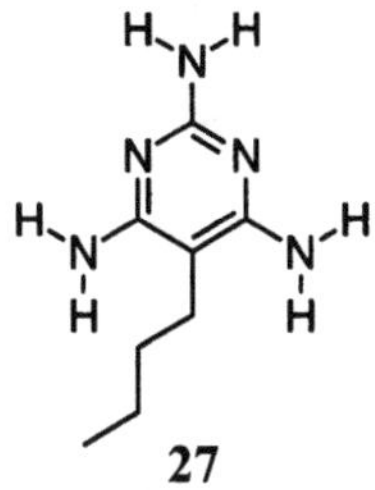

27

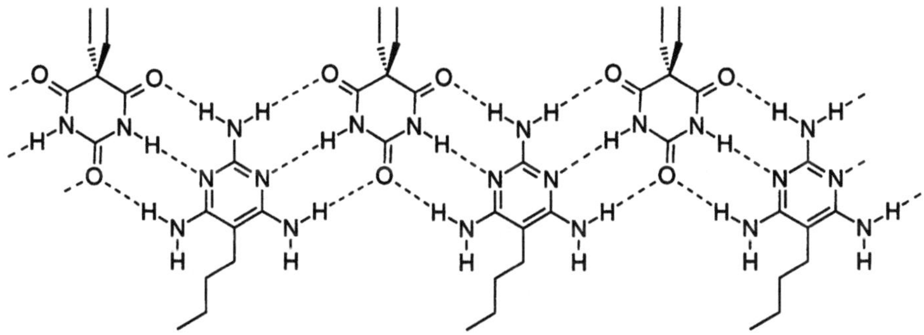

Fig. 21. Tapes formed by complementary monomers **25** and **27**

ality orientation. In monomers **28** and **29** (Fig. 22), the symmetry of the hydrogen bonding groups is destroyed. As a consequence, the monomers can only complement each other in a cyclic pattern. However, they are also self-complementary and as a consequence recrystallization of the 1:1 complex rendered crystals of compound **29** only, as seen in Fig. 22 [61].

2-Aminopyrimidine (**30**) has two hydrogen bond donors and two hydrogen bond acceptors, which are all used in its crystal structure pattern (Fig. 23) [62]. Etter and Adsmond recognized that carboxylic acids have the appropriate hydrogen bond functionality to form dimers with **30**. Several mono- and di-carboxylic acids were selected in this study to show the hydrogen bonding preference of **30** to carboxylic acids rather than to itself [51]. The ribbons formed in the crystal structure of **31**, which is a 1:1 complex formed by **30** and succinic acid, shows the extension of the motif formed by **30** (Fig. 23).

Our group has explored the hydrogen bond complementarity of carboxylic acids and 2-amidopyridines. A series of bis(amidopyridine) derivatives **32** were synthesized to explore the recognition properties of dicarboxylic acids **33** (Fig. 24). When the dicarboxylic acid is of the correct size, a discrete host-guest complex is formed (Fig. 24a). However, when the dicarboxylic acid is increased in length the hydrogen-bonding ribbon motif is observed (Fig. 24b–d) [18]. Furthermore, the relative orientation of the dicarboxylic and bis(amidopyridine) derivatives could be controlled by varying their relative length.

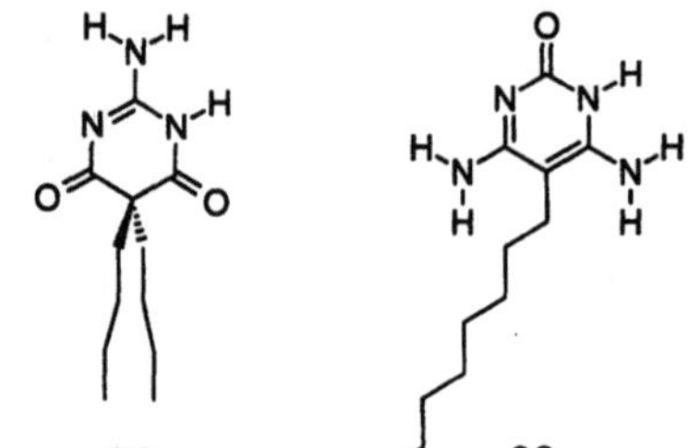

28 29

a

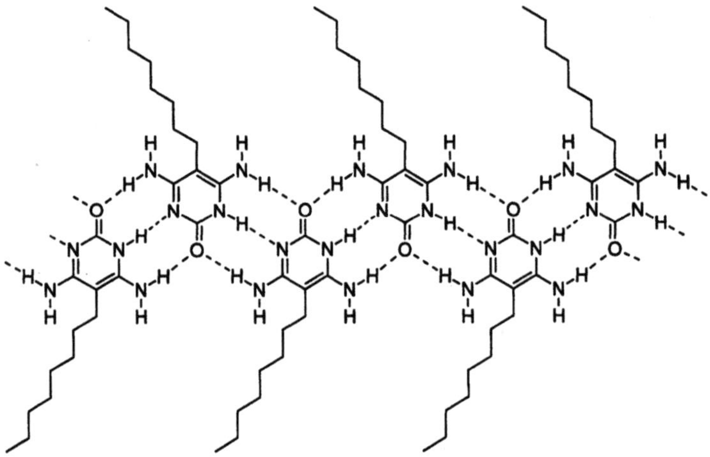

b

Fig. 22 a, b. Formulae of: **a** monomers; **b** tapes formed by self-assembly of **29**

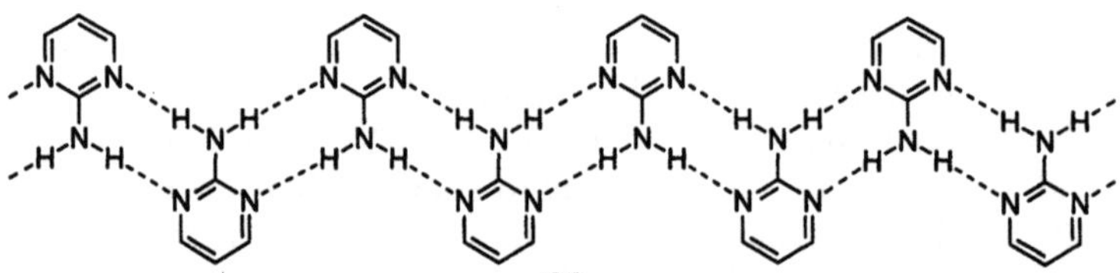

30

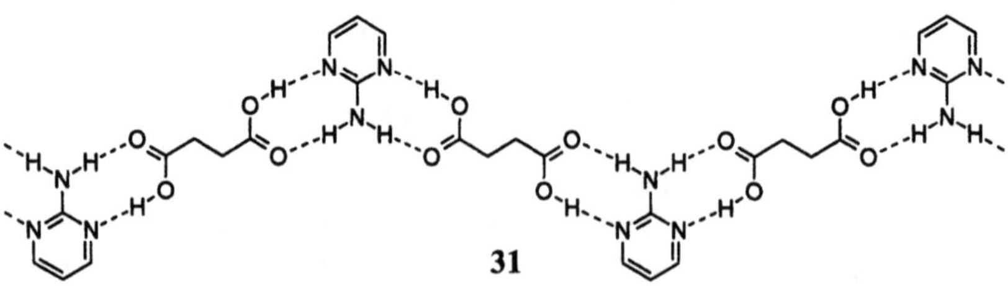

31

Fig. 23. Hydrogen bonding pattern in **30** and an extended pattern formed by **31**

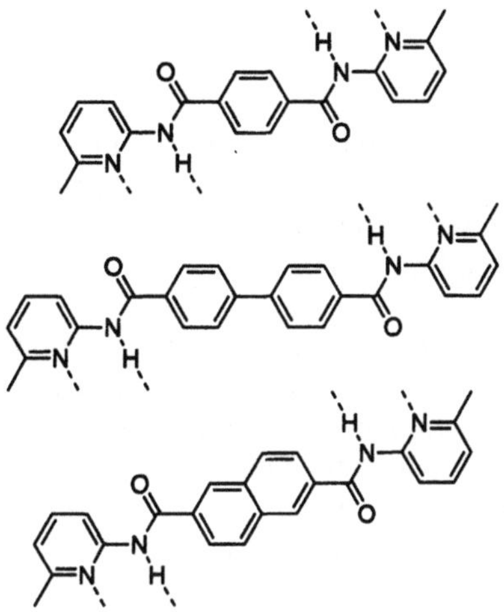

32

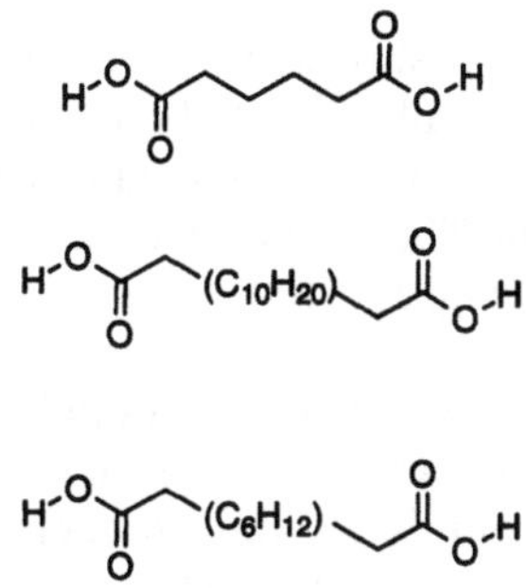

33

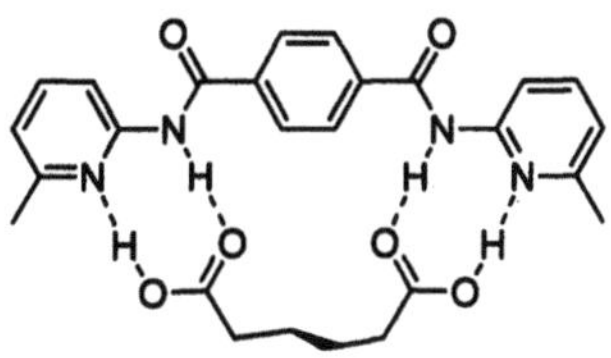

a

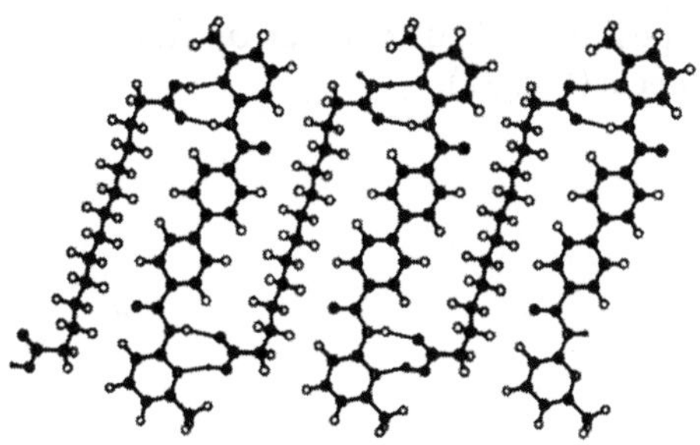

b

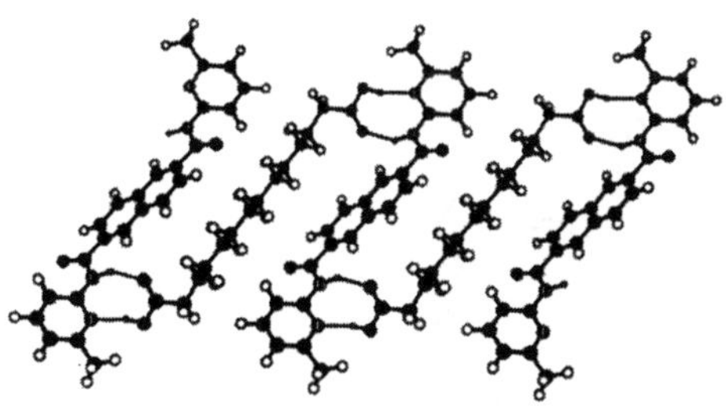

c

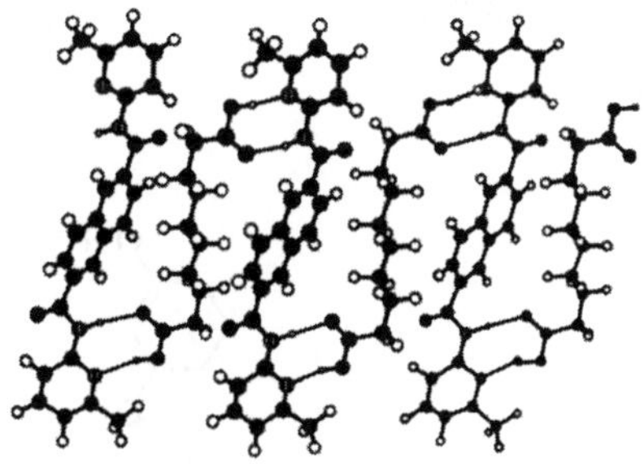

d

Fig. 24a–d. Hydrogen bonded networks formed by bis(amidopyridine) derivatives 32 and dicarboxylic acids 33. Examples of: **a** discreet host-guest complex; **b–d** ribbons

3.3
Other Strong Hydrogen Bonds

Etter et al. showed the formation of a cyclic hexamer when 1,3-hexanedione (**34**) is cocrystallized with benzene or deuterated benzene. The beauty of this work is that the hexamer aggregate is only formed when the guest solvent molecules are present. The crystal structure of 1,3-hexanedione forms a tape in the absence of guest (Fig. 25) [63].

Aoyama et al. have also developed the use of N-H···O hydrogen bonds to generate sheet structures [64]. They also take advantage of the π-stacking interactions of the molecules to control the supramolecular architecture of the solid. Compounds **35** and **36** cocrystallize to give a sheet held together by strong hydrogen bonding (Fig. 26). These controlled crystal lattices possess some selectivity in their inclusion of small molecules [65] and have even been shown to catalyze bimolecular Diels-Alder reactions within the pores [66] (more details are given in the article by Y. Aoyama in this volume).

Amidinium cations are interesting functional groups for the purpose of crystal engineering. Hosseini et al. have applied amidinium **37** in the design of tapes using the fumarate dianion **38** as a spacer [67]. In the crystal structure the tapes are linked together by fumaric acid spacers therefore propagating into sheets (Fig. 27).

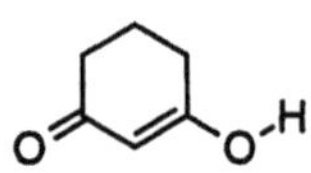

34

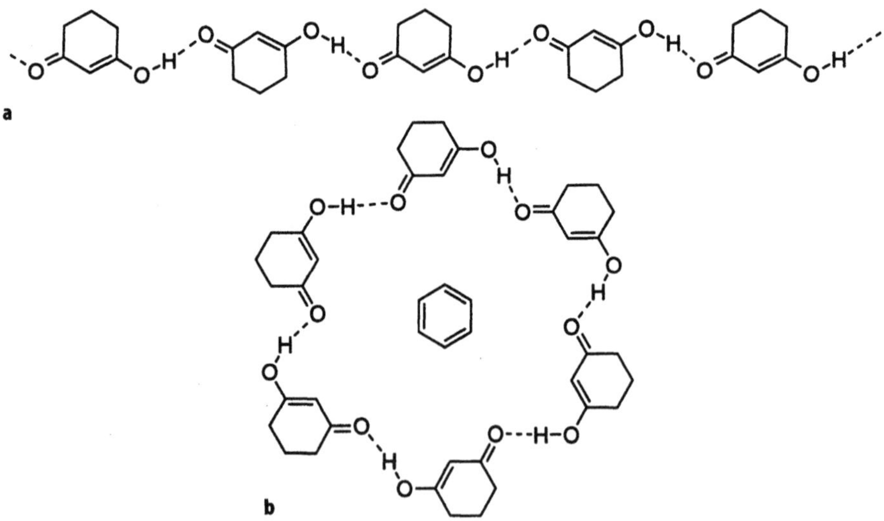

a

b

Fig. 25 a, b. Compound 34 forms: **a** tapes in the absence of guest; **b** cyclic hexamers in the presence of guest such as benzene

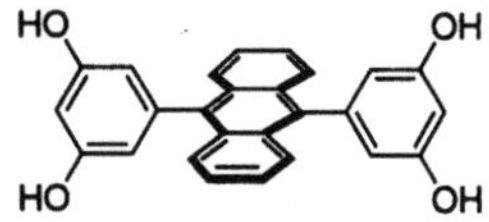

35

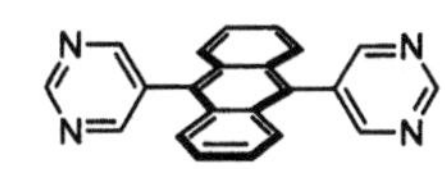

36

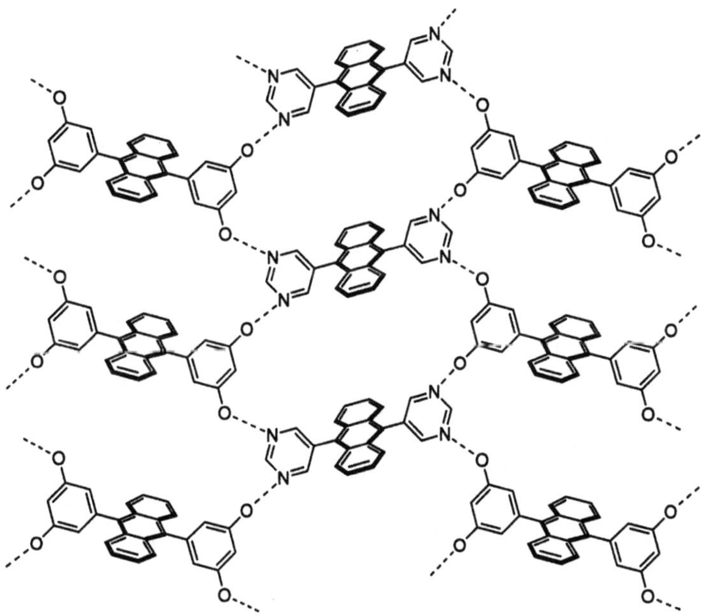

Fig. 26. Sheets formed in the crystal structure of **35** and **36** (hydrogens on oxygen have been omitted)

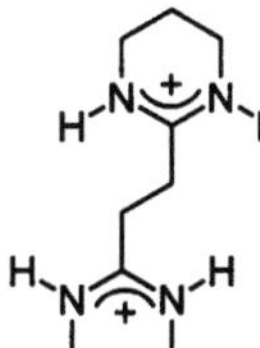

37

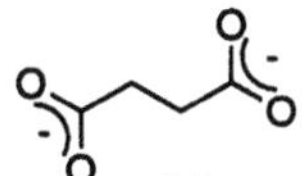

38

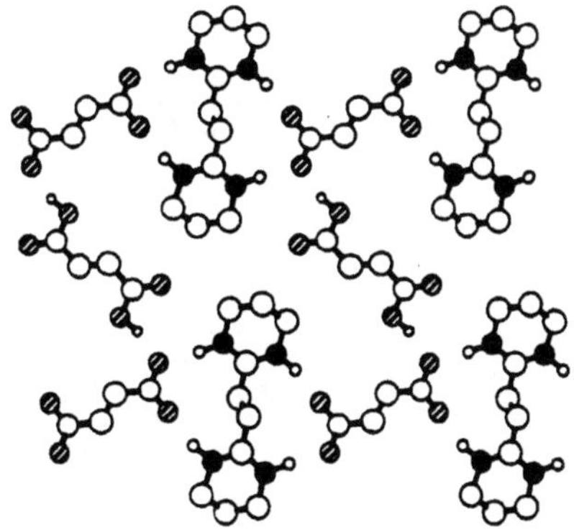

Fig. 27. X-ray structure of ribbons formed by **37** and **38** linked by fumaric acid [67]. (Reproduced with permission of the German Chemical Society (1997) from Angew Chem Int Ed Engl 36:102)

3.4 Weak Interactions

In addition to strong hydrogen bonds, the use of weak intermolecular interactions has been predicted to yield new and exciting results in the future of crystal engineering. Hydrogen bonds of type C-H···O(N) and O(N)-H···π as well as polarization induced iodo···nitro interactions have been examined and used in designing new solids [68].

The study of C-H···O hydrogen bonds has become increasingly important in our understanding of the packing of molecules in crystals [69]. In the past, observation of higher boiling [70] and melting [71] points in organic compounds have been ascribed to weak C-H···O hydrogen bond formation. In a survey of 113 published neutron diffraction organic crystal structures, Taylor and Kennard [72] concluded that C-H···O, C-H···N, and C-H···Cl interactions are more likely to be attractive than repulsive. In their study, typical C···O(N) distances occur in the range of 3–4 Å and the hydrogen bond angles are observed between 100–180°, but with a major cluster around 150–160° (see also the articles by J.P. Glusker and A.Nangia and G.R. Desiraju in this volume).

Desiraju has identified a series of recognition patterns which possess similar symmetry and directionality. The similarity in the formation of tapes by piperazine-2,5-dione (**39**), 1,4-benzoquinone (**40**) and 1,4-dicyanobenzene (**41**) can be observed in Fig. 28. The knowledge of the structure of **39** stabilized by N-H···O hydrogen bonds can serve as a guide in anticipating the formation of tapes by **40** and **41**. In the latter examples the directional properties of C-H···O and C-H···N hydrogen bonds are evident [68].

39

40

41

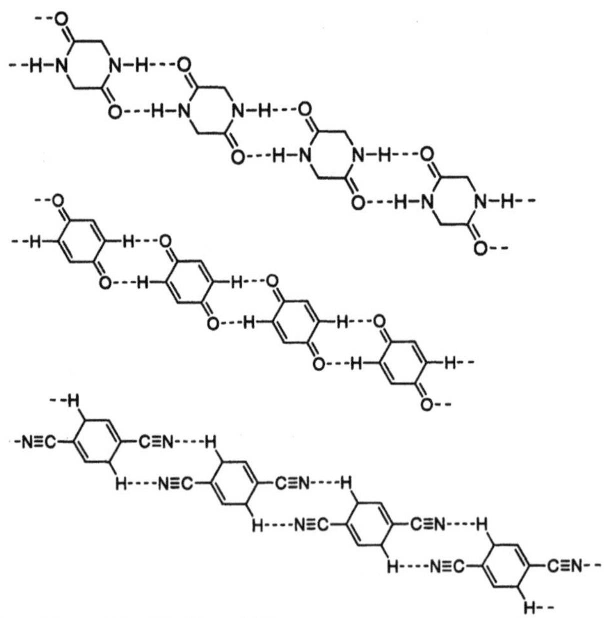

Fig. 28. Tape structures formed by **39**, **40** and **41**

The formation of C-H···O hydrogen bonds between the dimethylamino and nitro groups has been elegantly expressed by using *N,N'*-dimethylnitramine (**42**) as a prototype molecule. The 1:1 complexes of 4-nitrobenzoic acid with 4-(*N,N'*-dimethylamino)benzoic acid (**43**) [73] and 4-nitrobenzoic acid with 4-(*N,N'*-dimethylamino) cinnamic acid (**44**) [74] show a balance of strong and weak hydrogen bonding. In **43** and **44** the two acid molecules are linked by O-H···O hydrogen bonds to form dimers, as shown in Fig. 29. In addition, the dimers organize themselves to provide the right orientation to form C-H···O bonds between the NO_2 and the $N(CH_3)_2$ groups.

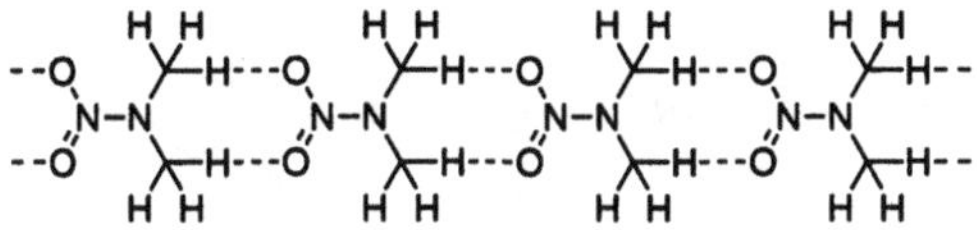

42

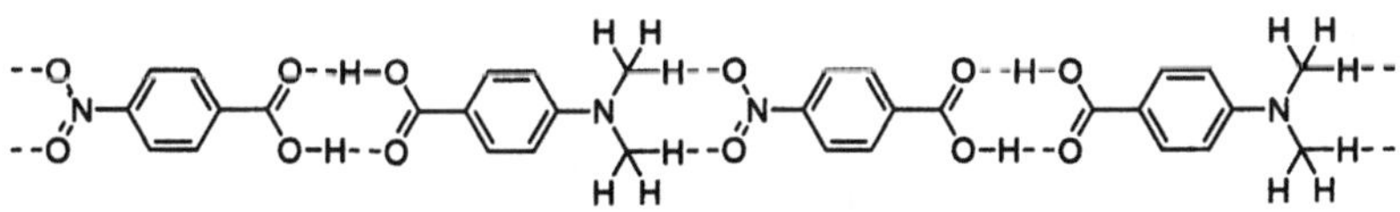

43

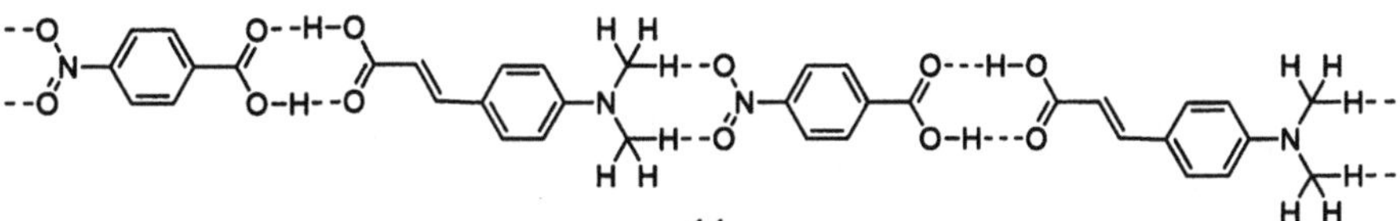

44

Fig. 29. C-H···O hydrogen bonding is evident in the tapes formed by **42, 43** and **44**

The observation of C-H···N hydrogen bonds between the C≡N groups and the aromatic H atoms in crystalline cyanocinnamic acids has led to the identification of 1,3,5-tricyanobenzene (**45**) as a candidate for the formation of sheets stabilized by C-H···N≡C bonds. Indeed, **45** forms a hexagonal network when crystallized in a 1:1 complex with hexamethylbenzene (Fig. 30a) [75]. The C···N distances in the sheets are 3.471(4) and 3.516(6) Å and C-H···N angles of 172(2) and 180°. In the same way, Desiraju et al. identified the presence of C-H···O hydrogen bond donors and acceptors in the same geometrical orientation in trimethylisocyanurate **46**. Not surprisingly, hexagonal networks form in its 1:1 complex with 1,3,5-trinitrobenzene (Fig. 30b) [76].

Weak polarization induced interactions of the C≡N···Cl type have also been used in crystal structure design. Cambridge Structural Database surveys show that this interaction has definite length (N···Cl 3.00–3.50 Å) and angle (C≡N···Cl 100–180°) attributes and therefore a certain structure-defining ability which can be used in crystal engineering. The structure of 2,3-dicyano-5,6-dichlorobenzenes (**47**) has been studied for the formation of molecular

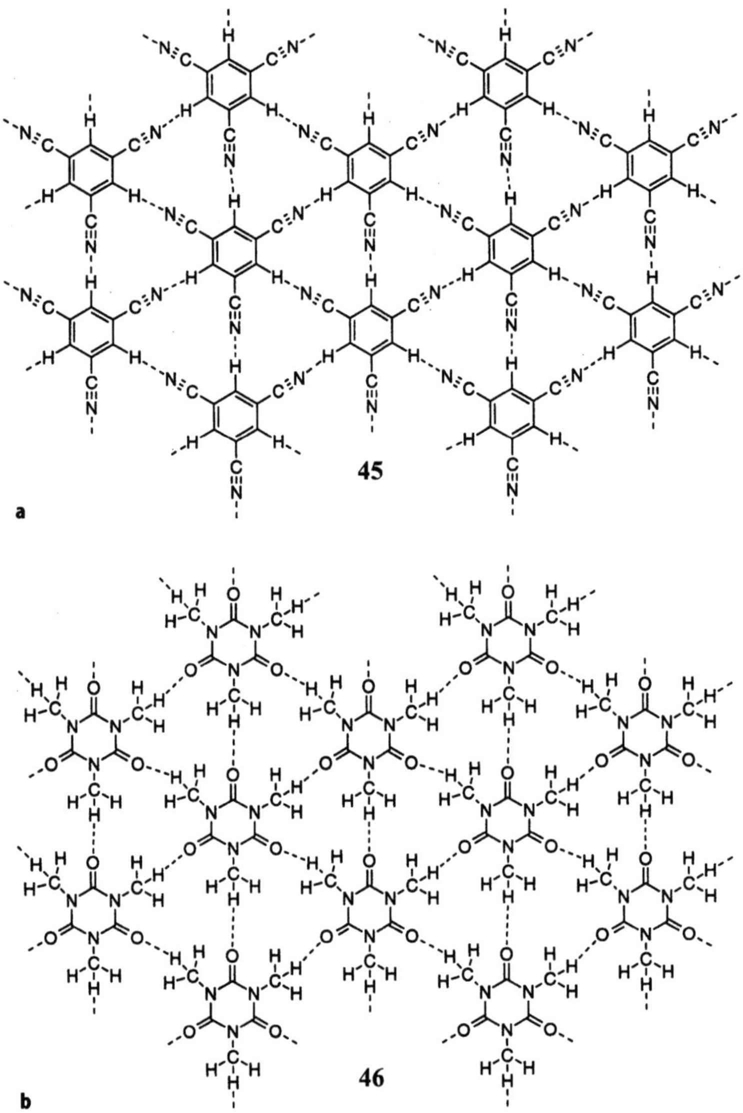

Fig. 30. Formation of sheets stabilized by C-H···N (**45**) and C-H···O (**46**) hydrogen bonds

tapes stabilized by C≡N···Cl interactions [77,78]. Eight compounds of this type have been studied crystallographically. The structure of the prototype 2,3-dicyano-5,6-dichlorobenzene consists of molecular tapes which are linked by a "dimeric loop" synthon (Fig. 31 a). When methoxy groups are in the 1,4-positions the same pattern is observed; however, as the bulk of the alkoxy groups is increased, the C≡N···Cl interactions become distorted to the point that hydrophobic interactions of the alkoxy chains are the major determinants of the struc-

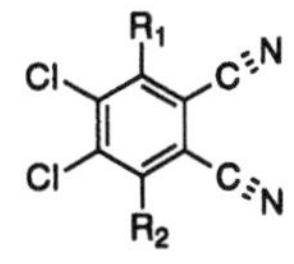

47

(1) $R_1 = R_2 = H$
(2) $R_1 = R_2 = OCH_3$
(3) $R_1 = R_2 = OC_2H_5$
(4) $R_1 = R_2 = n\text{-}OC_3H_7$
(5) $R_1 = R_2 = n\text{-}OC_4H_9$
(6) $R_1 = R_2 = n\text{-}OC_8H_{17}$
(7) $R_1 = OH, R_2 = n\text{-}OC_8H_{17}$
(8) $R_1 = R_2 = OH$

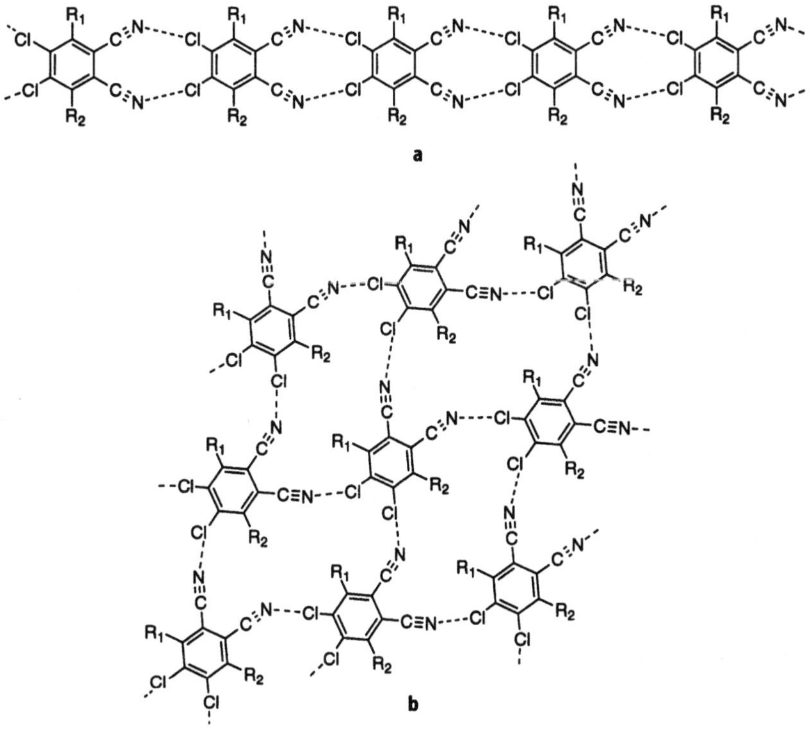

Fig. 31 a, b. C≡N···Cl mediated tapes are formed when R_1 and R_2 are small non-hydrogen bonding groups (1 and 2). **b** However, as the substituents get larger, sheets start forming

ture (Fig. 31b). In the case of the dihydroxy derivative, the Cl is unable to compete with the O-H groups and one can expect the O-H···N interaction to dominate in the crystal structure.

Pascal and Ho have also observed the formation of N···Cl donor-acceptor interactions in the formation of sheets by cyanuric chloride (**48**) (Fig. 32) [79]. They have also synthesized and studied molecular associations of receptors based on these donor-acceptor forces [80].

The iodo···nitro synthon has been used in the structures of 4-iodonitrobenzene (**49**), the 1:2 complex between 1,4-dinitrobenzene and 4-iodocinnamic acid (**51**) [81], and the 1:1 complex between 1,4-dinitrobenzene and 1,4-diiodobenzene (**50**) [82]. The target in the work was the formation of a linear tape

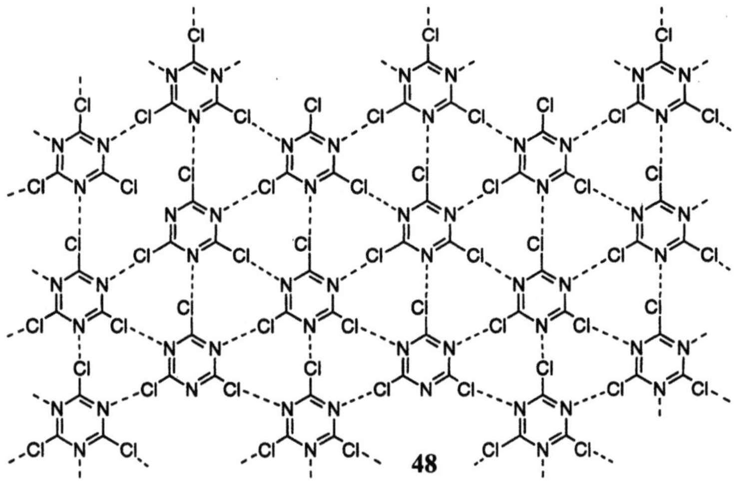

48

Fig. 32. The crystal structure of 48 forms a sheet through Cl···N interactions

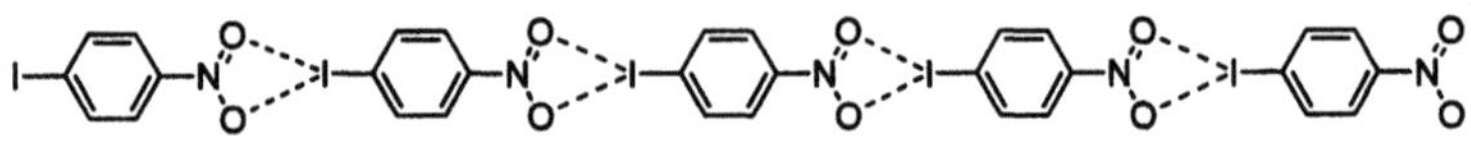

49

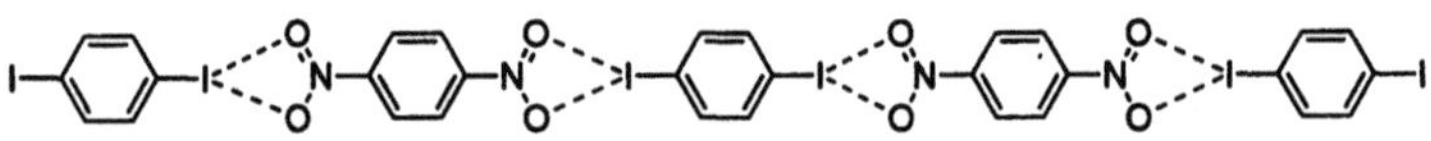

50

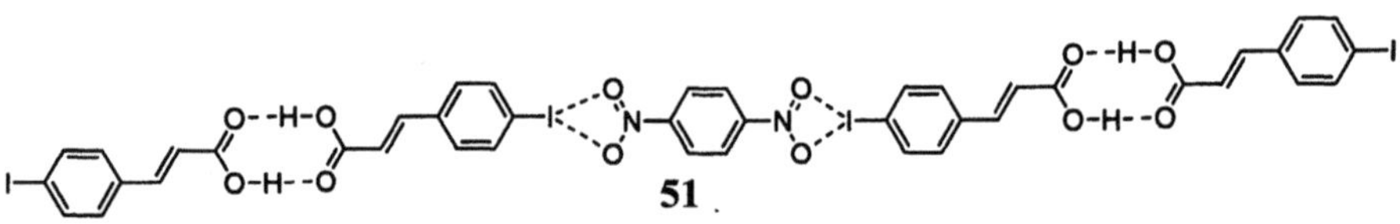

51

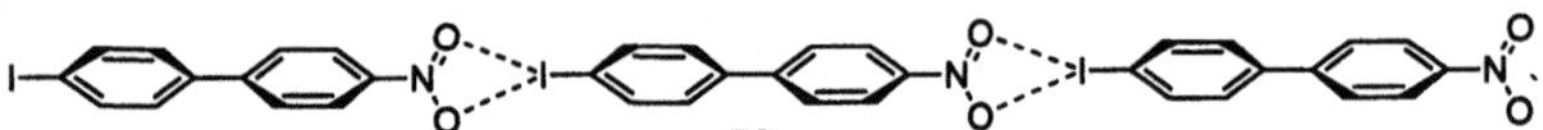

52

Fig. 33. The iodo···nitro synthon has been used in the design of tapes 49–52

based on the iodo···nitro polarization induced interaction. The iodo···nitro synthon has been separated by different spacers; in this case the carboxylic acid dimer has also been included. In order to design a non-centrosymmetric crystal, 4-iodo-4′-nitrobiphenyl (52) was prepared. The molecular chirality of the twisted biphenyl units and the parallel orientation of the ribbons lead to crystal chirality. This crystalline compound is second harmonic generation active. It has also been shown that this polarized interaction prevails in the structure of 52, due to the twisted biphenyl, which gives chirality to the structure [83] (Fig. 33).

4 Conclusion

In summary, we have shown that a wide range of molecular subunit shapes and orientations can be used to generate designed packing arrangements in the solid state. A key feature of these approaches is the exploitation of well defined and directional intermolecular interactions. In several cases, certain recognition motifs have been borrowed from solution based host-guest chemistry to impart the necessary control and specificity on multicomponent aggregation in the solid state. Both strong and weak hydrogen bonding interactions have been used extensively to produce ordered tapes, ribbons, and sheets. As a result, new solid materials with functional physical and chemical properties have been produced. These have included polar crystals with non-linear optical properties, as well as porous solids with well defined cavities that can selectively complex and even catalyze reactions of small organic molecules.

Acknowledgment. The portions of this work carried out at the University of Pittsburgh and Yale University were supported by the National Science Foundation (CHE 9213937).

5 References

1. Schmidt GMJ (1971) Pure Appl Chem 27:647
2. Desiraju GR (1989) Crystal engineering, the design of organic solids. Elsevier, Amsterdam
3. Lehn J-M (1988) Angew Chem Int Ed Engl 27:89
4. Dunitz JD (1991) Pure Appl Chem 63:177
5. Braga D, Grepioni F (1994) Acc Chem Res 27:51
6. Hulliger J (1994) Angew Chem Int Ed Engl 33:143
7. Vainshtein BK, Fridkin VM, Indenbom VL (1994) Structure of crystals, 2nd edn. Springer, Berlin
8. Desiraju GR (1995) The crystal as a supramolecular entity. Wiley, Chichester
9. Aakeröy CB, Seddon KR (1993) Chem Soc Rev 22:397
10. Zaworotko MJ (1994) Chem Soc Rev 283
11. Maddox J (1988) Nature 335:201
12. Wald G (1954) Sci Am 191:44
13. Lawrence DS, Jiang T, Levett M (1995) Chem Rev 95:2229
14. Fyfe MCT, Stoddart JF (1997) Acc Chem Res 30:393
15. Desiraju GR (1995) Angew Chem Int Ed Engl 34:2311
16. Leiserowitz L (1976) Acta Crystallogr B32:775
17. Leiserowitz L, Schmidt GMJ (1969) J Chem Soc 2372
18. Garcia-Tellado F, Geib SJ, Goswami S, Hamilton AD (1991) J Am Chem Soc 113:9265
19. Lady JH, Whetsel KB (1964) J Phys Chem 68:1001
20. MacDonald JC, Whitesides GM (1994) Chem Rev 94:2383
21. Panunto TW, Urbánczyk-Lipkowska Z, Johnson R, Etter MC (1987) J Am Chem Soc 109:7786
22. Herbstein FH (1996) 1,3,5-Benzenetricarboxylic acid (trimesic acid) and some analogues. Lehn J-M, Atwood JL et al. (eds) In: Solid state supramolecular chemistry: crystal engineering. Pergamon, New York, chap 3. See also: Herbstein FH (1987) In: Weber E (ed) Molecular inclusion and molecular recognition – clathrates I. Topics in Current Chemistry, vol 140. Springer, Berlin Heidelberg New York, p 107
23. Etter MC, Frankenbach GM (1989) Chem Mater 1:10

24. Zeegers-Huyskens T, Huyskens P (1991) Intermolecular forces – an introduction to modern methods and results. Springer-Verlag, Berlin Heidelberg New York
25. Atkins PW (1989) General chemistry. Scientific American Books, New York
26. Vinogradov SN, Linnell RH (1971) Hydrogen bonding. Van Nostrand Reinhold, New York
27. Etter MC (1990) Acc Chem Res 23:120
28. Etter MC (1991) J Phys Chem 95:4601
29. Jeffrey GA, Saenger W (1991) Hydrogen bonding in biological structures. Springer, Berlin
30. Jeffrey GA (1997) An introduction to hydrogen bonding. Oxford University Press, New York
31. Taylor R, Kennard O (1984) Acc Chem Res 17:320
32. Philp D, Stoddart JF (1996) Angew Chem Int Ed Engl 35:1155
33. Turi L, Dannenberg JJ (1993) J Phys Chem 97:7899
34. Emsley J (1980) Chem Soc Rev 9:91
35. McMahon TB, Larson JW (1982) J Am Chem Soc 104:5848
36. Taylor R, Kennard O (1983) Acta Crystallogr B39:133
37. Bernstein J, Davis RE, Shimoni L, Chang N-L (1995) Angew Chem Int Ed Engl 34:1555
38. Frankenbach GM, Etter MC (1992) Chem Mater 4:272
39. Bailey M, Brown CJ (1967) Acta Crystallogr 22:387
40. Alcala R, Martinez-Carrera S (1972) Acta Crystallogr B28:1671
41. Duchamp DJ, March RE (1969) Acta Crystallogr B25:5
42. Ermer O (1988) J Am Chem Soc 110:3747
43. Yang J, Marendaz J-L, Geib SJ, Hamilton AD (1994) Tetrahedron Lett 35:3665
44. Yang J (1996) PhD thesis, University of Pittsburgh
45. Valiyaveettil S, Enkelmann V, Müllen K (1994) J Chem Soc Chem Commun 2097
46. Eichhorst-Gerner K, Stabel A, Moessner G, Declerq D, Valiyaveettil S, Enkelmann V, Müllen K, Rabe JP (1996) Angew Chem Int Ed Engl 35:1492
47. Herbstein FH, Kapon M, Reisner GM (1987) J Inclusion Phenom 5:211
48. Kolotuchin SV, Fenlon EE, Wilson SR, Loweth CJ, Zimmerman SC (1995) Angew Chem Int Ed Engl 34:2654
49. Melendez RE, Sharma CVK, Zaworotko MJ, Bauer C, Rogers RD (1996) Angew Chem Int Ed Engl 35:2213
50. Ducharme Y, Wuest JD (1988) J Org Chem 53:5787
51. Etter MC, Adsmond DA (1990) J Chem Soc Chem Commun 589
52. Penfold BR, White JCB (1959) Acta Cryst 12:130
53. Zhao X, Chang Y-L, Fowler FW, Lauher JW (1990) J Am Chem Soc 112:6627
54. Gallant M, Viet MTP, Wuest JD (1991) J Org Chem 56:2284
55. Brienne M-J, Gabard J, Leclercq M, Lehn J-M, Cesario M, Pascard C, Cheve M, Dutruc-Rosset G (1994) Tetrahedron Lett 35:8157
56. Wang Y, Wei B, Wang Q (1990) J Crystallogr Spectrosc Res 20:79
57. Zerowski JA, Seto CT, Wierda DA, Whitesides GM (1990) J Am Chem Soc 112:9025
58. Zerkowski JA, MacDonald JC, Seto CT, Wierda DA, Whitesides GM (1994) J Am Chem Soc 116:2382
59. Zerkowski JA, Seto CT, Whitesides GM (1992) J Am Chem Soc 114:5473
60. Lehn J-M, Mascal M, DeCian A, Fischer J (1990) J Chem Soc Chem Commun 479
61. Lehn J-M, Mascal M, DeCian A, Fischer J (1992) J Chem Soc Perkin Trans 2 461
62. Scheinbeim J, Schempp E (1976) Acta Crystallogr B32:607
63. Etter MC, Urbańczyk-Lipkowska Z, Jahn DA, Frye JS (1986) J Am Chem Soc 108:5871
64. Aoyama Y, Endo K, Anzai T, Yamaguchi Y, Sawaki T, Kobayashi K, Kanehisa N, Hashimoto H, Kai Y, Masuda H (1996) J Am Chem Soc 118:5562
65. Endo K, Sawaki T, Koyanagi M, Kobayashi K, Masuda H, Aoyama Y (1995) J Am Chem Soc 117:8341
66. Endo K, Koike T, Sawaki T, Hayashida O, Masuda H, Aoyama Y (1997) J Am Chem Soc 119:4117
67. Felix O, Hosseini MW, DeCian A, Fischer J (1997) Angew Chem Int Ed Engl 36:102
68. Desiraju GR (1997) J Chem Soc Chem Commun 1475

69. Desiraju GR (1991) Acc Chem Res 24:290
70. Pauling L (1960) The nature of the chemical bond, 3rd edn. Cornell University Press, Ithaca
71. Dougill MW, Jeffrey GA (1953) Acta Crystallogr 6:831
72. Taylor R, Kennard O (1982) J Am Chem Soc 104:5063
73. Sharma CVK, Panneerselvam K, Pilati T, Desiraju GR (1992) J Chem Soc Chem Commun 832
74. Sharma CVK, Desiraju GR (1994) J Chem Soc Perkin Trans 2 2345
75. Reddy DS, Goud BS, Panneerselvam K, Desiraju GR (1993) J Chem Soc Chem Commun 663
76. Thalladi VR, Panneerselvam K, Carrell CJ, Carrell HL, Desiraju GR (1995) J Chem Soc Chem Commun 341
77. Reddy DS, Panneerselvam K, Pilati T, Desiraju GR (1993) J Chem Soc Chem Comm 661
78. Reddy DS, Ovchinnikov YE, Shishkin OV, Struchkov YT, Desiraju GR (1996) J Am Chem Soc 118:4085
79. Pascal RA Jr, Ho DM (1992) Tetrahedron Lett 33:4707
80. Xu K, Ho DM, Pascal RA Jr (1995) J Org Chem 60:7186
81. Thalladi VR, Goud BS, Hoy VJ, Allen FH, Howard JAK, Desiraju GR (1996) J Chem Soc Chem Commun 401
82. Allen FH, Goud BS, Hoy VJ, Howard JAK, Desiraju GR (1994) J Chem Soc Chem Commun 2729
83. Sarma JARP, Allen FH, Hoy VJ, Howard JAK, Thaimattam R, Biradha K, Desiraju GR (1997) J Chem Soc Chem Commun 101

Functional Organic Zeolite Analogues

Yasuhiro Aoyama

Institute for Fundamental Research of Organic Chemistry, Kyushu University, Hakozaki, Higashi-ku, Fukuoka 812-8581, Japan. *E-mail: aoyamay@ms.ifoc.kyushu-u.ac.jp*

CREST, Japan Science and Technology Corporation

There is much current interest in organic solid hosts, whose guest-binding properties are reminiscent of traditional inorganic zeolites. The basic design strategy is to assemble organic and metal-ion building blocks into a network by using directional intermolecular interactions such as hydrogen bonding and coordination. The network structures can be controlled by the geometrical and topological properties of the building blocks. When free from interpenetration, the resulting networks afford cavities or channels (occupying, in some cases, as much as 60–70% of the total volume) capable of selective guest binding. Some coordination and multiply hydrogen-bonded networks are robust enough to withstand the removal of included guests, thus sustaining as large as 10 Å guest-free channels. Less robust hydrogen-bonded networks undergo a transition to more dense structures upon guest removal; they are, however, flexible enough to readsorb the guests and restore the single-crystal structures of the host-guest adducts. Guest exchange also occurs, during which crystallinity is retained. Another important consequence of host-guest complexation is activation of trapped guests. Facilitated intracavity reactions, coupled with dynamic guest-binding behaviours (exchange of products and reactants as guests), suggest a potential use of the present type of microporous organic and metal-organic solids as catalysts. This is in fact demonstrated for the Diels-Alder and related ene reactions. The present stage of functional organic zeolite analogues and the problems and prospects associated therewith are discussed from both static and dynamic viewpoints.

Keywords: Zeolite analogue, Microporous material, Network, Pore, Catalysis.

1 **Introduction** 132

2 **Static Aspects** 133

2.1 Molecular Networks Using Supramolecular Building Blocks 133
2.2 Symmetry-Controlled 3D Nets 134
2.3 Layered 2D Nets 136
2.4 Interaction-Supported Channels 142
2.5 Modes of Guest Binding 147
2.6 Selectivities and Pore Size Control 148

3 **Dynamic Aspects** 150

3.1 Robust Networks and Maintenance of Permanent Voids 150
3.2 Flexible Networks and Induced-Fit Adjustment 151
3.3 Apohosts and Guest-Binding Selectivities 153

4 **Catalysis** 154

4.1 Criteria of Zeolitic Catalysis 154
4.2 Catalysis by a Hydrogen-Bonded Organic Solid 154
4.3 Immobilization of Soluble Metal Complexes 157

5 **Concluding Remarks and Future Prospects** 158

5.1 Manipulation of Pores 158
5.2 Solid Catalysts in Organic Transformations 158
5.3 Organic Zeolite Analogues as Enzyme Mimics in Water 159

6 **References** 159

1 Introduction

Zeolites are a class of microporous inorganic crystals composed of aluminosilicate tetrahedra, whose internal cavities are capable of not only reversible guest binding but also catalysing chemical reactions [1–5]. These unique properties challenge chemists to mimic some of the zeolitic functions by using organic or metal-organic components [6–9]. A great advantage of molecular solid materials is that they are spontaneously formed from molecular building blocks. The diversity and designability of organic structures as well as intermolecular interactions spur the research activity in this area [10–13].

There is no widely-accepted definition of organic zeolite analogues. The definition may depend on the functions expected for such materials. An obvious function is as selective adsorbents, which may be used in the separation, purification, or trapping and storage of particular chemicals [6, 9]. Another potential one is that of functional materials as either host-guest adducts or guest-free porous solids, which may exhibit unique electronic, optical or mechanical properties [14, 15]. Still, a third or a most important function is that of solid catalysts, which should be readily recovered; this is highly attractive from the environmental as well as resource utilisation viewpoints. In mimicking the key guest-binding properties of zeolites, a variety of lattice inclusion compounds were prepared, where guest molecules were included in well-defined host cavities [6–13]. However, "unusual" properties and catalytic activities of solid organic hosts remain almost completely unexplored. Heterogeneous catalysts have so far been exclusively inorganic materials including zeolites. In addition to the general problems in crystal engineering [16, 17], there seem to be a couple of fundamental questions associated with functional organic zeolite analogues. One is static or structural; is it really possible to maintain a large volume of permanent voids in the absence of any guests in organic crystals? The other is dynamic or kinetic; how do molecules diffuse in solid materials?

The object of this article is to review the present stage of mimicking the zeolitic functions, to highlight the achievement, although very preliminary, in constructing catalytic organic and metal-organic solids, and to discuss the problems and prospects associated therewith. The dynamic nature, especially the catalysis, is concerned, among others, with the author's own work. Several proposals will be made, whose generalisation awaits further testing in other systems.

2
Static Aspects

2.1
Molecular Networks Using Supramolecular Building Blocks

Porosity may be generated in crystals when the awkward shape of constituent molecules prohibits their close packing. Examples are such hosts as **1**–**4**. They are all rigid and have bulky substituents (**1** [18] and **2** [19]), inclined planes (**3** [20,21]) [22] or bent surfaces (**4** [23]) with or without polar groups capable of intermolecular hydrogen bonding. They all form lattice inclusion compounds. Indeed, the tetraarylporphyrins **3** and bile acid derivatives **4** constitute the most comprehensively studied clathrate hosts. While general packing modes and lattice patterns of the hosts are well conserved, the guest molecules are included in open voids and the guest-binding modes are often guest-dependent, especially when host-guest hydrogen-bonding comes into play [24].

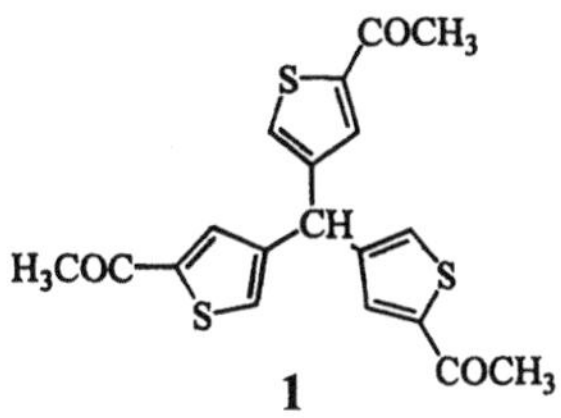

1

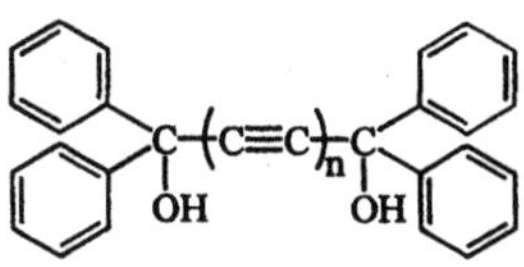

2

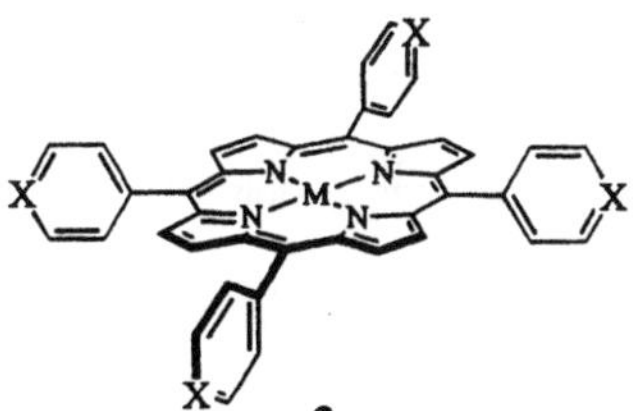

3

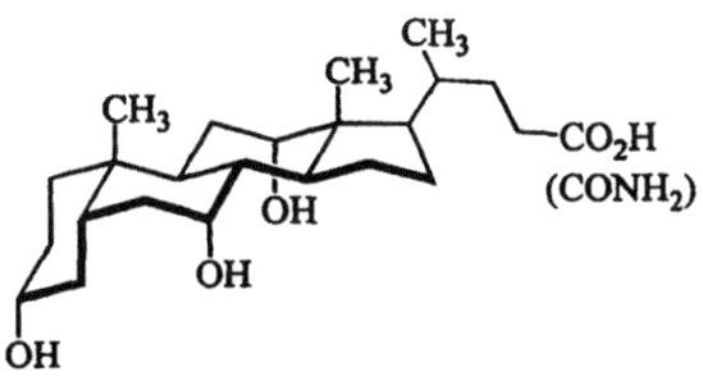

4

In the context of zeolite analogues, we are concerned about closed voids, which are more or less designable and hopefully modifiable systematically and are less sensitive to the included guests. A common strategy is to network rigid molecular building blocks [25] or tectons [26] by using directional intermolecular interactions such as hydrogen bonding, metal coordination, and others (e.g. X···X, CN···H, CN···X, and CX···π interactions; X=halogen) [27] as supramolecular synthons [28] or modules [29]. A robust n-dimensional (nD) module which competes favourably with van der Waals packing forces would reduce the crystal engineering problem from 3D to (3–n)D in the resulting nD-controlled material [30].

2.2 Symmetry-Controlled 3D Nets

Infinite 3D networks are found in many inorganic crystals. They are characterised by the symmetry elements of constituent atoms or ions. One approach to crystal engineering is to construct organic lattices having the same intermolecular topologies as the interatomic or interionic topologies of target inorganic lattices.

Diamond (**5**, Fig. 1), for example, consists of tetrahedral carbon atoms which are directly linked by covalent C-C bonds. If these C atoms are replaced by molecules which interact with each other in a tetrahedral fashion, we will have a molecule-based diamondoid lattice. The simplest remarkable example is ice (**6**) (Fig. 1), where oxygen atoms form a diamond-like framework with intermole-

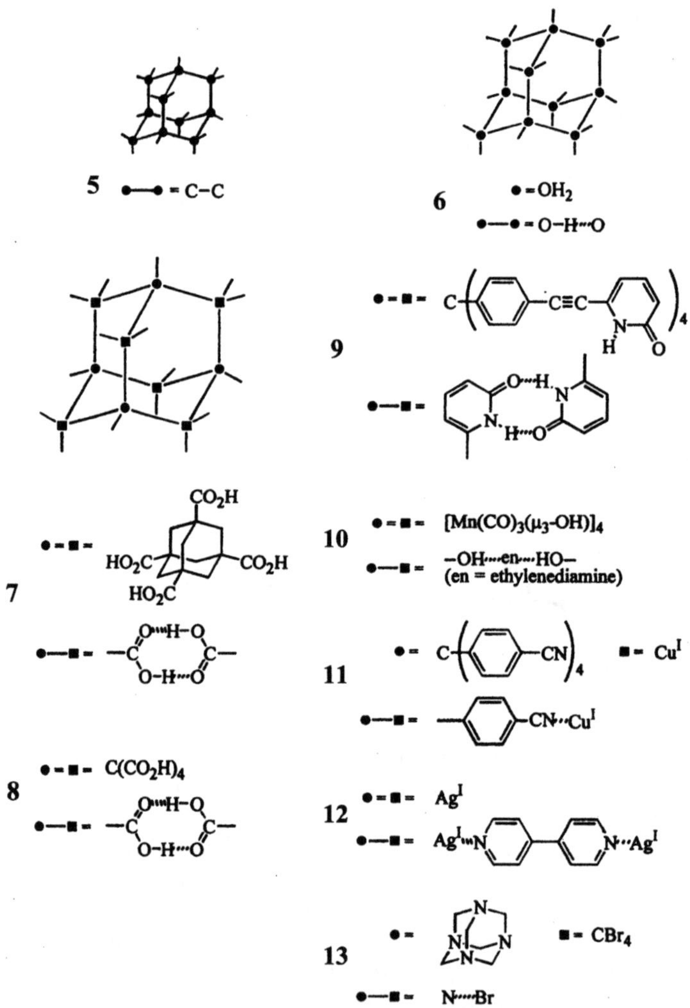

Fig. 1. Structure of diamond (**5**) and diamondoid lattices (**6**–**13**)

cular O-H···O hydrogen bonds in place of C-C covalent bonds as in diamond. The resulting ice is necessarily a porous and hence low-density material, although the parent diamond is the high-density hardest solid even known. Much larger diamondoid lattices, e.g. **7** [31], **8** [32], **9** [33], **10** [34], **11** [35], **12** [36] and **13** [37] (Fig. 1) are obtained by using tetrahedral organic and inorganic components such as 4,4′,4″,4‴-tetrasubstituted methane and 1,3,5,7-tetrasubstituted adamantane derivatives and Cu^I and Ag^I centres, respectively. Rod-like spacers or connectors can also be used. The resulting diamondoid huge cavities in **9** and **11** incorporate two molecules of well-ordered butyric acid and > 7.7 molecules of disordered nitrobenzene and counter anion (BF_4^-), respectively. Guest binding does not occur in other cases, where two or more networks overlap in such a way as to fill the cavities with each other. Such a self-inclusion can be prevented, however, by modifying the building blocks so that interpenetration or catenation of the networks becomes sterically less likely to occur [38, 39].

The PtS lattice contains both the square-planar (Pt) and tetrahedral (S) centres. A square-planar Pt complex $Pt(CN)_4^{2-}$, when interconnected with tetrahedral Cu^I centres ($CN\cdots Cu^I$), affords an infinite network similar to the PtS structure [40]. Tetra(4-pyridyl)- (**3**; X=N and $M=Cu^{II}$) and tetra(4-cyanophenyl) porphyrin (**3**; X=C-CN and $M=Cu^{II}$) provide a huge square-planar site. They are also assembled with Cu^I ($CuBF_4$) to give a similar structure **14** having porphyrin-walled channels in the two directions shown by the arrows (Fig. 2) [25].

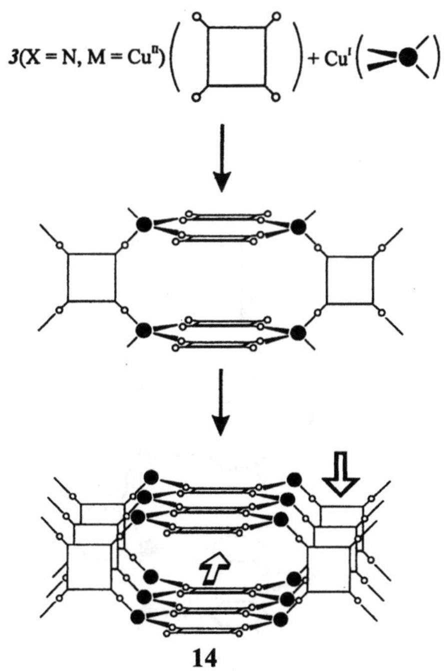

14

Fig. 2. PtS-related structure of $3 \cdot CuBF_4$ (X=N and $M=Cu^{II}$ in **3**). The counteranion is not shown for clarity. Identical channels (**14**) run in two directions shown by the *arrows*

A trigonal centre is found in the $ThSi_2$ lattice. A similar structure **16** is formed when a tritopic tricyano derivative **15** is connected with trigonal-bipyramidal Ag^I (AgOTf; Tf=CF_3SO_2) centres (Fig. 3) [41].

The large channels generated, ~15 Å in **15**·AgOTf and at least 10 Å for the free space (i.e. distance between van der Waals surfaces of porphyrins facing each other) in **3**·$CuBF_4$, are filled with solvent molecules and counter anions. It is not surprising that there is topological similarity in general between packing patterns of symmetric organic molecules and inorganic crystal structures. For example, 1,3,5,7-tetrahydroxyadamantane has four OH groups, each acting as a hydrogen-bond donor as well as an acceptor; the molecule as a whole behaves as an 8-connector and its crystal structure is topologically related to CsCl [42].

2.3 Layered 2D Nets

The channel is an important structural motif through which included guests may move. While, as noted above, porous 3D nets have a channel-like continui-

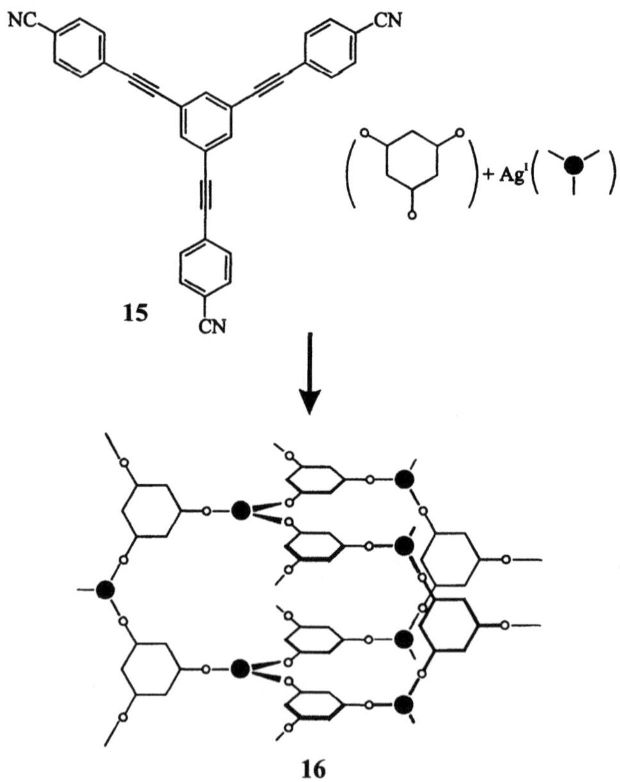

Fig. 3. $ThSi_2$-related structure of **15**·AgOTf. The counteranion in **16** is not shown for clarity

ty of the cavities, the channels can be designed in a more straightforward manner by letting porous 2D nets stack sheet by sheet. Pillared clays and graphite, in contrast to zeolites and diamond, represent typical layer structures. Porous 2D nets are obtained by assembling planar tri- or tetratopic building blocks directly or via connectors. Layered 2D nets may be classified according to the types of intersheet interactions.

When assembled with Cd^{II} ($Cd(NO_3^-)_2 \cdot 2H_2O$) as a linear (*trans*) or bent (*cis*) 2-connector, the above-mentioned tetra(4-pyridyl)porphyrin (**3**; X=N and M=Pd^{II}) forms a (4,2)-connected 2D net **17** (Fig. 4) [43].

The same Cd^{II} ion in $Cd(NO_3^-)_2$ acting in this case as a square planar 4-connector is used to assemble 4,4′-bipyridine **18** as a bidentate linear ligand to afford a (2,4)-connected 2D net **19** (Fig. 5) [44].

Tritopic tricyano derivative **20** [41] or **22** [45] and the Ag^I ion (AgOTf) in the trigonal-bipyramidal geometry form a (3,3)-connected 2D net composed of a hexagonal **21** or 12-annulene-like cyclic motif **23**, respectively (Figs. 6 and 7).

The resulting molecular sheets stack vertically to form channels incorporating solvent molecules and counter anions; the intersheet distances depend on the extent of coplanarity of the in-sheet aromatic moieties and are 4.68, 6.30,

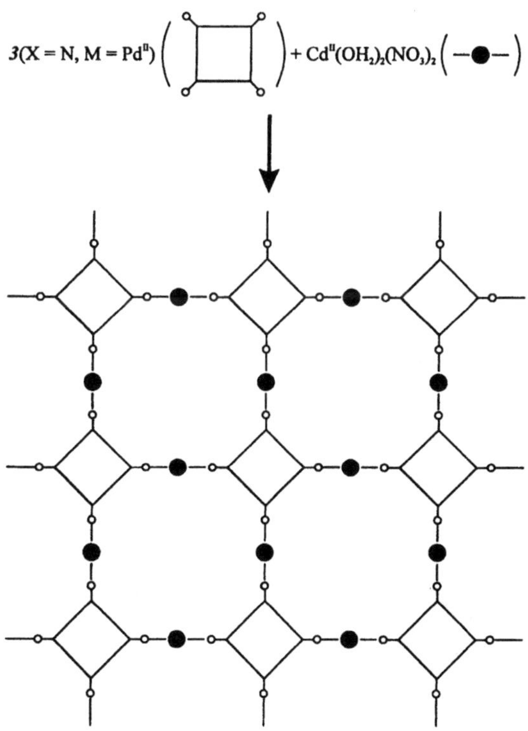

17

Fig. 4. 2D net formed by **3** · $Cd(OH_2)_2(NO_3)_2$ (X=N and M=Pd^{II}). The sheets (**17**) are layered to give intersheet channels

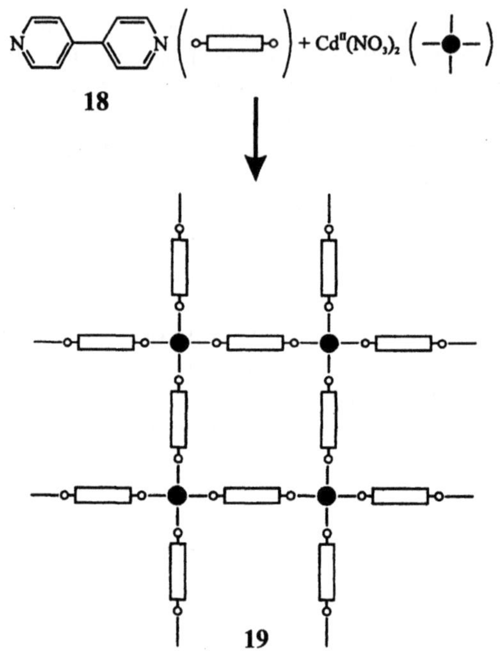

Fig. 5. 2D net formed by $2(\mathbf{18})\cdot Cd(NO_3)_2$. The sheets (**19**) are layered to give intersheet channels

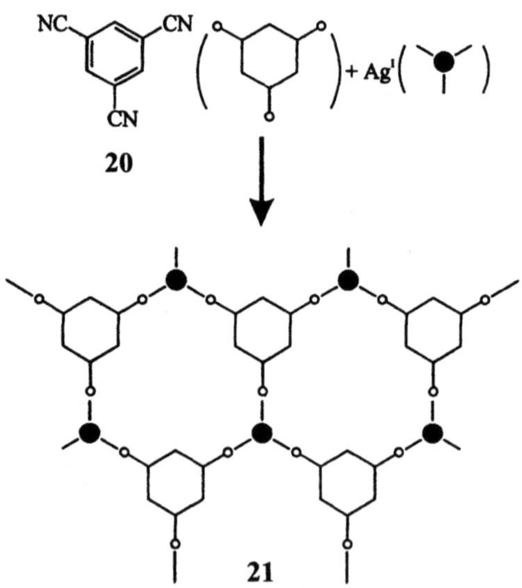

Fig. 6. Hexagonal motif formed by 20 · AgOTf. The counteranion is not shown for clarity. The sheets (**21**) are layered by stacking

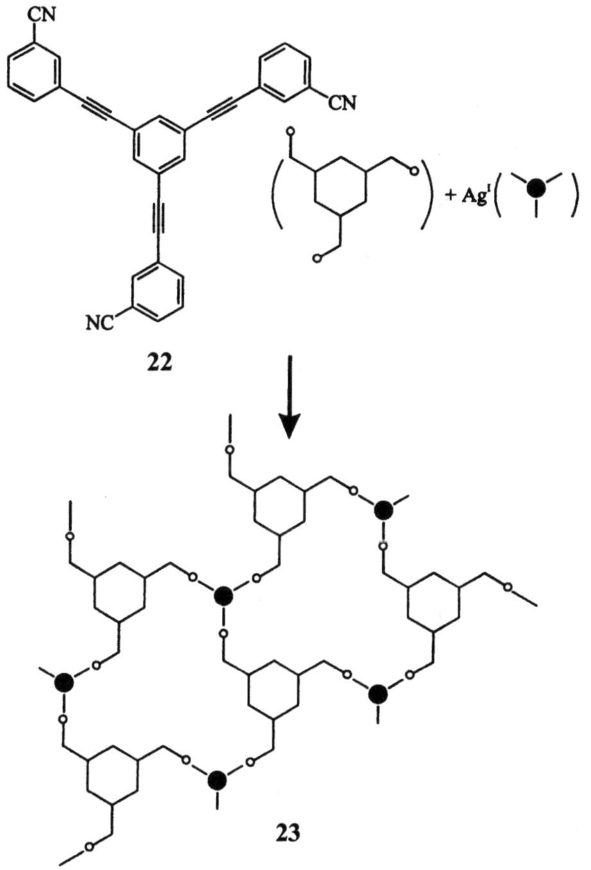

Fig. 7. 12-Annulene-like motif formed by **22**·AgOTf. The counteranion is not shown for clarity. The sheets (**23**) are layered to form intersheet channels

and 3.55 Å in the porphyrin (**3**·$2Cd(NO_3)_2$·$8.6H_2O$), bipyridine [2(**18**)·$Cd(NO_3)_2$·2(*o*-$C_6H_4Br_2$)] and tricyano (**20**·AgOTf) systems, respectively.

Trimesic (benzene-1,3,5-tricarboxylic) acid (**24**) forms a hydrogen-bonded hexagonal motif **25** with a 14 Å-hole (Fig. 8). It stacks in the vertical direction as above. An extensive interpenetration of the networks, however, leaves no channel capable of guest binding [46] (see also the article by R.E. Meléndez and A.D. Hamilton in this volume). Such an interpenetration or polycatenation can be elegantly prevented by building a covalently-closed macrocyclic ring **26** with a 9 Å-hole [47] or an open dimer **27** [48]. The use of a bulky amine as a spacer also leads to a non-interpenetrating and hence guest-binding motif **28** with a 12.7 Å-hole [49]. The cyclohexyl groups in the amine partially cover the guest-binding cavity, thereby inhibiting interpenetration of the networks.

Anthracene-diresorcinol derivative **29** [50–52] is an in-plane tetraol, having an inclined or orthogonal anthracene spacer (Fig. 9). It forms a hydrogen-bon-

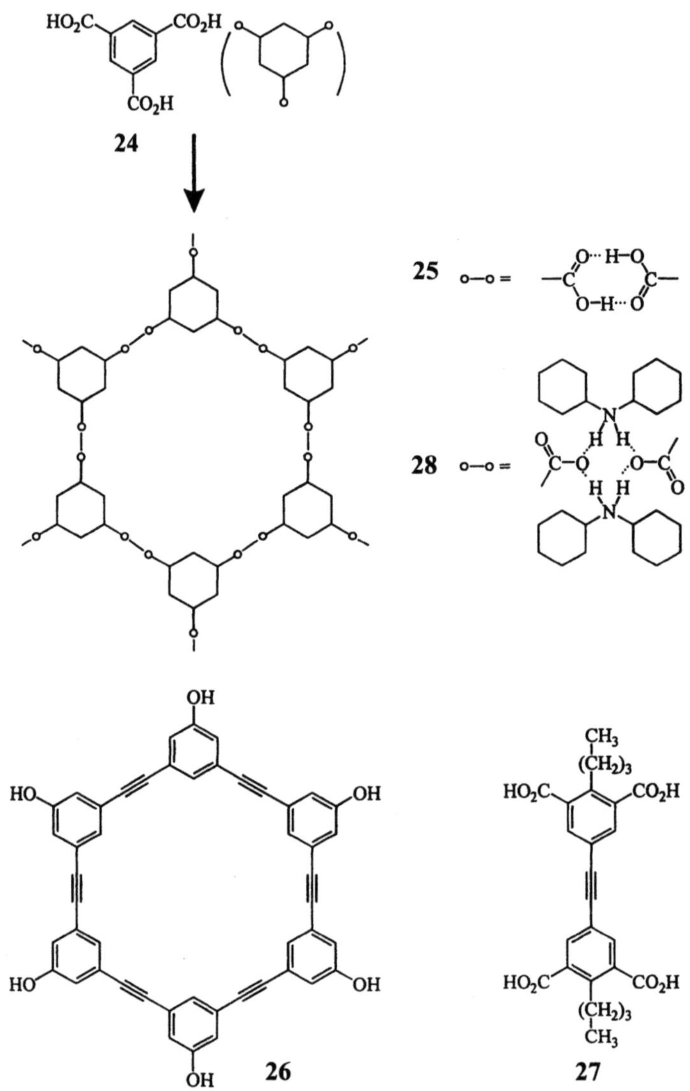

Fig. 8. Hexagonal motifs formed by **24** and its covalently-linked derivatives

ded (O-H···O-H) 2D net **30.** There is no interpenetration of the nets. Their layer structure is somehow different from the previous cases. The presence of the orthogonal spacers sticking out of the sheet prevents close packing thereof. Actually, the sheets are layered in a staggered manner (**31** for the top view of neighbouring sheets) with an intersheet distance of ~7 Å. This gives rise to a "depth" of the big cavities (~ 10 Å height for the free space, referring to **30**), which incorporate two guest molecules (ketone or ester), antiparallel with each other and roughly parallel with respect to the anthracene rings, via host-guest hydrogen bonding (O-H···O-H···OC).

Anthracene-monoresorcinol **32** as a 1D counterpart forms a similar hydrogen-bonded polyresorcinol chain **33** (Fig. 9). The chains are assembled via intercalation of the orthogonal anthracene substituents to give a pseudo 2D net **34** [53]. The resulting small cavity (~ 3.5 Å height for the free space) accommodates one guest molecule via hydrogen bonding.

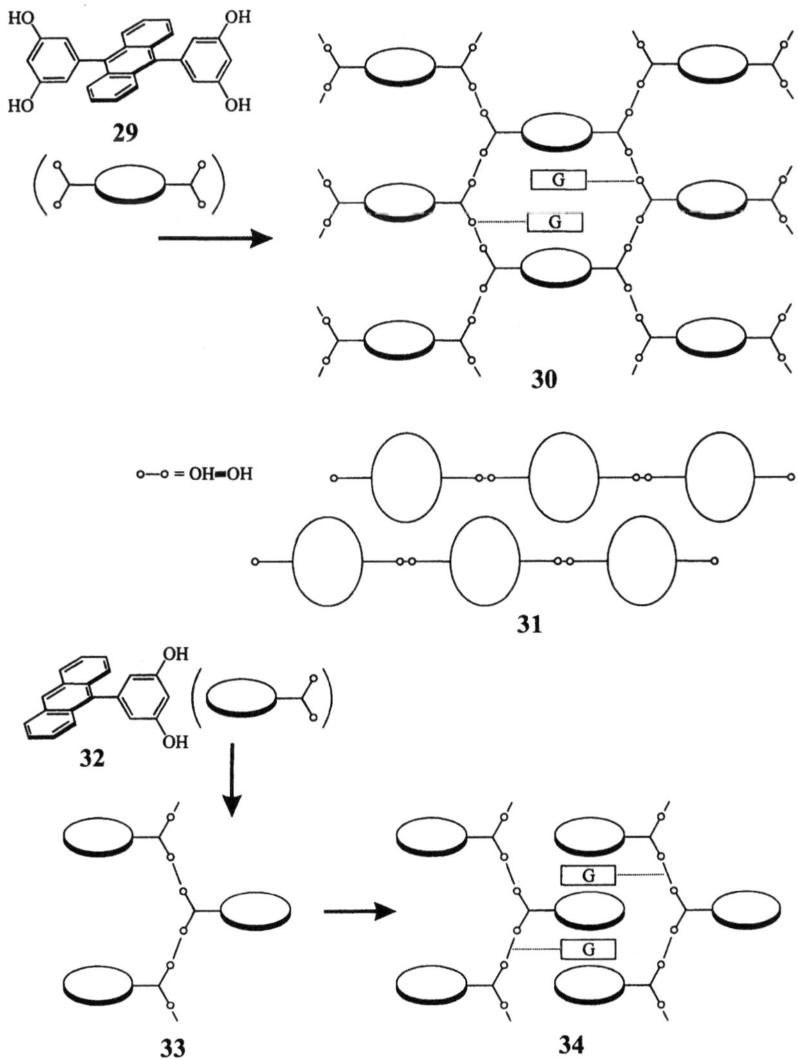

Fig. 9. 2D net formed by **29** (side view in **30** and top view of two adjacent sheets in **31**) or **32** (side view in **34**). G stands for the guest bound via hydrogen bonding

2.4 Interaction-Supported Channels

When excess such interactions are available, the sheets can be held together via intersheet interactions. 2,4-Diaminotriazine can form a hydrogen-bonded tape motif. A tetraphenylmethane derivative of this heterocycle **35** [54] forms a 2D net **36** via hydrogen bonding of a well known type **37** (Fig. 10). The sheets are then linked via another set of hydrogen bonding to give a 3D molecular polymer **38**. One amino group not participating in the network formation is used to hook the guest molecules in the large cavity (11.8 Å diameter for the free space). All the hydrogen bonding capacity of the diaminotriazine ring is thus fulfilled hierarchically, i.e. formation of a sheet, stack of the sheets, and guest binding (see also the article by R.E. Meléndez and A.D. Hamilton in this volume).

Another example is a tetraresorcinol derivative of porphyrin. The corresponding diresorcinol derivative **39** affords a hydrogen-bonded (O-H···O-H) 2D net (**40** for the top view of the sheet in Fig. 11) [55] similar to that (**31** in Fig. 9) for

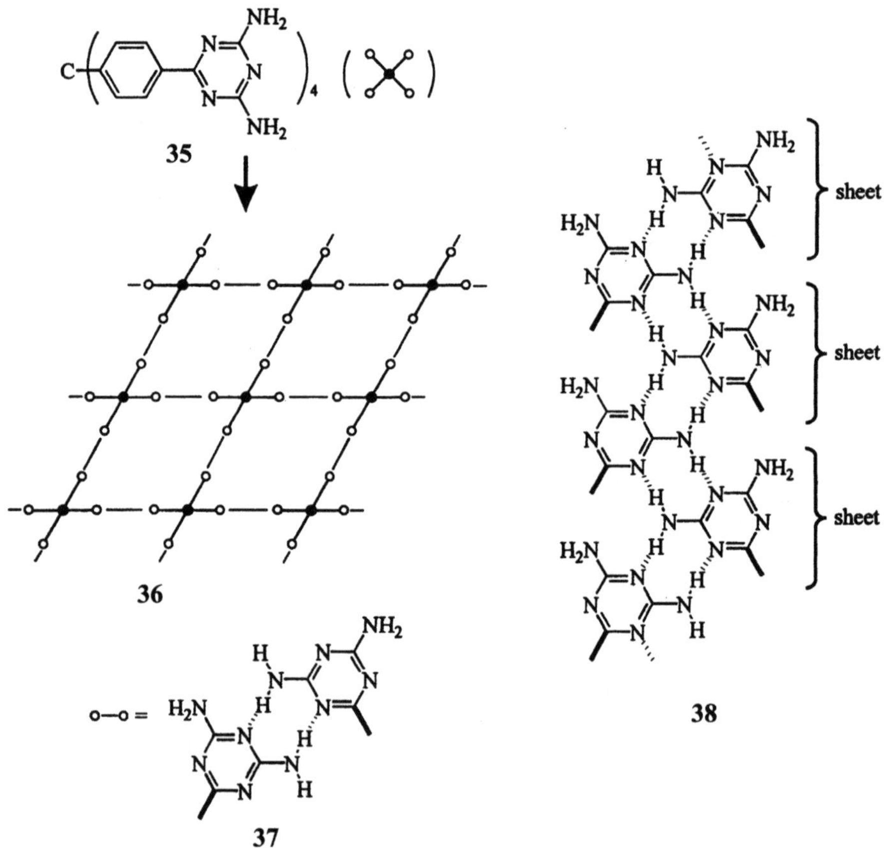

Fig. 10. 2D net (**36**) formed by **35** and intersheet hydrogen bonding to form layered 3D net (**38**)

the anthracene analogue. The excess hydrogen bonding capability is used in guest binding (O-H···O-H···OC) as in the anthracene case above. The sheet formation for the tetraresorcinol derivative **41** occurs in such a way as depicted in **42**. In this case, intersheet hydrogen bonding (O-H···O-H···O-H···O-H) is sterically allowed and gives rise to a continuous channel (cross-sectional area 33 Å^2) incorporating solvent ($7C_6H_5CN$) molecules [56].

A set of vertical hydrogen bonding is also used successfully to assemble cyclic peptides having alternating l- and D-amino acid moieties into nanotubes [57].

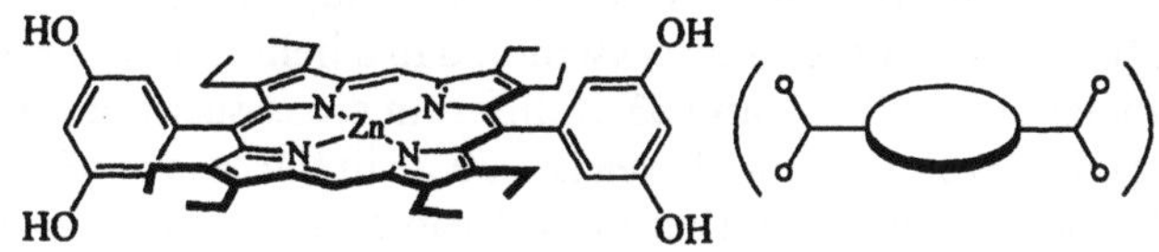

39

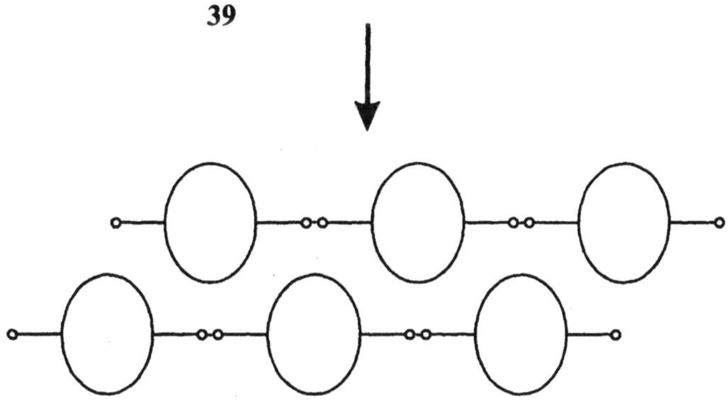

40

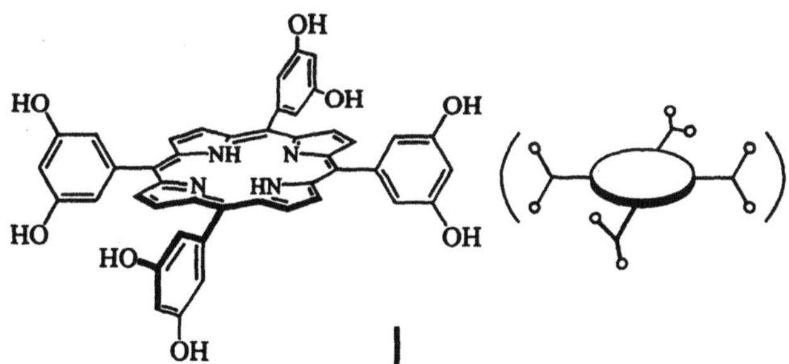

41

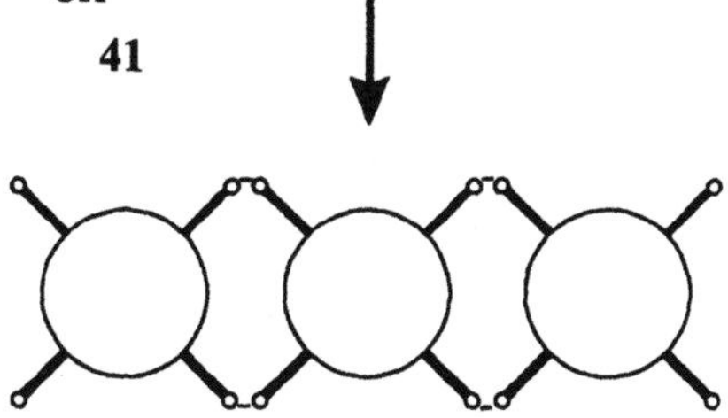

42

Fig. 11. Top view of 2D net formed by **39** or **41**

In the presence of a strong but weakly coordinating base such as triethylamine, fully deprotonated trimesic acid (**43**) (Fig. 12) acts as a hexamonodentate unit. It is assembled with the 6-coordinate Zn^{II} ion ($Zn(NO_3)_2$) formally in a 2D fashon **44** and then into a rigid metal-coordinated 3D network **45** [58]. The basic unit in adduct **43** · 2Zn · (NO_3) · (H_2O) · 5(CH_3CH_2OH) is a 10-membered macrocyclic ring motif **46**. Highly mobile ethanol and water molecules reside in the extended channel having a 14 Å cross-section. Coordination networks can often be further cross-linked via hydrogen bonding between metal ligands [59, 60].

A chiral bicyclic diol **47** and its analogues such as **50** and **52** form spiral hydrogen-bonded spines ···O-H···O-H···O-H···O-H··· (**48**) around a threefold screw axis (Fig. 13) [61]. The molecules are assembled in a helical manner to give well-defined intrahelix canals **49, 51,** and **53** of different cross-sections) supported by six such spines. The cross-sectional area is ~ 20 Å^2 in the case of **49** and a variety of guest molecules are included therein to form a class of helical tubulate in-

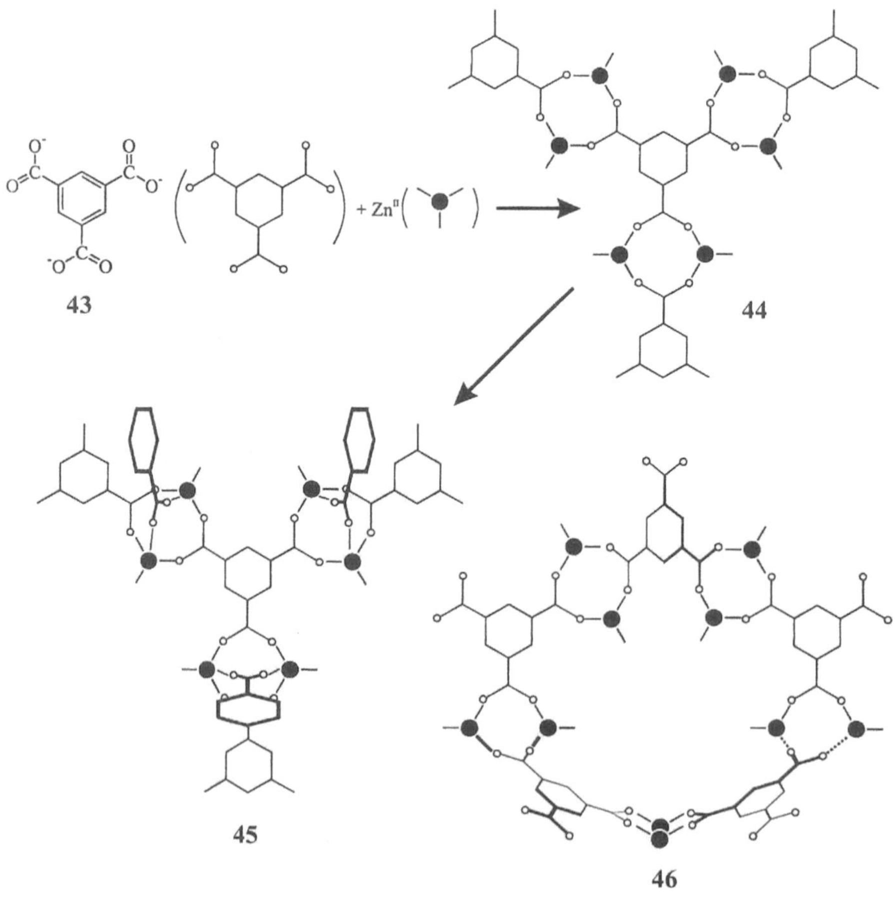

Fig. 12. 3D net formed by **43**·2Zn·(NO_3). The counteranion is not shown for clarity

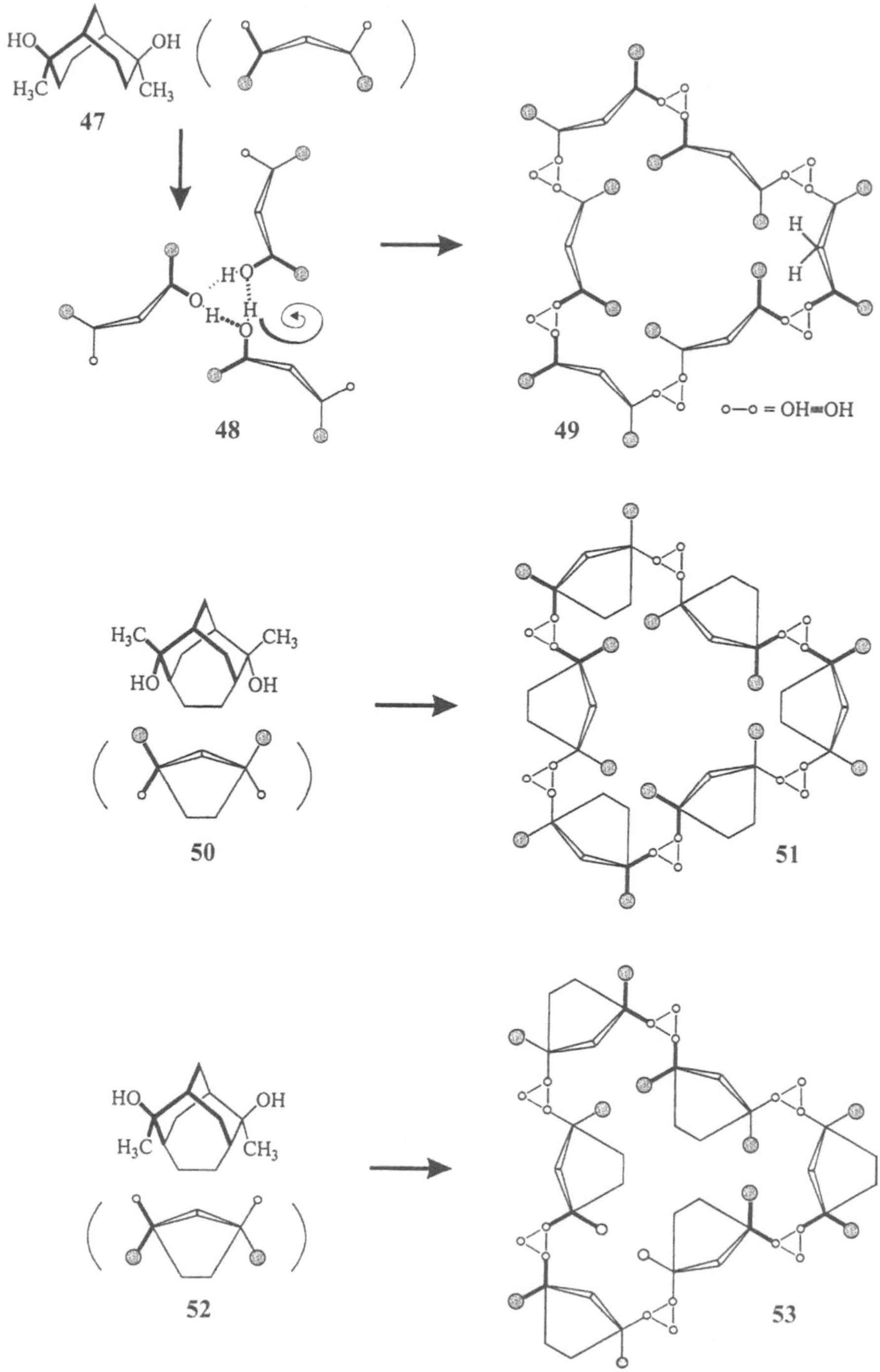

Fig. 13. Helical canals formed by chiral bicyclic diols **47, 50,** and **52**

clusion compounds (see also the article by R. Bishop and I.G. Dance in volume 149 of this series).

The molecular sheets can remarkably also be linked by covalent bonds. An alkane- or arenesulfonate anion (RSO_3^-) **54** and the guanidinium cation **55** form an interesting inorganic 2D net **56**, in which the organic moieties Rs are sticking out of the sheet (Fig. 14). The sheets are intercalatively layered in an R-dependent manner (**57** for the top view) [62]. A simple use of an alkane- or arenedisulfonate **58–62** (Fig. 14) results in covalent linkage of the neighbouring sheets (**63**) [30].

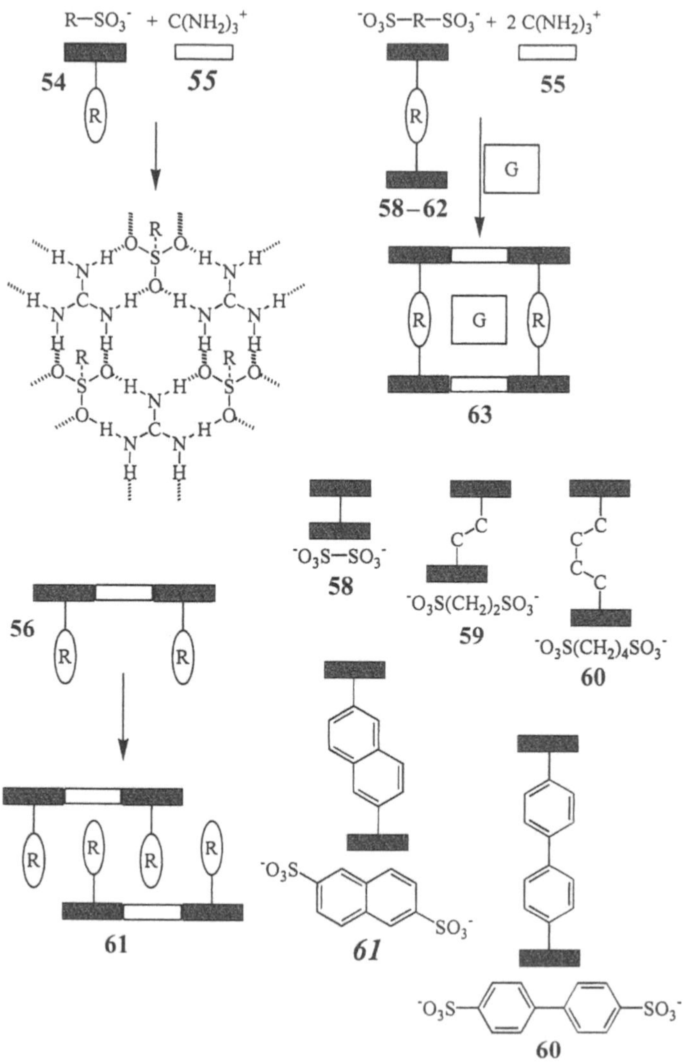

Fig. 14. Assembly (**57**) and covalent linkage (**63**) of 2D net formed by guanidinium alkanesulfonate (arenesulfonate) salts

2.5
Modes of Guest Binding

The percent volume occupied by host or guest provides a measure of the porosity of a network. In this respect, many of the networks such as **11** (Fig. 1), **23** (Fig. 7), **30** (Fig. 9), **38** (Fig. 10), **42** (Fig. 11), and **45** (Fig. 12) are highly porous, i.e. $\geq 50\%$ of the total volume of the adduct is occupied by the guest or only $\geq 50\%$ of that is occupied by the host framework. Some of the guest-binding channels have a cross-section as wide as 10 Å. For comparison, typical pore sizes of zeolites (molecular sieves 4 A, 5 A, and 13X) fall in the range 4–13 Å and void volumes are generally $< 50\%$ [63].

For the stoichiometry, selectivity, mobility and catalysis it is of crucial importance if there is host-guest interaction. In many cases of multiple hydrogen bonds such as those between carboxyl groups (**1** and **2** in Fig 1), pyridones (**3** in Fig. 1), or 2,4-diaminotriazines (**38** in Fig. 10), all or most of the hydrogen-bonding capacity is consumed to build a robust network. In most metal-organic solids whose crystal structures have been determined, the metal centres are coordinatively saturated [58, 64]; there is no guest-metal interaction expected unless guest/ligand exchange occurs. In the absence of strong host-guest (metal-guest) interaction, guest (solvent) molecules included in a large cavity or channel are, not surprisingly, highly disordered. Thus, it is remarkable that two-thirds by volume of what is definitely a crystal (adduct **11** · $x(C_6H_5NO_2)$ ($x \geq 7.7$) is essentially a liquid [35]. It is also likely that the host-guest stoichiometry is primarily controlled on a size basis. Host **35** (Fig. 10), for example, forms such adducts as **35** · 5(dioxane), **35** · $10(CH_3CN)$, **35** · $10(HCO_2H)$ and **35** · $21(H_2O)$ [54].

The simple hydrogen bond between OH groups is coordinatively unsaturated. Two or one such free protons (O-H···O-H) capable of hydrogen-bonding fixation of a polar guest are available for the diresorcinol or monoresorcinol host **29** or **32** (Fig. 9), respectively. This gives a 1:2 **29**:guest or 1:1 **32**:guest stoichiometry for various polar guests having different sizes [52, 53]. In fact, when crystallized from ethyl acetete, host **29** only affords a 1:2 adduct. On the other hand, when crystallized from a mixture of ethyl acetate and benzene, the host gives rise to a ternary adduct **29** · 2(ester) · 2(benzene) [50]. Thus the stoichiometry of the polar host-guest interaction is very strict and the void space still available is better filled with a hydrocarbon guest. The formation of such a ternary adduct containing polar and nonpolar guests simultaneously bound in a cavity provides a basis of its application as a solid catalyst to the bimolecular Diels-Alder reactions [65].

An analogous diresorcinol derivative of Zn^{II}-porphyrin (**39**) generates a similar 2D net **40** (Fig. 11), having cavities much wider than those (**30** in Fig. 9) of the anthracene host **29**. The porphyrin host forms adduct **39** · 4THF [55]. Two molecules of the guest are hydrogen-bonded to the host as above, while the remaining two are coordinated to the central Zn^{II} ions. Cooperation of coordination and hydrogen bonding would be a potential tool for assembling different types of guests in a cavity.

The host-guest hydrogen bonding can also be applied to achieve a particular alignment of host and guest components. Quinones as strong electron-acceptors

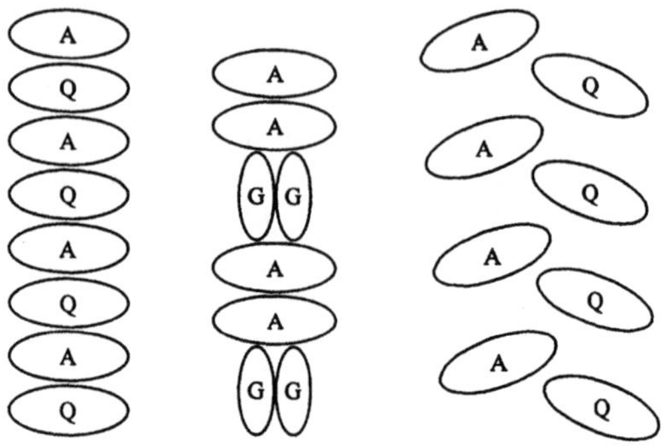

64 **65** **66**

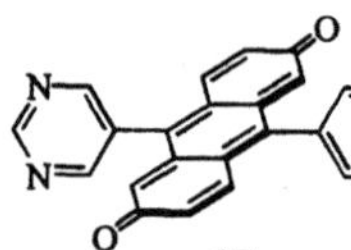

67

Fig. 15. Formation of various columns in host-guest adducts; A, Q, G represent anthracene ring, quinone moiety, and guest species, respectively

are intriguing guests. Duroquinone, for example, can be inserted into the narrow cavities in the pseudo 2D net of anthracene-monoresorcinol host (**34** in Fig. 9), thereby generating a system of closely packed alternating anthracene-quinone (A-Q) columns (**64** in Fig. 15) [53].

Use of a less bulky guest such as ethyl acetate induces a dimeric lattice pattern **65**, which exclusively emits excimer fluorescence from the face-to-face dimeric anthracene units [53]. Segregated anthracene (A) and anthraquinone (Q) columns (**66**) are found in the charge-transfer molecular crystals of adduct **29**·2 (anthraquinone) [66]. This is also the case for the 1:1 complex of the diresorcinol derivative of anthracene (**29**) and the dipyrimidine derivative of anthraquinone (**67**) as a specific hydrogen-bond donor and an acceptor, respectively [66].

2.6
Selectivities and Pore Size Control

Selectivity is a particularly important aspect of inclusion phenomena. Although a number of highly enantioselective [67, 68], regioselective [44] or functional group-selective [69] solid-state complexations are known, it is not easy to understand fully the origin of such a selectivity. Even when host-guest complexation is primarily driven by a particular interaction such as hydrogen bonding, the actual affinity of a guest markedly depends on how other binding forces, especially the cavity packing, cooperate with the essential host-guest interaction.

Detailed studies [52, 53] on the selectivities of hosts **29** and **32** (Fig. 9) reveal that the hydrogen bonded network is somehow adjustable to smaller guests but there is a rather strict upper limit for the size of guests able to be accommodated. As noted above, two guest (ketone or ester) molecules (R_1COR_2; R_1, R_2 = alkyl, aryl or alkoxy) are bound in each cavity of the 2D network **30** with their carbonyl groups hydrogen-bonded to the host in the sheet and their alkyl, aryl or alkoxy groups extending into the neighbouring sheets. In reference to the top view **31** (Fig. 9) of the sheets, the affinity of a guest would then be governed by the intersheet distance. This in fact results in a remarkable chain-length selectivity. The optimal chain-length in R_1 and R_2 is 4. Thus, 4,4-ketones such as 5-nonanone, benzophenone and propyl or isobutyl benzoate constitute a class of highest affinity guests. The shorter one (≤ 3) with a reduced affinity is still able to be accommodated, but the longer one (≤ 5) is hardly incorporated; the (4,4) vs (5,5) selectivity being 5-nonanone/6-undecanone $\cong$ 100/0. Ethyl benzoate (a 3,4-ketone) is one of the suitable guests, while its alkyl/aryl exchanged isomer, phenyl propionate (a 2,5-ketone) shows almost no affinity [52].

A similar cooperation of hydrogen bonding and hydrophobic/van-der-Waals cavity packing may explain the preferential binding of alcohol from an aqueous alcohol solution [70].

What is remarkable about zeolite pores is that various pore sizes are available. Pore size control is an important target of designed organic zeolite analogues. There are a number of promising approaches. One is concerned with thickness control. The huge cavity of height ~10 Å provided by the diresorcinol 2D net **30** (Fig. 9) is more than enough to incorporate two molecules of a simple guest, e.g. alkyl benzoate. The branching in the alkyl moiety allows better packing of the cavity and hence affinity-enhancing, i.e. isopropyl > ethyl and isobutyl > propyl [52]. Monoresorcinol host **32** (Fig. 9), on the other hand, forms a 1:1 adduct in which the guest must be inserted in a much thinner (~3.5 Å) cavity (**34**), which fits for an aromatic ring having a π-electron thickness of ~3.5 Å as well as a linear alkyl group. This results in a remarkable thickness or linear/branched selectivity for the ketone guests. Competition, for example, between diethyl ketone and diisopropyl ketone or between ethyl phenyl ketone and isopropyl phenyl ketone results in an exclusive incorporation of the former [53]. The branched isopropyl group is too thick to be incorporated in the cavity without breaking the hydrogen-bonded polyresorcinol chains which are essential for maintaining the cavity. Attempted crystallization of host **32** from isopropyl benzoate affords a quite different 1:2 (host to guest) adduct having no network structure.

The size and shape of the pores may be systematically changed by modifying the hosts. This approach, however, often fails since even an apparently slight structural change could dramatically alter the crystal structure. A family of chiral bicyclic diols including **47**, **50**, and **52** (Fig. 13) are remarkable in this respect, since the key spiral spine motif **48** is well conserved in the family, while the size of the internal canal depends on the stereochemistry at the OH-bearing carbon, i.e. **50** vs **52**, and the presence or absence of an alkyl bridge i.e. **47** vs **52** [61]. The canal cross-sectional areas (e.g. **49**, **51**, and **53**) can be changed in a range from 0 to 35 Å^2 and a great variety of guest molecules with different sizes, including those as wide as ferrocene and as long as squalene, can be included in tuned intrahelix canals.

The tuning or engineering of the pore sizes is more straightforward in the case of covalently-constructed diguanidinum alkane- or arenedisulfonate crystals (**63**), where the height of an alkane or arene moiety represents that of the cavity (Fig. 14). The cavity or gallery heights increase with respect to the alkane- or arenedisulfonates in the order dithionate (**58**, 3.0 Å, no guest included) < ethane-1,2-disulfonate (**59**, 5.5 Å, no guest included) < butane-1,4-disulfonate (**60**, 8.3 Å, $2CH_3CN$) < naphthalene-2,6-disulfonate (**61**, 9.5 Å, benzonitrile) < biphenyl-4,4′-disulfonate (**62**, 8.9 Å, $2CH_3OH$; 11.5 Å, *m*-xylene; 11.1 Å, toluene and styrene). Guest molecules shown in parentheses are included in the cavities [30].

3 Dynamic Aspects

3.1 Robust Networks and Maintenance of Permanent Voids

Solvent molecules trapped in the cavity or channel can be removed when the adduct is heated. This is also the case for volatile guests hydrogen-bonded to the host or labile metal ligands such as alcohol and water. In many cases, cavities collapse when imprisoned guests escape and the host lattices undergo a transition to a more condensed structure. However, there are a number of recent cases where guests can be taken out without significantly affecting the lattice structures of the hosts so that guest-removed cavities are left more or less intact. Evidence mostly comes from X-ray diffraction which refers to a similarity in the powder patterns for the guest-removed material and the starting adduct, although at present there seems to be no report which rigorously demonstrates the presence of significant internal voids by the direct single-crystal X-ray diffraction method. Desorption of the guests often occurs in a stepwise manner; each step is well-characterised when followed by the gravimetric, calorimetric and spectroscopic analyses (see also the article by M.R. Caira in this volume).

Examples of porous coordination networks are those involving tritopic tricyano and tricarboxylate ligands **15** (Fig. 3), **22** (Fig. 7) and **43** (Fig. 12). Thus, the guest benzene can be completely or partially removed respectively from the benzene solvates of adducts **15** · AgOTf [71] and **22** · AgOTf [45], while maintaining the host lattices. Similarly, most of the ethanol and water molecules occupying a 14 Å channel in adduct **43** · $2Zn^{II}$ · (NO_3) · (H_2O) · $5(CH_3CH_2OH)$ is removed without destruction of the porous framework [58]. The loss of metal-bound ligands leaves vacant coordination sites for the metal centres. This in turn bring about unique selectivities in the subsequent guest-readsorption processes. Dehydration of adduct **43** · $3Co^{II}$ · $12(H_2O)$ yields monohydrate **43** · $3Co^{II}$ · $12(H_2O)$ having channels of a 4 × 5 Å pore size [64]. Destruction of an extensive hydrogen-bonded network present in the dodecahydrate may force the metal centres to form a cluster. This would sterically prevent interpenetration of the coordination networks in a similar manner as discussed in Sect. 2.3. In a 4,4-bipyridine adduct of Co^{II}, adduct $2Co^{II}$ · 3(**18**) · $4(NO_3)$ · $4(H_2O)$, there is no strong interaction between water and the framework [72]. Dehydration of this material leaves 3 × 6 Å channels.

The hydrogen bond is generally much weaker than the metal-ligand interaction. The structural integrity of multiply hydrogen bonded 3D nets may be shown by host (tecton) **38** (Fig. 10) [54]. Up to 63% of the guest can be removed from adduct **38**·5(dioxane). The resulting material remains optically transparent and undergoes a small but systematic contraction in the cell parameters, indicating that the network remains essentially the same after most of the guest having escaped the cavity. The remarkable robustness of the present network is due to the huge number (16) of hydrogen bonds which participate to maintain a single cavity. This number is equivalent to the number of hydrogen bonds supplied by a host; 16, 8, 4, and 2 for tetra(diaminotriazine) **38**, tetracarboxylic acids **7** and **8** (Fig. 1), tetraol **29**, and diol **32** (Fig. 9), respectively.

In this respect, it may be surprising that a bicyclic but simple diol **47** (Fig. 13) affords a porous guest-free host with empty canals [61]. The unique spiral spine motif **48** may play an important role. This is, however, not the sole origin of the robustness of the helical network, since a similar network but with a larger internal void (**51**) provided by an analogous host **50** collapses into a layer structure in the absence of guest. It is suggested that the methyl groups and the methylene moieties pointing into the canal sterically interfere with each other (as shown in **49**) to protect the structure against collapse in a similar manner as the stone blocks support an arched bridge.

When immersed in a guest liquid or kept in contact with a guest vapour, the guest-free apohost readily readsorbs the guest. Guest exchange also occurs. The exchange of nonfunctionalised aliphatic and aromatic molecules makes no change greater than 0.4 Å for any axis in the original cell parameters for adduct **15**·AgOTF·2(benzene) (Fig. 3) [41]. When single crystals of adduct **43**·n(HCO_2H)·m(dioxane) (cf. Fig. 12) are suspended in dioxane or acetonitrile, complete guest exchange occurs rapidly to produce single crystals of **43**·5(dioxane) or **43**·10(CH_3CN) [54]. In a similar manner, single crystals of the dioxane adduct, when suspended in water, give rise to single crystals of water adduct **43**·21(H_2O). In all the cases, the network remains the same and the unit cell parameters show reasonably small and systematic changes. For metal coordination networks, exchange of counter anions occur readily [35, 73, 74].

Although the detailed mechanism of guest exchange is not clear, especially with respect to the timing of sorption/desorption and the modes of lattice diffusion, there is little doubt that the exchange is indeed a solid-state intracavity event [43, 52, 54, 58]. Recrystallization mechanisms are ruled out, since the hosts are essentially insoluble in the media used, the morphology of the crystals is essentially unaffected, less than one-hundredth at best of the total volume of the crystal, and the guest exchange is much faster (minutes) than the recrystallization which may take days.

3.2
Flexible Networks and Induced-Fit Adjustment

Simple hydrogen-bonded networks, especially lower-dimensional ones, do not seem to be robust enough to sustain guest-free cavities. Volatile guests such as ethyl acetate can be readily removed from adduct **29**·2(guest) (cf. Fig. 9) to give

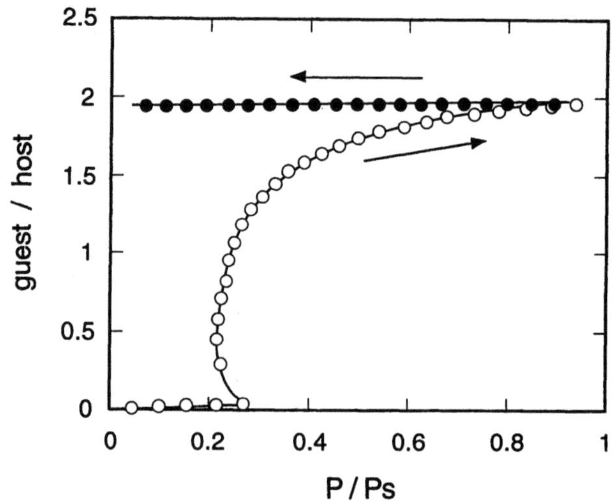

Fig. 16. Binding isotherm (guest/host ratios plotted against pressure of the guest) for gaseous ethyl acetate with host **29** at 25 °C. The pressure is increased (→) from 0 up to the saturation vapour pressure of the guest (P_s) for the sorption (*open circles*) and then decreased (←) to 0 for the desorption (*filled circles*)

polycrystalline guest-free apohost [52]. There is at least no evidence in support of the presence of a significant void there; the cavity must have collapsed somehow, due to a conformational (anthracene-resorcinol dihedral angle) change in the molecule, a closer approach of the sheets, and possibly partial breakdown of the hydrogen bonds. Nevertheless, apohost **29** in the solid state readily binds various guests including hydrocarbons and haloalkanes (-arenes) not only as liquids and gases but also as solids in some cases. The complexable solid guests include benzophenone, benzoquinone and cyclohexanediol [52].

The solid-liquid complexation occurs instantaneously, while showing no apparent change in the size and shape of the host. There are two remarkable aspects in the present solid-state complexation. One is stoichiometry which is strictly the same as that in recrystallization, i. e. 1:2 (host to guest) in most cases. This is also true for the solid-solid complexation. If a 1:3 (host to guest) mixture is used, for example, the 1:2 adduct results and one equivalent of excess guest remains simply as such. The other is guest-induced structure change; the host-guest adducts resulting from solid-liquid, solid-gas or solid-solid complexations exhibit almost the same X-ray powder diffractions (XRPD) as those of their single-crystalline specimen obtained by recrystallization.

The binding isotherm for gaseous ethyl acetate with apohost **29** is shown in Fig. 16, where guest/host molar ratios are plotted against pressure of the guest in the vapour phase up to its saturation vapour pressure at 25 0 °C. The sorption curve is remarkably sigmoidal [75]. In conjunction with XRPD evidence above, this indicates that the transition from the tense guest-free apohost with collapsed cavities to the relaxed host-guest complex with a single-crystal structure occurs in a highly cooperative or allosteric manner (Eq. 1).

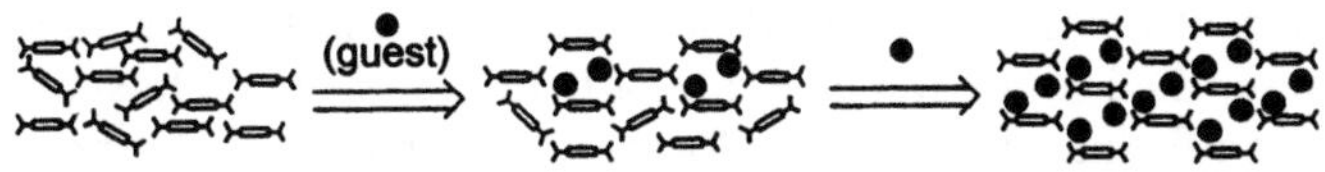

(1)

In fact, the resulting complex is so "stable" that desorption of the guest therefrom does not occur practically under the conditions, i.e. at 25 0°C (Fig. 16). The solid-gas complexation, even with non-pulverized large pieces (1 × 1 × 1 mm) of the apohost, is relatively rapid, having a typical half-life of minutes. The rates of guest-uptake are roughly linear with respect to the pressure (concentration) of the guest in the gas phase. Extrapolation into the liquid phase suggests that the solid-liquid complexation should occur instantaneously, as observed. Guest exchange also takes place at least in some cases [51]. When immersed in methyl benzoate, single crystals of adduct **29** · $2(C_6H_5CO_2CH_2CH_3)$ are converted to adduct **29** · $2(C_6H_5CO_2CH_3)$, whose XRPD is identical with those of the corresponding single crystalline sample. Curiously enough, the exchange in the reverse direction, i.e. from **29** · $2(C_6H_5CO_2CH_3)$ to **29** · 2 $(C_6H_5CO_2CH_2CH_3)$ does not occur, even though ethyl benzoate shows a higher affinity to the host than methyl benzoate.

The characteristic guest-binding (removal, addition and exchange) properties observed for host **29** are also exhibited by monoresorcinol host **32** (Fig. 9) which has much more compact cavities and seem to be generally shared by other hydrogen-bonded hosts including diol **47** (Fig. 13) and steroidal bile acid derivatives [76]. Such a dynamic and cooperative nature may be an important aspect of hydrogen-bonded flexible networks. Furthermore, as far as guest binding is concerned, the presence of permanent void is by no means a prerequisite of organic zeolite analogues. We will come back to this important point later.

3.3
Apohosts and Guest-Binding Selectivities

The guest-free apohosts are "unsaturated" not only with respect to the space but also in some cases to the functional moieties involved. In the Ag^I-networked host **15**· AgOTf (cf. Fig. 3) having a huge aromatic ligand, the metal centres are coordinatively saturated. The host shows affinity to aromatic guests, which may allow host-guest π-π interactions [71].

In the Zn^{II}-networked host material **43** · $2Zn^{II}$ · (NO_3) · $0.5(H_2O)$ · (CH_3CH_2OH) (cf. Fig. 12), on the other hand, the metal centres have vacant coordination sites, due to loss of volatile ethanol/water ligands. This host now shows high affinities to alcoholic guests, e.g. ethanol, which may readily coordinate with the metal [58]. Other guests such as THF, methyl ethyl ketone, acetonitrile and acetone, having an appropriate shape and size for inclusion, are excluded. This is also the case for a porous apohost **43** · $3Co^{II}$· (H_2O), having a 4 × 5 Å channel. This host is highly specific to NH_3 having lone-pair electrons again required for metal coordination [64].

Apohost $2Co^{II}$ · 3(**18**) · $4(NO_3)$ has a similar pore size (3 × 6 Å) and coordinatively saturated Co^{II} centres. Typical zeolite guests such as CH_4, O_2 and N_2 can be reversibly bound in the internal channels under high pressure conditions [72].

The desorption of the included guest occurs readily and there is no significant hysteresis. In marked contrast to the case of hydrogen-bonded host **29** (Fig. 16), there is no significant structural/conformational change in the host network upon sorption/desorption of the guest. This is taken as another sort of evidence for the robustness of metal-coordination networks.

Water is a potential guest of zeolites (molecular sieves). Many hosts [77], including those described above, bind water as a guest. Apohost **29** (Fig. 9), for example, incorporates ~16 molecules of water. When immersed in an aqueous solution of such guests as ethanol, alkyl acetate and cyclohexanol (-diol), the host preferentially picks up two molecules of the guest together with 3–5 molecules of water [78]. Thermogravimetry for adduct $\mathbf{29} \cdot 2(CH_3CH_2OH) \cdot 5(H_2O)$ indicates that water escapes the cavity even at room temperature and completely at $< 50\,°C$, while ethanol is desorbed only at $> 100\,°C$. These results, coupled with IR evidence, suggest that it is the guest (ethanol) molecules that are hydrogen-bonded to the host and vapour-like water molecules fill the void space left. On the other hand, treatment of the apohost with an aqueous salt (e.g. NaCl) solution results in exclusive incorporation of the water.

4 Catalysis

4.1 Criteria of Zeolitic Catalysis

The great advantage of solid catalysts is that they are readily recovered and, in principle, used repeatedly. For the catalysis to be zeolitic, there seem to be a couple of criteria. (1) The catalysis should not be a simple surface phenomenon; in other words, internal cavities should make an essential contribution to the catalysis. (2) The catalysis should be specific in some way to the solid state of the catalyst; at least the heterogeneous catalysis should be superior to the corresponding homogeneous one, when available, in terms of either rate or selectivity. The facile guest exchange and the availability of protic (in hydrogen-bonded networks) or metal sites (in coordination networks) suggest their potential application as zeolite-like acid catalysts. However, there is only scattered information at best [44]. An apparent general problem is that the stable single-crystalline materials suited for structural studies do not usually have vacant coordination sites required for the catalysis.

4.2 Catalysis by a Hydrogen-Bonded Organic Solid

A prototype of catalysis that meets the above criteria is that of host **29** (Fig. 9) in the Diels-Alder reaction between acrolein and 1,3-cyclohexadiene (Eq. 2) [65].

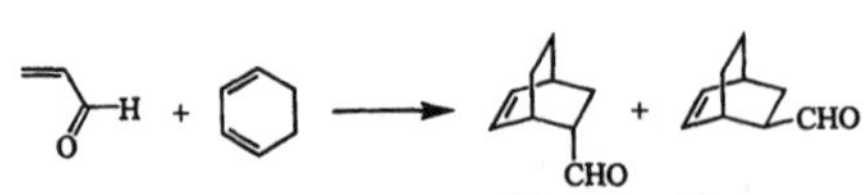

68$_{endo}$ **68**$_{exo}$ (2)

In the absence of any catalyst, the reaction is ~90% endo-selective and is very slow with a half-life of $\tau = 500$ h at 25 °C. The reaction is catalyzed by the solid state of the host (**29**) almost equally when the latter is used as pulverized powders (with a turnover rate constant of 0.33 h^{-1}) or as nonpulverized pieces of an approximate size of $1 \times 1 \times 1$ mm (0.20 h^{-1}). The catalyzed reaction gives rise to the endo product (**68_{endo}**) with a 96% selectivity. In reference to the criteria shown above, the present heterogeneous catalysis is approximately an order of magnitude more efficient than the corresponding homogeneous catalysis exhibited by soluble resorcinol and shows a remarkably small size effect for the solid catalyst.

When immersed in a mixture of reactants, the host readily forms a 1:2:2 adduct **29**·2(acrolein)·2(diene) in seconds. The isolated adduct undergoes a gradual intracavity Diels-Alder reaction with an approximate half-life of 2 h. The resulting product adduct **29**·2(**68_{endo}**), when dipped in the reaction mixture, undergoes a facile product/reactant exchange to regenerate the reactant adduct **29**·2(acrolein)·2(diene) in minutes. These results, coupled with the negligibly small size effect for the catalyst, suggest that the principal catalytic sites are the internal cavities and not the surface of the solid catalyst.

The plausible mechanism (Fig. 17) involves (1) simultaneous binding of two reactants in a cavity, (2) facilitated (~250-fold as compared with the spontaneous process) but still rate-determining intracavity reaction which is preorganized and hence stereoselective, and (3) product/reactant exchange resulting in turnover of the catalyst. In light of the crystal structure for an analogous adduct **29**·2(ethyl acrylate)·(diene), there is no doubt that the proximity of dienophile and diene is responsible for the acceleration of the present bimolecular process in the cavity. However, it should not be overlooked that the dienophile hydrogen-bonded to the host (O-H···O-H···O=C-C=C) must be activated electronically, although separation of an overall effect into steric and electronic factors is not easy.

The effect of hydrogen bonding becomes clearer when one examines a unimolecular process which should be free from proximity effects. Host **29** also catalyzes the intramolecular ene reaction of an unsaturated terpenoid aldehyde cytronellal (**69**) to isopulegol (**70**) (Eq. 3) [79].

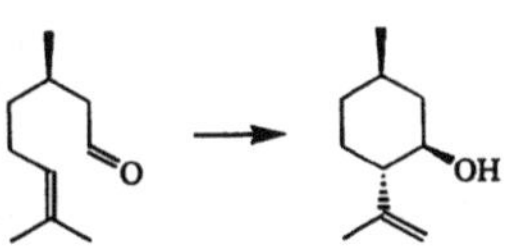

69 **70**

(3)

In the initially formed 1:2 adduct **29**·2(**69**), substrate **69** is hydrogen-bonded to the host with its alkenyl group in an extended conformation. Nevertheless, the present cyclization reaction in the cavity is ~100-fold faster than the spontaneous process. The acceleration should be primarily ascribed to the host-substrate hydrogen bonding.

The Diels-Alder and related ene reactions provide a mechanistic prototype of the zeolitic catalyses played by microporous organic solids. From a practical

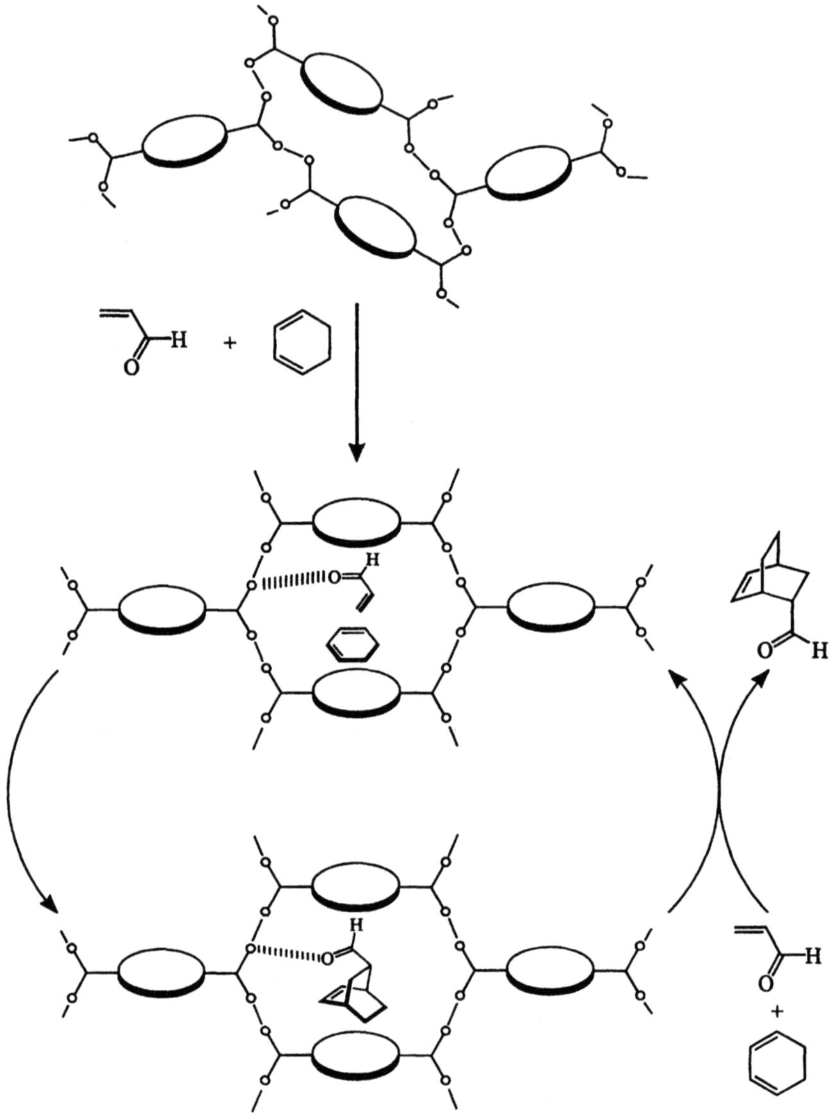

Fig. 17. Suggested mechanism for the Diels-Alder reaction catalyzed by host **29** as a microporous solid. Each cavity actually binds two molecules of both reactants. Only one molecule of each is shown here for clarity

point of view, there are a couple of problems. One is efficiency. The activity of the present hydrogen-bonded network host **29** is much smaller than those of soluble Lewis acid catalysts conventionally used in organic syntheses. The other is generality. Although the acrolein reaction goes in a catalytic manner, this is not the case for otherwise closely related alkyl acrylates. For example, as noted above, ethyl acrylate forms a similar ternary adduct **29** · 2(dienophile)·(diene), which undergoes a facilitated and highly stereoselective (98%) intracavity Diels-Alder reaction [65]. The product, however, resists being exchanged by the reactants, resulting in a product inhibition of the catalysis. While the essential difference between the acrolein and ethyl acrylate systems is not clear, the diminished exchangeability of the bulkier acrylate product is not unexpected. The hydrogen-bonded network in host **29** is flexible enough to adjust to the guest structures. This may be a good aspect as far as guest binding is concerned. However, such an induced-fit "stabilization" of the adduct may make the desorption of included guest more difficult (Fig. 16). An efficient turnover of a catalyst is based on a good balance of sorption and desorption. It is likely that for this purpose we need more rigid and less guest-sensitive networks.

4.3 Immobilization of Soluble Metal Complexes

Hydrogen-bond donor atoms (O and N) are potential ligands for metal ions. This suggests that the hydrogen-bonded network is convertible to a metal-coordinated one via H^+/M^{n+} exchange. Treatment of apohost **29** with $TiCl_2(OCH(CH_3)_2)_2$ affords an insoluble amorphous solid formulated as $\mathbf{29}^{4-}\cdot 2[TiCl(OCH(CH_3)_2]$ (Ti-host) with concomitant liberation of one molecule of HCl and isopropanol ($\mathbf{29}+2[TiCl_2(OCH(CH_3)_2)_2]\rightarrow\mathbf{29}^{4-}\cdot 2[TiCl(OCH(CH_3)_2]+HCl+(CH_3)_2CHOH$) ($\mathbf{29}^{4-}$ is the deprotonated tetraanion of **29**). This formulation suggests that the hydrogen-bonded network (O-H···O-H) in apohost **29** has been replaced by a metal-organic network (O^-–Ti^{4+}–O^- where $Ti^{4+}=Ti^{4+}Cl(OCH(CH_3)_2)$. When dipped in liquid ethyl acetate, alkyl benzoate, alkyl acrylate or acrolein as a guest, Ti-host binds 4 moles (i.e. 2 moles on each 4-coordinate Ti^{4+} centre) of the guest. Gas adsorption studies indicate that the guest binding to Ti-host is much more reversible than to apohost **29** (Fig. 16). Hydrocarbon guests such as benzene, *p*-xylene and 1,3-cyclohexadiene (4–6 moles) are also accomodatable.

Ti-host remarkably catalyzes the highly endo-selective (>99%) acrolein-1,3-cyclohexadiene Diels-Alder reaction (Eq. 2) [80]. The half-lives of this reaction in the absence and presence (3 mol%) of a catalyst are $\tau=500$ h (no catalyst), 50 h (apohost **29** as an insoluble catalyst), 1 h ($TiCl_2(OCH(CH_3)_2)_2$ as a soluble catalyst) and ~5 min (Ti-host as an insoluble catalyst). Thus, as a Lewis acid Ti-host shows a much higher activity than its soluble counterpart. As a solid catalyst, it allows easy product-catalyst separation, recovery of the catalyst without deactivation and hence its repeated use. In addition, Ti-host is also capable of catalyzing the Diels-Alder reaction between ethyl acrylate and 1,3-cyclohexadiene, which the apohost **29** fails to catalyze because of the lack of desorption of the product from the cavities.

A soluble Ti^{4+} Lewis acid can thus be immobilized with a known hydrogen-bonded supporting network as a microporous multi-ligand. This simple strategy would in principle be applicable to various organic networks and metal ions.

5
Concluding Remarks and Future Prospects

5.1
Manipulation of Pores

Zeolites, in a sense, are 3D-networked covalent polymers. The attempts to mimic their structural integrity and unique properties led to a number of important discoveries which are summarised as follows. (1) Significant permanent voids can be sustained by using multiply-bonded coordination and organic networks. Rigid pores may be used directly as low-dielectric materials and also as the sites of selective guest binding. Functionalised pores may also be an important future concern. The construction of robust networks has something to do with the essence of crystal engineering. It provides a powerful strategy for precisely designing ordered materials on the basis of predictable intermolecular interactions which compete favourably with unpredictable van der Waals packing forces. (2) Simple hydrogen-bonded networks are not so robust as to maintain guest-free cavities. They are, however, remarkably flexible, dynamic and adjustable. Guest molecules readily diffuse in to open the channels. The high cooperativity of organic network materials may find their unique applications as on-off switchable devices. (3) For the catalysis to occur, porous materials should have vacant coordination sites. Such sites also bring about selectivities in the guest binding. Simple hydrogen-bonded systems have this capacity, i.e. capability of forming additional hydrogen bonds. Vacant metal coordination sites can be generated by the loss of labile ligands from preformed saturated adducts [58, 64] or by immobilising soluble metal complexes with a preformed hydrogen-bonded network [80]. In this way, the catalytic zeolite analogues are becoming a reality. Two areas of particular interest are shown below. It should also be noted that the crystal integrity becomes less pronounced upon introduction of vacant coordination sites.

5.2
Solid Catalysts in Organic Transformations

The catalytic activity of the immobilized Ti complex suggests a bright future in the use of solid metal-organic catalysts in fine organic synthesis, where soluble metal complexes and organometallic derivatives have been extensively used. In principle, there can be various combinations of organic networks as microporous polymeric ligands and metal complexes as catalytic sites. An especially interesting area is the use of chiral networks. In view of the importance of asymmetric transformations, the abundance of chiral soluble catalysts, and the nonexistence of chiral zeolites, catalytic zeolite analogues may claim their maximal significance in the chirality control.

Catalysis by metal-organic solids may also be applied to redox reactions. An especially intriguing target would be manipulation of hydrocarbon transformations. Many metal-organic networks so far reported in fact contain redox-active transition metals (Cu^{II}, Pd^{II}, Co^{II} and so on). Metalloporphyrins are potential redox catalysts and have been incorporated in a variety of porous structures. As for the guests, there are good examples of hydrocarbon incorporation.

The use of efficient solid catalysts may play an essential role in the construction of waste-free and workup-free molecular transformations which are friendly to the environment and resource-saving.

5.3 Organic Zeolite Analogues as Enzyme Mimics in Water

From the environmental point of view, the most ideal solvent is water. This is also the medium where enzymatic reactions take place. There are a couple of questions in the use of water for the zeolite analogues. One is as for the maintenance of structures; can a hydrogen-bonded or metal-coordinated network survive in an aqueous environment? The other is with respect to the efficiency of catalysis; even when the structure is maintained, can a guest be bound to the host via either hydrogen bonding or coordination to the metal in preference to the water? Both questions arise from the very polar nature of water, which acts as a strong hydrogen-bond former as well as a potent ligand to a metal ion, especially a Lewis acid centre.

The strength of the metal-coordination networks against hydrolysis may be proved by the fact that many of them are prepared by crystallisation from an aqueous medium. Use of water-tolerable Lewis acids such as lanthanoid metals may be an interesting extension. The water-tolerance of a hydrogen-bonded network is also proved. For example, host **35** (Fig. 10), when immersed in water, not only maintains its hydrogen-bonded network but actually incorporates a large number (~21) of water molecules in each huge cavity and still retains single-crystallinity. In a similar manner, host **29** (Fig. 9) adsorbs ~16 water molecules. Furthermore, this host preferentially binds highly hydrophilic guest molecules such as ethanol, ethyl acetate and acetonitrile in water via host-guest hydrogen bonding. This suggests that cavity-bound polar guests can be activated via hydrogen bonding in water in a similar manner as in enzymatic catalyses. In fact, there is a remarkable apparent relevance between microporous organic solids and enzyme proteins in the maintenance of structures and functions.

Acknowledgements. The support of our research activities by CREST (Core Research for Evolutional Science and Technology) of Japan Science and Technology Corporation and also by Grant-in-Aids from the Ministry of Education, Science, and Culture of the Japanese Government is greatly acknowledged. I am also grateful to my capable co-workers, especially Dr. K. Endo (CREST, Kyushu University) and Dr. K. Kobayashi (Tsukuba University) for their collaboration.

6 References

1. Breck DW (1974) Zeolite molecular sieves, structure, chemistry, and use. Wiley, New York
2. Herron N (1991) In: Atwood JL, Davies JED, MacNicol DD (eds) Inclusion compounds, vol 5, chap 3. Academic Press, Oxford
3. Suib SL (1993) Chem Rev 93:803

4. Hölderich W, Hesse M, Näumann F (1988) Angew Chem Int Ed Engl 27:226
5. Thomas JM (1988) Angew Chem Int Ed Engl 27:1673
6. Weber E (1987) Top Curr Chem 140:3
7. Zaworokto (1994) Chem Soc Rev 23:283
8. Bishop R (1996) Chem Soc Rev 25:311
9. Zimmerman SC (1977) Science 276:543
10. Schmidt GM (1971) Pure Appl Chem 27:647
11. MacNicol DD, McKendrick JJ, Wilson DR (1978) Chem Soc Rev 7:65
12. Lehn JM, Atwood JL, Davies JED, MacNicol DD, Vögtle F (eds) (1996) Comprehensive supramolecular chemistry, vol 6. Pergamon Press, Oxford
13. Atwood JL, Davies JED, MacNicol DD (eds) (1984) Inclusion compounds, vols 1–3. Academic Press, London; (1991) Inclusion compounds, vols 4, 5. Oxford University Press, Oxford
14. Baughman RH, Galvão DS (1993) Nature 735:365
15. Evans KE, Hutchinson IJ (1991) Nature 353:124
16. Kitaigorodskii AI (1973) Molecular crystals and molecules. Academic Press, New York
17. Desiraju GR (1989) Crystal engineering: the design of organic solids. Elsevier, Amsterdam
18. Pang L, Brisse F (1994) Can J Chem 72:2318
19. Toda F (1991) In: Atwood JL, Davies JED, MacNicol DD (eds) Inclusion compounds, vol 4, chap 4. Oxford University Press, Oxford
20. Byrn MP, Curtis CJ, Goldberg I, Hsiou Y, Khan SI, Sawin PA, Tendick SK, Strouse CE (1991) J Am Chem Soc 113:6549
21. Byrn MP, Curtis CJ, Goldberg I, Hsiou Y, Khan SI, Sawin PA, Tendick SK, Terzis A, Strouse CE (1993) J Am Chem Soc 115:9480
22. Weber E, Czugler M (1988) Top Curr Chem 149:45
23. Miyata M, Sada K (1996) In: Lehn JM, Atwood JL, Davies JED, MacNicol DD, Vögtle F (eds) Comprehensive supramolecular chemistry, vol 6, chap 6. Pergamon Press, Oxford
24. Golberg I, Krupitsky H, Stein Z, Hsiou Y, Strouse CE (1995) Supramol Chem 4:203
25. Abrahams BF, Hoskins BF, Michail DM, Robson R (1994) Nature, 369:72
26. Su D, Wang X, Simard M, Wuest JD (1995) Supramol Chem 6:171
27. Desiraju GR (1997) J Chem Soc Chem Commun 1475
28. Desiraju GR (1995) Angew Chem Int Ed Engl 34:2311
29. Russel VA, Ward MD (1996) Chem Mater 8:1654
30. Russel VA, Evans CC, Li W, Ward MD (1997) Science 276:575
31. Ermer O (1988) J Am Chem Soc 110:3747
32. Ermer O, Eling A (1988) Angew Chem Int Ed Engl 27:829
33. Simard M, Su D, Wuest JD (1991) J Am Chem Soc 113:4696
34. Copp SB, Subramanian S, Zawarotko MJ (1992) J Am Chem Soc 114:8719
35. Hoskins BF, Robson R (1990) J Am Chem Soc 112:1546
36. Carlucci L, Ciani G, Proserpio DM, Sironi A (1994) J Chem Soc Chem Commun 1994:2755
37. Reddy DS, Craig DC, Rae D, Desiraju GR (1993) J Chem Soc Chem Commun 1993:1737
38. Ermer O, Lindenberg L (1991) Helv Chim Acta 74:825
39. Copp SB, Subramanian S, Zaworotko MJ (1993) J Chem Soc Chem Commun 1993:1078
40. Gable RW, Hoskins BF, Robson R (1990) J Chem Soc Chem Commun 1990:762
41. Gardner GB, Venkataraman D, Moore JS, Lee S (1995) Nature 374:792
42. Reddy DS, Craig DC, Desiraju GR (1995) J Chem Soc Chem Commun 1995:339
43. Abrahams BF, Hoskins BF, Robson R (1991) J. Am Chem Soc 113:3606
44. Fujita M, Kwon YJ, Washizu S, Ogura K (1994) J Am Chem Soc 116:1151
45. Venkataraman D, Gradner GB, Lee S, Moore JS (1995) J Am Chem Soc 117:11,600
46. Duchamp DJ, Marsh RE (1969) Acta Crystallogr B25:5
47. Venkataraman D, Lee S, Zhang J, Moore JS (1994) Nature 371:591
48. Kolotuchin SV, Felton EE, Wilson SR, Loweth CJ, Zimmerman SC (1995) Angew Chem Int Ed Engl 23:2654
49. Melendez RE, Sharma CVK, Zaworotko MJ, Bauer C, Rogers RD (1996) Angew Chem Int Ed Engl 24:2213

50. Kobayashi K, Endo K, Aoyama Y, Masuda H (1993) Tetrahedron Lett 34:7929
51. Aoyama Y, Endo K, Kobayashi K, Masuda H (1995) Supramol Chem 4:229
52. Endo K, Sawaki T, Koyanagi M, Kobayashi K, Masuda H, Aoyama Y (1995) J Am Chem Soc 117:8341
53. Endo K, Ezuhara T, Koyanagi M, Masuda H, Aoyama Y (1997) J Am Chem Soc 119:499
54. Brunet PB, Simard M, Wuest JD (1997) J Am Chem Soc 119:2723
55. Kobayashi K, Koyanagi M, Endo K, Masuda H, Aoyama Y (1998) Chem Eur J 4:417
56. Bhyrappa P, Wilson SR, Suslick KS (1997) J Am Chem Soc 119:8492
57. Hartgerink JD, Granja JR, Milligan RA, Ghadiri MR (1996) J Am Chem Soc 118:43
58. Yaghi OM, Davis CE, Li G, Li H (1997) J Am Chem Soc 119:2861
59. Wu LP, Yamamoto M, Kuroda-Sowa T, Maekawa M, Fukui J, Munakata M (1995) Inorg Chim Acta 239:165
60. Munakata M, Wu LP, Yamamoto M, Kuroda-Sowa T, Maekawa M (1996) J Am Chem Soc 118:3117
61. Ung AT, Gizachew D, Bishop R, Scudder ML, Dance IG, Craig DC (1995) J Am Chem Soc 117:8745
62. Russel VA, Etter MC, Ward MD (1994) J Am Chem Soc 116:1941
63. Davis ME, Robo RF (1992) Chem Mater 4:756
64. Yaghi OM, Li H, Groy TL (1996) J Am Chem Soc 118:9096
65. Endo K, Koike T, Sawaki T, Hayashida O, Masuda H, Aoyama Y (1997) J Am Chem Soc 119:4117
66. Aoyama Y, Endo K, Anzai T, Yamaguchi Y, Sawaki T, Kobayashi K, Kanehisa N, Hashimoto H, Kai Y, Masuda H (1996) J Am Chem Soc 118:5562
67. Toda F, Tanaka K, Miyahara I, Atsutsu S, Hirotsu K (1994) J Chem Soc Chem Commun 1994:1795
68. Korkas PP, Weber E, Czugler M, Náray-Szabó G (1995) J Chem Soc Chem Commun 1995:2229
69. Marjo CE, Bishop R, Craig DC, O'Brien A, Scudder ML (1994) J Chem Soc Chem Commun 1994:2513
70. Toda F, Tanaka K, Mak TCW (1985) Bull Chem Soc Jp, 58:2221
71. Gardner GB, Kiang Y-H, Lee S, Asgaonkar A, Venkataraman D (1996) J Am Chem Soc 118:6946
72. Kondo M, Yoshitomi T, Seki K, Matsuzaka H, Kitagawa S (1977) Angew Chem Int Ed Engl 36:1725
73. Yaghi OM, Li H (1996) J Am Chem Soc 118:295
74. Yaghi OM, Li H (1995) J Am Chem Soc 117:10,401
75. Dewa T, Aoyama Y (unpublished results)
76. Scott JL (1995) J Chem Soc Perkin Trans 2 495
77. Hawkins SC, Bishop R, Dance IG, Lipari T, Craig DC, Scudder ML (1993) J Chem Soc Perkin Trans 2 1729
78. Aoyama Y, Imai Y, Endo K, Kobayashi K (1995) Tetrahedron 51:343
79. Sawaki T, Endo K, Kobayashi K, Hayashida O, Aoyama Y (1998) Bull Chem Soc Jpn 70:3075
80. Sawaki T, Dewa T, Aoyama Y (submitted for publication)

Crystalline Polymorphism of Organic Compounds

Mino R. Caira

Department of Chemistry, University of Cape Town, Rondebosch 7700, South Africa.
E-mail: xraymino@psipsy.uct.ac.za

Crystal polymorphism is encountered in all areas of research involving solid substances. Its occurrence introduces complications during manufacturing processes and adds another dimension to the complexity of designing materials with specific properties. Research on polymorphism is fraught with unique difficulties due to the subtlety of polymorphic transformations and the inadvertent formation of pseudopolymorphs. In this report, a summary of thermodynamic, kinetic and structural considerations of polymorphism is presented. A wide variety of techniques appropriate to the study of organic crystalline polymorphism and pseudopolymorphism is then surveyed, ranging from simple crystal density measurement to observation of polymorphic transformations using variable-temperature synchrotron X-ray diffraction methods. Application of newer methodology described in this report is yielding fresh insights into the nature of the crystallization process, holding promise for a deeper understanding of the phenomenon of polymorphism and its practical control.

Keywords: Crystal polymorphism, Pseudopolymorphism, Crystallization.

1 Introduction 164

2 Crystal Polymorphism – Theoretical Principles and Practical Implications 165

2.1 Background – The Role of Polymorphism in the Production of Materials 165
2.2 Crystallization and Polymorphic Transformations – Thermodynamic and Kinetic Considerations 166
2.3 Polymorphism – Structural Considerations 171

3 Methodology for the Study of Crystal Polymorphism 177

3.1 Review of Preparative Methods 177
3.2 Review of Investigative Methods 180
3.3 Polymorphic Systems and Polymorphic Changes – Case Studies . . . 198

4 Towards Control of Polymorphism 199

4.1 The Need for Polymorphic Control 199
4.2 Current Strategies and Prognosis for Crystal Engineering 200

5 References 204

Topics in Current Chemistry, Vol. 198

1
Introduction

The protean nature of a chemical substance, reflected in its ability to crystallize in different structural arrangements (polymorphs), has since its discovery [1] been a source of both fascination and frustration for chemists. At a given temperature and pressure, only one polymorphic form of a substance is thermodynamically stable, all other forms being metastable. Since the rate of transformation of metastable polymorphs to the stable one may be slow, it is quite common to encounter several polymorphs of a single compound under normal laboratory conditions. Organic compounds tend to form different polymorphs owing to weak, non-directional intermolecular interactions which exist in the solid state. When a compound can be isolated in different polymorphic modifications, each with the potential of possessing unique properties (solubility, density, melting point, enthalpy of fusion, chemical reactivity, electrical conductivity, to name a few), the chemist, pharmaceutical chemist, or chemical engineer is presented with a degree of flexibility of choice for a particular application. Balanced against this flexibility, however, are the considerable practical difficulties that can arise, both in ensuring reproducible preparation of a specific polymorph and, during the lifetime of its application, preventing its spontaneous transformation to an undesirable form. Since free energy differences between polymorphic forms of a given substance are generally around a few kJ mol^{-1} [2] and the process of crystallization is affected by many physical parameters (e.g. nature of the solvent, cooling and stirring rates, temperature, pressure, presence of impurities), minor variations in preparative conditions can tip the balance in favour of crystallization of a polymorph which is not necessarily the thermodynamically stable one. This element of unpredictability in the outcome of the crystallization process has serious implications for solids design in crystal engineering [3], where the required specificity of molecular organization in the crystalline state is crucial.

Various aspects of organic crystalline polymorphism and its occurrence in the fine chemicals, pharmaceuticals and other industries have been the subjects of several recent reviews. With varying degrees of overlap, these reviews can be roughly grouped into the following, according to their focus: thermodynamic and kinetic aspects [4–7], structural aspects [2, 3, 8–17], methodology [18–27], the crystallization process [28–34] and polymorphic control [35, 36]. The reader is referred to the above for a comprehensive view of what is a rather pervasive phenomenon in chemistry and whose pursuit is currently enjoying an upsurge of interest from solid-state researchers [5].

This report describes some recent developments in the understanding of the thermodynamic, kinetic and structural aspects of organic crystal polymorphism with an emphasis on the application of newer methodology used for its study, since this is one of the areas in which significant progress has been made in recent years. Numerous examples of polymorphic systems are described to illustrate the applications of both older and newer techniques for their investigation. These include studies of pseudopolymorphism manifested by hydrates and solvates of the parent organic molecule. Finally, the crucial question of control in polymorphism is briefly addressed with a view to illustrating current strategies and their implications for the design of solids.

2 Crystal Polymorphism – Theoretical Principles and Practical Implications

2.1 Background – The Role of Polymorphism in the Production of Materials

Many of the inconsistencies encountered in product performance in the chemical, chemical engineering, pharmaceutical, food and related industries can be attributed to polymorphism. An important example is inconsistent behaviour of drug substances upon dissolution which may have a direct influence on bioavailability. This arises because different polymorphic forms of the same drug may have solubilities which differ by an order of magnitude [24]. Inadvertent production of the 'wrong' polymorph at the crystallization stage following synthesis or at any of the intermediate processing stages can therefore result in pharmaceutical dosage forms which are either ineffective or toxic [37,38]. Spontaneous polymorphic transformations mediated by solvents is common [4] and liquid preparations of metastable drugs frequently lose their effectiveness due to precipitation of less soluble, thermodynamically more stable polymorphs or pseudopolymorphs. A case in point is the antiprotozoal agent metronidazole benzoate which, when stored as an aqueous suspension below 38 °C is metastable, leading to precipitation and growth of the insoluble monohydrate [39,40].

Dunitz and Bernstein [5] have recently documented several cases of "vanishing" polymorphs. These are usually metastable forms which, despite their thermodynamic instability, may have crystallized preferentially due to more rapid nucleation. Such metastable forms may persist and be used for many years before being "displaced", when a thermodynamically more stable form is prepared. Attempts to regenerate the original polymorph are frequently met with failure. Specific compounds with such a history include e.g. 1,2,3,5-tetra-*O*-acetyl-*β*-D-ribofuranose, benzocaine picrate and xylitol. This disturbing phenomenon extends to pseudopolymorphs. A previously known monohydrate of the antibiotic ampicillin has not been obtained since the appearance of the trihydrate [24]. A possible explanation for this behaviour is that after minute particles of the stable polymorph enter the environment, they eventually become widely disseminated ("planetary seeding" [5]) and serve as nuclei promoting crystallization of their own kind exclusively.

Manufacturing processes including crystallization scale-up, drying, heating, compression and milling can induce polymorphic transformations [24] and it follows that careful quality control is necessary at all stages to monitor undesirable changes. Systematic investigation of a compound to determine whether it is prone to polymorphism, as well as the nature of the polymorphism (enantiotropic or monotropic) [23], is routine practice in pharmaceutical pre-formulation studies. Identification of the different polymorphic forms of a drug substance, determination of their chemical and physical properties, thermodynamic stabilities, and temperatures and rates of interconversion are essential for ensuring drug preparations with reproducible behaviour [24]. Already, legislation requiring drug manufacturers to provide information relating to the occurrence (or apparent absence) of polymorphism in their products has been introduced [41]. Demonstrating the absence of a tendency to polymorphism is not easy; most substances when investigated for a sufficiently long time will reveal more than one polymorph [42].

Successful preparation of crystals of organic compounds having special properties (e.g. second-harmonic generation, metallic conductivity) may hinge on polymorphism, only one polymorph of the compound in question displaying the desired property. Bernstein has recently described representative systems which clearly illustrate the relationship between a polymorphic structure (a crystal architecture characterised by well-defined molecular interactions) and the unique physical properties which that structure confers on the solid material [13].

These remarks serve to emphasise some of the more important practical implications and consequences of polymorphism. Overcoming the problems encountered requires a deeper understanding of the processes of nucleation, crystal growth and polymorphic transformation. Several recent studies relating to these topics are reviewed in the next section.

2.2 Crystallization and Polymorphic Transformations – Thermodynamic and Kinetic Considerations

Crystallization of a specific polymorph from a melt, solution or vapour, commences with nucleation, i.e. the formation of a critical "embryonic" nucleus which is the structural blueprint for subsequent development and growth of the macroscopic crystal. The factors determining nucleation rate (e.g. the associated Gibbs free energy of activation, molecular volume, interfacial energy) generally differ for polymorphs of the same substance [4]. Since, in a supersaturated solution, nuclei of all possible polymorphs of the dissolved substance may be imagined to exist [36], the outcome of crystallization is kinetically complicated by competitive nucleation processes. Thermodynamic considerations of polymorphic crystallization include Ostwald's law of stages [4, 43], according to which, at high supersaturation, the first form which crystallizes is the thermodynamically least stable (most soluble) form. This form subsequently dissolves and transforms into a more stable one. The cycle continues until only the thermodynamically stable (least soluble) polymorph remains. The practical implication is that it should be possible to isolate the different polymorphs of a given compound at different levels of solution supersaturation and hence exercise some control over the crystallization process.

As regards polymorphic transformations in general, two types are distinguished, namely enantiotropic and monotropic [23]. These can be described in terms of the Gibbs free energy G, which has a minimum value for the thermodynamically stable phase of a polymorphic system and larger values for metastable phases and is such that the polymorph with the higher entropy will tend

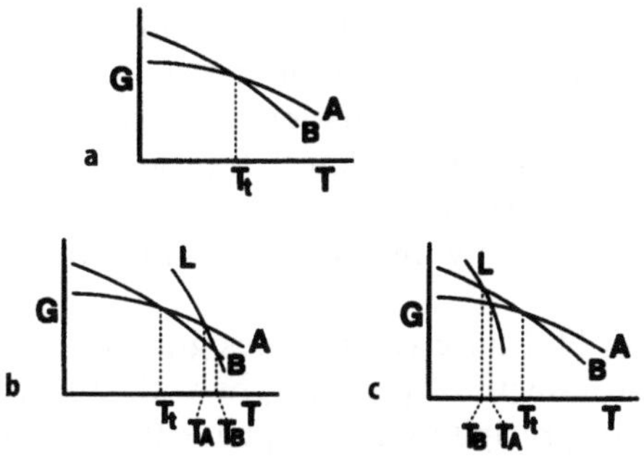

Fig. 1a–c. Gibbs free energy vs temperature for: **a** a dimorphic system, exhibiting; **b** enantiotropy; **c** monotropy

to become the stable form at higher temperature (species B in Fig. 1a). Above T_t (the transition temperature), B is the stable polymorph while A is metastable and vice versa at temperatures below T_t. In an enantiotropic system (Fig. 1b), the free energy curve for the common liquid phase L intersects the A and B curves at $T > T_t$. In this case, the lower melting form (A) is stable at $T < T_t$, the higher melting form is stable at $T > T_t$, and the transition between the two forms is in principle reversible. Since transition temperatures in practice are often in the range 20–200 °C, one practical implication of enantiotropy is that conversion of one polymorph into another may be favoured during routine manufacturing processes [24]. On the other hand, for a system displaying monotropy (Fig. 1c), curve L intersects those for A and B below T_t and the higher melting form (A) is always the thermodynamically stable one. Thus, below the melting point, only one form is stable and the other metastable. In practice, if a desired metastable polymorph is obtained during manufacture, it can revert to the stable polymorph under suitable conditions (e.g. in suspension, via solvent-mediation, or during compression). It follows that to prepare a specific polymorph and be aware of its possible fate during handling, it is advantageous to know the transition temperatures and thermodynamic stabilities of all the forms that may appear in the system [24].

The general considerations above highlight the importance of nucleation and the role of environmental conditions (e.g. solvent, temperature) in the crystallization of polymorphs as well as their interconversions. These areas continue to be the subject of intense interest especially in the context of polymorphic control in crystallization.

Some fundamental aspects of the nucleation process have been investigated by molecular dynamics (MD) methods. In a recent review [44] the advantages and limitations of molecular cluster models in simulating the dynamics of nucleation and phase changes have been discussed. In this approach, molecular dynamic simulations are correlated with experimental nucleation rates extracted from electron diffraction patterns of molecular supersonic jets. The dynamics of freezing of ammonia, CCl_4 and water, and the phase transformations of *t*-butyl chloride have been analysed. A useful feature of the MD computational approach is visual representation of phase transformations. Figure 2 illustrates MD-derived images of a crystalline cluster of 188 molecules of *t*-butyl chloride at various stages of freezing. MD simulations show that when sufficiently supercooled, the tetragonal phase spontaneously transforms to a lower temperature, ordered monoclinic phase. Diffraction patterns computed from the MD molecular packing were consistent with experimental neutron powder patterns for this phase.

Other investigators [45] have recently designed a numerical model to describe nucleation and growth of polymorphs with the aim of calculating the temporal sequences of precipitation and phase transformation in metastable solutions of polymorphic substances. Another group has recently modelled the formation and aggregation of polymorphs in continuous precipitation [46]. Consideration was given to the simultaneous growth and agglomeration of two different polymorphs as well as the case of nucleation of a single polymorph which subsequently transforms into a second one. The results indicated that the ratio of the nucleation rates, the ratio of the growth rates, and the aggregation tendencies determined polymorphic product composition as well as particle size distributions. This study is important since simultaneous precipitation of different polymorphs is encountered frequently in industrial crystallizations.

A mathematical phase-field model for the kinetics of isothermal polymorphic crystallization has recently been proposed [47], according to which crystallization involves rapid relaxation of the metastable state followed by nucleation and growth of the polycrystalline phase. Computer simulations were used to obtain results which could be tested experimentally using X-ray scattering experiments. Growth rates of different polymorphic polymers have also been investigated [48]. Simultaneous development of spherulites of different polymorphs occurs at different rates under isothermal conditions. From observation of interspherulitic boundaries between the α- and γ-forms of polypivalolactone,

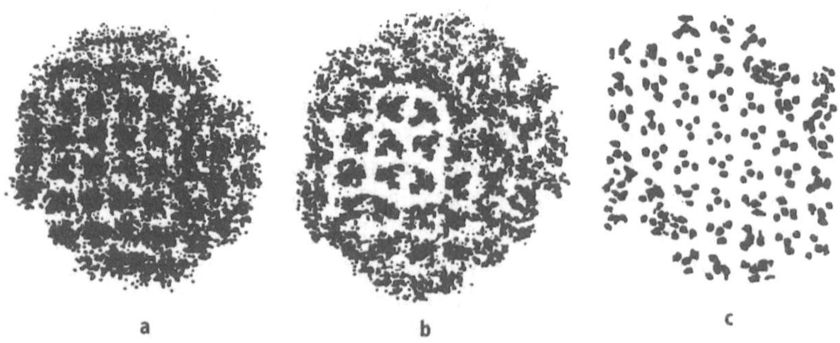

Fig. 2a–c. Images of a crystalline cluster of *t*-butyl chloride molecules at various stages of cooling, looking down the threefold molecular axis: **a** orientationally disordered tetragonal phase at 130 K; **b** nucleus of monoclinic phase growing in tetragonal phase at 80 K; **c** ordered monoclinic phase at 50 K after transformation. Surface molecules tend to be disordered at all temperatures. (Reprinted with permission from [44], copyright 1995 American Chemical Society)

their relative growth rates could be determined. The thermodynamics and kinetics of crystallization of large molecules from solution have been discussed [49]. Modelling of the crystallization of protein molecules indicated that the concepts and numerics of colloid stability theory are appropriate.

Recent studies of polymorphic transformations in organic crystals mediated by melt, solution and interface have been reviewed [4]. Interface-mediated transformation has only recently been recognized as a distinct mode of polymorphic transformation. It involves nucleation and growth of a new polymorph through mass transfer across the interface connecting single crystals of two different polymorphs and it differs from solid-solid transformations in that a microscopic solution layer is required as an interface. Transformations mediated by melt, solution and interface are usually more rapid than solid-solid transitions; for the latter, the activation energy is larger due to the fact that nucleation and growth of the new phase within a second phase involve diffusion and structural rearrangement at the reaction interface. The fundamental thermodynamic relationships governing polymorphic solid state transitions have been reviewed [6]. It has also been pointed out that the mechanisms of polymorphic transitions in molecular crystals are largely unknown, though order-disorder transitions are understood in reasonable detail [5]. An important technique for studying solid-solid transformations is thermal analysis and a review of the basic thermodynamic principles for interpreting thermal analysis data for both polymorphic and pseudopolymorphic systems has appeared [23]. Kinetic and thermodynamic aspects of the thermal decomposition of inclusion compounds (a special class of pseudopolymorphs) have also recently been discussed [50].

Theoretical and experimental studies of the role of solvent on polymorphic crystallization and phase transformations abound in the literature of the last few years and some pertinent examples are described here. For solvent-mediated transformations, the driving force is the difference in solubility between different polymorphs. An important earlier paper on the kinetics of such phase transformations [51] described a model featuring two kinetic processes in solid to solid phase changes via a solution phase, namely dissolution of the metastable phase and growth of the stable one.

The effect of solvent on the crystallization of polymorphs has recently been investigated [43] using as a model compound the antibacterial sulphathiazole whose four known polymorphs are well characterised. The study, whose express intention was to test the Ostwald law, involved crystallization of the pure polymorphic forms of the drug, solubility measurements and crystallizations from various solvent systems. Systematic variation of supersaturation was employed in an attempt to crystallize each of the four forms of sulphathiazole, as predicted by the Ostwald law. Solubility studies showed Form I to be the most soluble form, followed in order by Forms II, IV and III. The supersaturation crystallizations using acetone, acetone-$CHCl_3$ (3:2), *n*-propanol and water, revealed that only the acetone-$CHCl_3$ system yielded results in accord with theory, Forms I, III and IV being isolated from it by varying the supersaturation. Crystallization from *n*-propanol, for example, yielded only Form I at all supersaturation levels. Thus, the finding that some solvents selectively favour the crystallization of a particular form (or forms) indicated that, while supersaturation is an important factor determining crystallization of polymorphs, the solvent may play a dominating role which is not thermodynamic in nature, but rather kinetic. The proposed mechanism, namely selective adsorption of different solvent molecules on specific faces of particular polymorphs, is consistent with that suggested earlier [52] for the effects of additives and solvents on crystal morphology. Such specific adsorption might result in inhibition of nucleation of certain polymorphs or retardation of their growth, allowing other, thermodynamically less favoured, polymorphs to crystallize instead.

The mechanism of a solvent-mediated transformation may change with complete change of solvent, or more subtly, when a gradual change in polarity is effected by dilution of the original solvent with another. The effects of the solvent systems water [53] and ethanol/water [54] on the crystallization of L-histidine polymorphs have been investigated. In aqueous solution at the isoelectric point, it was found that both the A and B polymorphs precipitate with a nearly constant ratio over a wide concentration range. However, slow transformation from B to A (which does not occur in the absence of solvent) was observed and pure A could eventually be isolated. This transformation involves smooth growth of the stable A polymorph and dissolution of the metastable form B. The ratio of the rate constants for the appearance of A and the dissolution of B indicated a growth-controlled mechanism for the transformation. In subsequent experiments investigating the effect of added ethanol, however, it was found that the fraction of polymorph A in precipitates decreased rapidly with increasing volume fraction of ethanol in the mixture, and pure B could be obtained when this fraction was 0.4. An explanation for the change in growth mechanism with added ethanol, based on the decreased concentration of polymorph A, was proposed [54].

This type of behaviour is not confined to polymorphs but may extend to pseudopolymorphic forms such as hydrates and solvates. A recent case of solvent-mediated phase transformation involved polymorphic and pseudopolymorphic forms of thiazole carboxylic acid [55], where the transformation is again sensitive to the composition of the mixed solvent. Three forms of the compound are known, an anhydrous form, a 0.5 hydrate, and a 1.5 hydrate. In 50–80% solutions (% = vol.% MeOH-H_2O), transformation of the 1.5 hydrate to the 0.5 hydrate was observed while transformation to the anhydrous form occurred in 85–100% solutions. No transformation occurred in 0–30% solutions. Detailed study of a solvent-mediated polymorphic transition has also been carried out for the antiulcerative agent cimetidine [56] for which seven polymorphic forms are known. An important feature of this study was the systematic use of seed crystals to induce crystallization at different supersaturation ratios.

The possibility of relating solvent effects to polymorphic crystallization at the molecular level may be realized when the individual crystal structures of the polymorphs are known. An analysis of this kind was carried out for the anti-inflammatory drug piroxicam [57] which was found to crystallize as the α-polymorph from proton donor and basic solvents, but as the β-polymorph from non-polar solvents. On the assumption that crystallization of the drug requires release of solvent molecules from acceptor and donor sites on the molecule, it was possible to reconcile the observed intermolecular hydrogen bonding arrangements in the crystals (infinite chains in α-, cyclic dimers in β-) with probable sites of solvation on the piroxicam molecule in solvents of different polarities. In this way, formation of the α- and β-polymorphs from different solvent systems could be rationalised. Combination of polymorphic crystal structural data with appropriately detailed solvation models could be a useful adjunct to the existing methods of rationalising or predicting the outcome of polymorphic crystallization from solution.

2.3 Polymorphism – Structural Considerations

Dunitz has recently developed the theme of the crystal as an ordered supramolecular entity [2]. From this perspective, different polymorphic modifications of a given compound may thus be regarded as "supramolecular isomers". In discussing the possible structural arrangements occurring in crystals of polymorphs, a convenient distinction may be made [3] between rigid molecules (e.g. planar chloro-aromatics) and those with conformational flexibility (e.g. *N*-benzylideneanilines). In the former case, different polymorphic structures frequently display common features such as similar intermolecular directional contacts, layer stacking and unit cell dimensions which are related in a simple way. Conformational polymorphism, the existence of different conformers of a flexible molecule in the various crystal structures [58], may be expected to occur when the conformational energy minima differ by less than about 8 kJ mol^{-1}. If the energy barriers separating these minima are sufficiently low, these conformers may co-exist in solution and slight variations in crystallization conditions may lead to their individual isolation as conformational polymorphs [11]. Systems of this type have been exploited to study both the influences of the crystalline environment on molecular conformation as well as the properties of molecules which depend strongly on conformation [13]. An example of conformational polymorphism in which the structural arrangements are dictated by non-directional van der Waals forces only is shown in Fig. 3. The molecule in question is probucol, a drug used to control blood-cholesterol levels. Here, intermolecular hydrogen bonding between hydroxyl groups in the crystals is prevented owing to intramolecular steric crowding of these groups by neighbouring *t*-butyl substituents. The molecules adopt distinctly different conformations in the two polymorphs [59], the more symmetrical conformation approaching point symmetry C_{2v}.

Figure 4 shows representative hydrogen bonded (N-H···O, C-H···O) layers of planar nitrofurantoin molecules occurring in the α- and β-polymorphs [60] which are triclinic and monoclinic, respectively. In this system, the molecular conformations in the two polymorphs are indistinguishable but the symmetries of their intermolecular hydrogen bonding schemes differ significantly. The common molecular conformation shown here occurs in five modifications (two polymorphs and three pseudopolymorphs) of this compound [61]. The three-dimensional crystal structures of the polymorphs result from close stacking

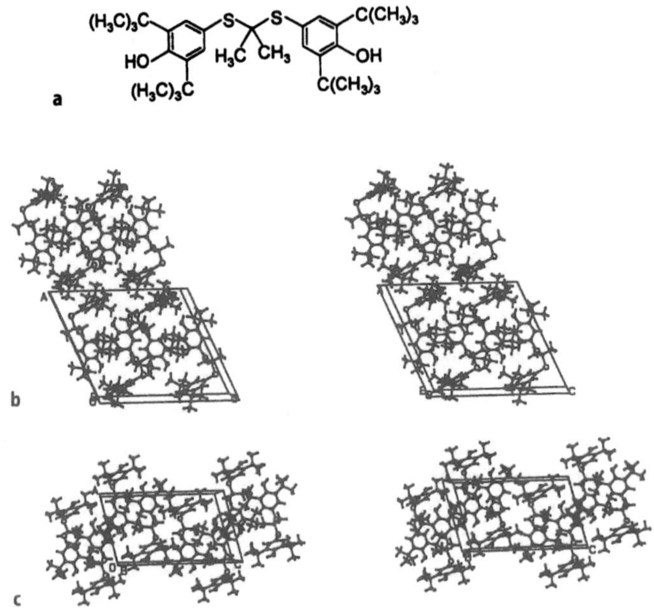

Fig. 3. **a** Molecular structure of probucol; **b** stereoview of the crystal structure of Form I; **c** stereoview of the crystal structure of Form II. (Adapted from [59] with permission)

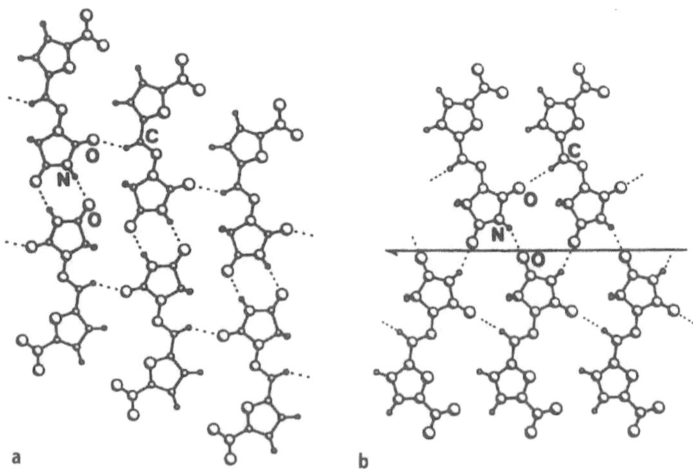

Fig. 4 a, b. Hydrogen bonded layers in nitrofurantoin polymorphs: **a** α-form; **b** β-form (Adapted from [60] with permission)

(~3.2 Å) of the respective layers shown in Fig. 4. This example conveys an idea of the variations in polymorphic structural arrangements that are possible with molecules containing hydrogen bonding functionalities.

During the last few years, the development of graph set analysis [16] has greatly facilitated the visualisation and comparison of polymorphic structures. Here, hydrogen bonding networks are classified as belonging to one of four distinct patterns, each specified by a designator (G in general): intramolecular (S), chains (C), rings (R), or other finite patterns (D). Further specification of the number of donor (d) and acceptor (a) atoms as well as the total number of atoms (n) comprising the pattern yields a concise and informative description of the hydrogen bonding arrangement, $G_d^a(n)$. A simple illustration of the use of graph set descriptors is given in Fig. 5 for a polymorph of thalidomide (space group C2/c, Z=8) [62]. Alternating hydrogen bonded ring motifs exist in this polymorphic structure as a result of bifurcated hydrogen bonding involving the N-H group. In contrast, the other known racemic modification of thalidomide (space group $P2_1/n$, Z=4) [63] contains only centrosymmetric dimers of the type $R_2^2(8)$. Application of graph set analysis to three polymorphs of iminodiacetic acid [64] has led not only to facile comparison of the crystal structures, but has also provided a basis for concise description of the polymorphic transformations occurring in that system. Thus, e.g. transformation of one polymorph of the acid into another is simply described as a conversion of $R_2^2(4)$ into $R_4^2(8)$. It has been pointed out that graph set analysis of hydrogen bonded systems also provides a means for seeking systematic correlations between the resultant patterns that demonstrate "hydrogen bond pattern functionality" [16]. This concept may be useful in the prediction of crystal structures as well as for the design of materials with desired supramolecular features. Extension of the graph set approach to encompass intermolecular atom-atom interactions other than hydrogen bonding for the classification of polymorphic crystal structures is envisaged as a natural and desirable development and more widespread use of this classification can be expected in future.

Elucidation of detailed polymorphic structural features and structural changes accompanying polymorphic transformations relies heavily on the single crystal X-ray diffraction technique. The inability to produce single crystals in

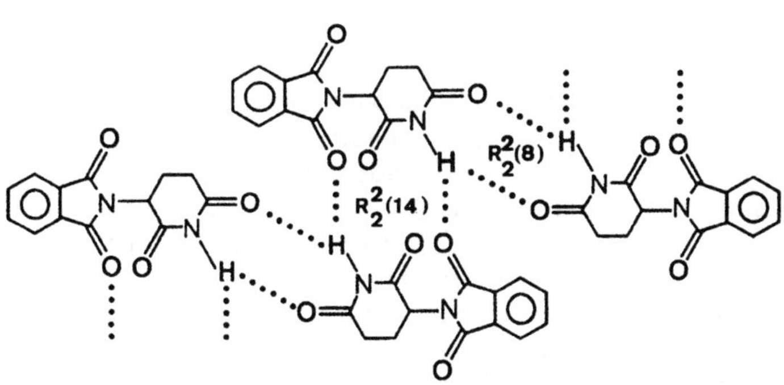

Fig. 5. Graph set notation for hydrogen bonded ring motifs in a polymorph of thalidomide

the laboratory (especially for metastable polymorphs) has hindered progress in research on polymorphism. If, however, a polymorph is available in a powdered crystalline form and a computational method exists for predicting possible three-dimensional crystal structures, then the latter can be used to generate computed X-ray powder patterns for comparison with the experimental pattern. While the scope of the present review does not permit detailed exposition of the approaches and algorithms employed in crystal structure prediction, a summary of some recent developments, especially as they relate to polymorphism, is in order. The studies of polymorphism and crystal structure prediction are indeed two facets of the same topic [65]. Additional powerful motivations for predicting crystal structures include a better understanding of the crystallization process as well as the design of new solid materials. The reader is referred to several recent treatments of crystal structure prediction [66–70]. Comments on some of these methods and their applications follow.

In the atom-atom potential (AAP) method as implemented in the program PROMET [66], symmetry operations (e.g. a screw axis, centre of inversion) appropriate to the chosen space group, are applied to a molecule in fixed conformation (possibly optimised previously by a molecular mechanics calculation) to generate molecular clusters. Following a search for the most stable clusters on the potential hypersurface (calculated using empirical atom-atom potentials), translation is applied to generate one or more periodic structures. Packing energies are computed and acceptable structures are optimised. This is a very useful means of generating a series of polymorphic structures for a given molecule. In this approach, consideration is usually given to only the most populous space groups (93% of organic molecules being confined to 18 space groups [71]) thereby running a small risk of an incorrect choice. If an experimental X-ray powder pattern of the material is available, the correct crystal structure may be identified by comparison with computed powder patterns from the candidate structures generated by PROMET. In a study using this procedure [72] several literature cases were selected, for each of which the crystal structure of one polymorph had been determined, and mention was made of the existence of a second polymorph whose crystals were unsuitable for complete structural elucidation by X-ray analysis. Only unit cell and space group data were available for these undetermined polymorphs. Each starting molecular conformation was assumed to be the same as that in the corresponding fully characterised polymorph and was submitted to the PROMET procedure. In all cases, satisfactory structures were generated, with predicted unit cell data in good agreement with the experimental ones and with packing energies in the expected ranges. These results represent authentic crystal structure prediction, assisted by partial X-ray data. The method is therefore an alternative to direct methods of structural solution and also implies that, provided cell and space group information can be acquired, full structure determination without diffraction data is feasible. The authors of this study are less optimistic as regards true ab initio crystal structure prediction for numerous reasons, among them that packing energies for polymorphs of the same compound are always very similar, rendering the choice of the correctly computed structure difficult in the absence of other data, and that the occurrence of polymorphs containing molecules in different conformations is common. The point was also made that the correct energy ordering of polymorphic structures by computational methods may bear no relation to the experimental situation, where kinetic factors may determine which polymorph actually crystallizes under given conditions.

The AAP approach, with some variations, has recently been applied with varying degrees of success to the prediction of several other structures including, in order of increasing molecular complexity, high pressure solid phases of benzene [73], the three polymorphs of sulfanilamide [74], and the low and high temperature phases of poly(*p*-hydroxybenzoic acid) [75].

An ab initio molecular packing analysis procedure (program mpa) which avoids any prior assumption of space group symmetry has recently been described [76]. Only the molecular structure and the force field are required as inputs and the program finds intermolecular energy minima for packed arrangements for any given number of molecules comprising the asymmetric unit. Space group symmetry operations are predicted in this procedure and successful application to the crystal structures of urea and benzene were reported. It is significant that the energy minimisation for the urea structure converged to the correct space group ($P\bar{4}2_1m$) which has a frequency of occurrence of only 0.39 %.

Quantum mechanical methods have also been applied to crystal structure prediction. A recent example involved the use of ab initio crystal field methods with the SM (supermolecule) model and the PC (point charge) model applied to the three known polymorphs of glycine [77]. Comparison of the optimised structures with published X-ray structures for these forms indicated that the quantum-mechanically based SM model employing a 15-molecule cluster produced results in better agreement with experiment than the PC model which describes the crystal environment purely electrostatically.

Despite the varying degrees of success attainable by such computational methods, a survey of the recent literature on this subject seems to indicate that crystal structure prediction by theoretical methods is more rapid, has greater chances of success, and consumes far fewer computing resources when coupled with other techniques which provide additional experimental data for the crystal in question. (Examples of such combined studies are discussed in Sect. 3.2). At the same time, it can be argued that successful ab initio crystal structure prediction (i. e. assuming only the molecular structure as given) by whatever means possible in the future, would represent a very significant advance in the understanding of the fundamentals of the crystallization process. Regarding the feasibility of crystal structure prediction in general, some philosophical and technical points have been discussed, together with excellent practical recommendations for a programme of experimental and theoretical studies for elucidating the basic principles of organic solid-state chemistry [78]. This is seen as a prerequisite to the solution of the problem of crystal structure prediction.

Some new insights into the nature of organic crystal polymorphism have been gleaned from a recent systematic analysis [79] of data for polymorphic structures retrieved from the Cambridge Structural Database. A total of 345 crystal structures were reduced to 163 clusters (a cluster referring to a group of two or more polymorphs of the same compound). These clusters comprised 147 with 2, 13 with 3, and 3 with 4 partners. Differences in molecular properties (P) (e.g. density, molecular volume, packing coefficient, AAP-calculated packing energies and other thermodynamic properties) were computed between cluster members (i, j) as $\Delta P = P_j - P_i$ or $\Delta P = 100(P_j - P_i)/P_j$ $(j > i)$ for a total of 204 data points. Both monovariate and bivariate statistical analyses were performed on the data, yielding several revealing trends and correlations. Histograms of differences in properties between polymorphic pairs, shown in Fig. 6a–c, indicate respectively that differences in crystal packing energies, densities and lattice vibrational entropies for polymorphs are rather small while Fig. 6d reveals that an appreciable proportion (actually 18%) of polymorphs have $Z' > 1$ (where Z' is the number of molecules in the crystal asymmetric unit). Some further important conclusions drawn from this study are as follows: both calculated and experimental values for the relative stability of crystal polymorphs are currently subject to large uncertainties; polymorphs with $Z' > 1$ are as stable, or even more stable than those with $Z' = 1$; higher crystal density is, as expected, found to

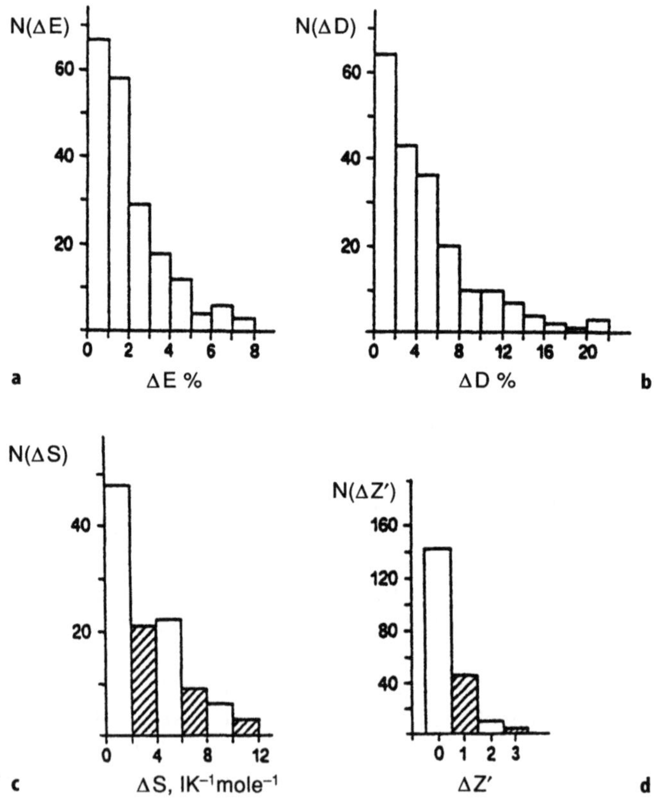

Fig. 6a–d. Histograms of differences in properties between polymorphs: **a** ΔE (packing energy, %); **b** ΔD (density, %); **c** ΔS (lattice-vibrational entropy, J K^{-1} mol^{-1}); **d** ΔZ′ (no. of molecules in the asymmetric unit). (Reprinted with permission from [79], copyright 1995, American Chemical Society)

correlate with higher packing energy. Finally, from the observation that 24 % of the polymorphic pairs analysed comprised both a centrosymmetric and a non-centrosymmetric partner, it was concluded that the link between molecular properties and centrosymmetry of crystals is evidently weak. The reader is referred to this seminal study [79] for the full exposition of the above conclusions and their justification, as well as further perceptive observations on the phenomenon of polymorphism.

3 Methodology for the Study of Crystal Polymorphism

3.1 Review of Preparative Methods

Research on the polymorphism of a new molecular entity normally commences with experimental screening which can indicate the occurrence of more than one crystalline form of the substance. An inexpensive method of such testing is hot stage microscopy (HSM), which has been used very extensively and effectively by a leading proponent [80] for many years to provide preliminary indications of the presence of crystalline polymorphic and pseudopolymorphic (solvated), as well as glassy (amorphous) forms, all of which may have practical utility. Pseudopolymorphic forms are molecular adducts containing solvent of crystallization and have been classified [37] as stoichiometric solvates and non-stoichiometric inclusion compounds possessing channel, layer or cage (clathrate) structures. Procedures for detecting the existence of multiple forms by HSM have been outlined [37, 81]. These include, for example, the observation of solid-solid transformation upon heating the substance, observation of transformation (spontaneous or mechanically induced) following the freezing of a melt, and detection of gas evolution as bubbles from pseudopolymorphs immersed in silicone oil during heating (indicative of a solid$\rightarrow$gas + solid transformation). Once the existence of multiple forms is established, practical methods for the preparation of specific forms on a larger scale may be explored. Frequently, recrystallization of the compound from solvents or solvent mixtures spanning a wide polarity range is effective in producing several of the different forms in sufficient quantity for complete characterisation by the analytical methods to be discussed. Most pseudopolymorphs are prepared by crystallization of the parent organic compound from the respective solvent, whereupon the latter becomes incorporated in the new crystal. Recrystallization from a mixed solvent system may yield a pseudopolymorph containing either or both solvents. Exposure of the parent organic compound to vapours may also result in the formation of pseudopolymorphs (as occurs e.g. when anhydrous drugs react with atmospheric water to form hydrates). This is exemplified by the drug indomethacin, for which four polymorphic forms (I–IV) have been identified [82]. Crystallization of the commercially available Form I from over fifty solvents yielded Forms I, II, IV and mixtures of these, as well as a number of pseudopolymorphs [83]. To ensure subsequent reproducibility in the preparation of specific forms by crystallization, careful attention to the details of solvent purity, degree of solution agitation, temperature, supersaturation and the rate of cooling of the solution is necessary. Detailed methods for isolating metastable polymorphs from the melt or from solution have been reviewed [81].

Desolvation of a pseudopolymorphic form by controlled heating leads to the formation of a polymorph or a mixture of polymorphs of the parent compound, hence providing an additional route to isolation of such species. For example, for indomethacin quoted above, twelve pseudopolymorphs with the general formula indomethacin · (solvent)$_n$ ($n = 0.2-1.1$) were isolated [83] and their products of desolvation were characterised. Whereas mixtures of Forms I and II resulted from heating the benzene, CCl_4, $CHCl_3$ and toluene pseudopolymorphs, pure Form II was obtained from desolvation of the acetone pseudopolymorph. The most soluble (and pharmaceutically desirable) polymorph, Form IV, was obtained in a pure state by desolvation of the pseudopolymorph containing methanol.

A technologically very important and potentially beneficial feature of some polymorphs obtained in this way is the possibility that they may acquire altered rheological or other properties (flowability, texture, particle size distribution, compressibility) when compared with the same polymorphic crystalline forms obtained by direct crystallization from solution. Two examples of considerable pharmaceutical relevance may be cited, namely those of lactose and paracetamol. The lactose used as an excipient in pharmaceutical tablets and capsules is α-lactose monohydrate. It was found that thermal dehydration of this species or desiccation of α-lactose containing methanol yielded a stable product with superior binding properties and excellent flowability [84]. Tablets prepared by compaction of the stable product had an overall porosity nearly equal to those of tablets prepared with the original materials. The very poor compression abilities of paracetamol prompted an investigation of solvation/desolvation as a process for preparing pure paracetamol with improved properties [85]. A crystalline hemisolvate of paracetamol was prepared by cooling a hot saturated solution of the drug in dioxane. Desolvation of this species yielded pure paracetamol with significantly improved technological properties including flowability, die filling, and hardness/pressure profile. A crucial point is that this improvement was not attributable to the production of a new polymorphic form of the drug; the X-ray powder diffraction pattern of the desolvated material was the same as that of the commercially available, monoclinic polymorph of paracetamol. However, detailed examination with scanning electron microscopy revealed that desolvation produces material with an unusually porous, sintered-like texture which lends itself to compression more readily than other forms of the drug. Interestingly, solvation/desolvation using other oxygen-donor solvents (e.g. acetone, cyclohexanone) yielded paracetamol in the same (monoclinic) form but failed to produce paracetamol crystals with the required texture. An earlier review on polymorphism [37] lists several drug pseudopolymorphs whose desolvation leads to significant particle-size reduction (or micronization) of the ensuing polymorphic crystallites. This usually results in improved dissolution and tabletting properties of the polymorph.

Mechanical grinding and compression of compounds represent another possible route to polymorphs. In the former case, the local pressures induced by the mechanical stress may initiate the transformation of the original polymorph into another crystalline form. From an industrial viewpoint, grinding and compression are attractive processes, being relatively inexpensive and requiring no solvents. In addition, complete polymorphic conversion may be effected in very short times (several minutes in some cases). Recent examples include the production of polymorphic Form I of sulphathiazole by planetary ball-milling of Form III [86], the transformation of metastable Form I of caffeine into the stable Form II by either grinding or compression [87] and production of Form I of probucol by manual trituration of Form II [59]. The role of grinding in the design of pharmaceutical dosage forms has been investigated and the effects of increased specific surface areas and enhanced solubilities following mechanical action have been noted [88]. The same group reported the effects of environmental temperature and compression energy on the polymorphic transformations of the antidiabetic chlorpropamide [89] and discussed the relation between the polymorphic transformation pathway during grinding and the physicochemical properties of bulk powders of cephalexin, chloramphenicol palmitate and indomethacin [90]. Local temperature variations during grinding may also play a role in effecting polymorphic transformation and it is desirable to separate the two influences. This was recently achieved with cortisone acetate by cryogrinding at 78 K, during which the monoclinic form transformed to an orthorhombic form in ten minutes purely by mechanical effects [91].

Prolonged mechanical grinding of a crystalline compound may produce material in an amorphous (or glassy) state which, due to the lack of long-range internal order of the constituent molecules, displays a broad melting temperature range and a diffuse X-ray diffraction pattern. A discussion of polymorphism without reference to amorphism would represent serious neglect of an important aspect of phase behaviour. The amorphous state represents the thermodynamically least stable form of the compound, which consequently has a tendency to revert to a more stable form. The amorphous material is also the most soluble form of the compound and this property is used to advantage in pharmaceutical preparations in cases where the solubility of the crystalline form of the drug is low, leading to poor systemic absorption. A well known example is the antibacterial novobiocin acid, for which the solubility of the amorphous (and therapeutically active) form is ten times that of the crystalline (inactive) form [92]. However, use of an amorphous form in a suspension may require addition of another component to suppress spontaneous transformation to a thermodynamically more stable form. For novobiocin acid preparations, the additives methylcellulose and polyvinylpyrrolidone are successful in this respect. It seems likely that the amorphous state may attain even greater practical significance in view of the recent reference to "amorphous polymorphism" [14], i.e. the existence of more than one distinct amorphous phase of the same substance. This phenomenon, which has been studied by computer simulation, evidently occurs in substances where the thermodynamic behaviour of the liquid state exhibits liquid-liquid phase separation or a tendency towards it.

In concluding this discussion of the methods of preparing polymorphs, several examples may be quoted of observations or procedures which, owing to their novelty or confinement to only one particular compound, cannot be described as general. However, since systematic study of the factors which might be involved in effecting or affecting polymorphic transformation in these cases could in principle find wider application, brief mention of such examples is made here. The effect of explosion on polymorphs of chitin and chitosan has recently been studied [93]. Explosion of α-chitin resulted in no change, but explosion of hydrated chitosan of low crystallinity yielded a product with increased crystallinity as well as a small amount of the anhydrous phase. The effect of an electric field on polymorphic transformation in certain classes of compounds may be a phenomenon warranting systematic investigation. Strong temperature-dependence of the $\beta \rightarrow \alpha$ polymorphic transition of isotactic polypropylene (containing additives) has been observed when the system is exposed to an electric field [94]. The effects of neutron-irradiation on the kinetics of polymorphic phase transitions for several inorganic compounds have been reported [95], but similar studies using organic compounds as substrates are lacking. It has, however, been demonstrated that ionizing radiation can induce a polymorphic transformation in an organic compound. A polycrystalline sample of the β-polymorph of a diacetylene nitroxide has been reported to undergo transition to the α-polymorph on irradiation with CuKα or ^{60}Co sources [96]. X-ray powder diffraction was used to monitor this phase tranformation.

Finally, the role of serendipity in producing polymorphs may be mentioned. Two recent cases involved unsuccessful attempts to produce molecular complexes by reaction of two components in solution, resulting instead in the precipitation of crystals of one component in a desirable polymorphic form. Attempts to grow crystals of methotrexate by various techniques failed [97]. However, tetragonal crystals of the compound were obtained from a solution containing methotrexate and thymidine, prepared for the purpose of obtaining a co-crystal of these components. Similarly, attempted complexation between 5-sulphamethoxydiazine and p-aminosalicylic acid failed, the solution of these species in a 1:1 molar ratio producing instead large crystals of Form II of the sulphonamide [98]. Interestingly, this polymorph of the drug is the biologically most active form and is usually crystallized from ethanol followed by rapid cooling of the solution to −12 °C. This yields small particles with little or no geometrical form [99]. A possible explanation for the crystallization of Form II during attempts to produce the complex has been discussed [98].

3.2 Review of Investigative Methods

Having outlined the methods of preparation of polymorphs and pseudopolymorphs, this report now focuses on the methodology used to study these forms. The discussion commences with a survey of some well established and still widely used techniques, each of which is illustrated by one or more applications. This is followed by a survey of newer methodology which is being used to probe organic crystal polymorphism.

A wide spectrum of analytical techniques may be used to characterize polymorphs and pseudopolymorphs in terms of their structure, spectral energies, thermodynamic stabilities, kinetics of transformation and solubility behaviour.

The choice of analytical and physicochemical methods for the characterization of polymorphs is dictated by the need to measure properties which ultimately depend on the different internal arrangements of the same molecules in these phases. When pseudopolymorphs are also considered, the range of suitable analytical techniques is significantly broadened owing to the presence in the crystal of the solvating molecule and the possibility of analysing the physical and chemical changes which may accompany both formation and decomposition of pseudopolymorphs.

Some remarks on the use of white- and polarized-light microscopy serve as an appropriate introduction to this section. With the advent of more sophisticated analytical methods, the use of microscopy has, to some extent, fallen into neglect. This is unfortunate, since investment of relatively little time and effort in studying a crystal of a polymorph or pseudopolymorph by microscopy can be invaluable, enabling one to assess the overall quality of a recrystallization, to detect crystal faults, fractures and macroscopic inclusions, and obtain preliminary information which will facilitate subsequent X-ray examination. Detection of different crystal habits (acicular, tabular, bladed, plate-like, prismatic) using white-light microscopy is not necessarily indicative of polymorphism since crystal habit depends on crystallization conditions and may vary widely for a given polymorph. However, this method is useful for distinguishing polymorphs having different colours in reflected or transmitted light. Pseudopolymorphs which undergo pseudomorphosis (i.e. loss of solvent on removal from their mother liquor) tend to form opaque, microcrystalline masses which are also discernible by ordinary microscopy. Addition of a polarizing attachment allows distinction between optically isotropic crystals and anisotropic crystals as well as the measurement of refractive indices [100]. Optically isotropic crystals belong to the cubic system and have a single value for their refractive index. The vast majority of crystalline organic compounds are optically anisotropic, having multiple refractive indices and displaying numerous optical effects which may be used to differentiate polymorphic forms. Anisotropic crystals reveal themselves by producing variable interference colours as well as regular extinction of plane-polarized light on rotation of the microscope stage. They may be uniaxial (characterized by two principal refractive indices and belonging to the trigonal, tetragonal or hexagonal systems) or biaxial (with three refractive indices and belonging to the triclinic, monoclinic or orthorhombic systems). Measurement of these refractive indices is certainly a means of identifying a polymorph unequivocally, but is seldom done for this purpose. The uniaxial or biaxial nature of the crystal is easily determined from observation of the respective characteristic interference figure when the crystal is viewed with condensed (conoscopic) light. Taken together, extinction directions, crystal morphology and uniaxial or biaxial character can facilitate the identification of a new polymorph, as exemplified by the following case from our laboratory. Carbamazepine commonly crystallizes in the monoclinic system with a prismatic habit [101]. Microscopic examination of crystal batches obtained by recrystallization of the drug from a wide range of solvents confirmed the predominance of this form. However, crystals obtained from tetrahydrofuran were acicular, yielding extinction parallel to the needle-axis and presenting a uniaxial interference figure.

Measurement of the interfacial angles of a crystal section both microscopically and using an optical goniometer yielded values of 120±1°. Thus, a new polymorph belonging to either of the trigonal or hexagonal systems was unequivocally identified. This was confirmed by subsequent X-ray photography which revealed a trigonal space group [102].

Owing to its vast depth of field, scanning electron microscopy (SEM) is widely used for observing the texture, morphology and surface features of both powders and large single crystals of polymorphs. The high vacuum used for sample observation precludes study of pseudopolymorphs containing volatile solvents but SEM micrographs of polymorphs are useful for purposes of identification provided that crystallization conditions and SEM sampling methods are carefully controlled.

Crystal density is an important technological parameter since it affects the flow properties of bulk solids. Due to the different bulk densities of polymorphs and their abilities to retain solvent, different isolation strategies are required in industry [24]. Furthermore, if mixtures of different solid phases are present in a sample (e.g. in a powdered pharmaceutical formulation), differences in component densities may lead to a heterogeneous product during processing due to phase segregation. Distinguishing polymorphs of the same compound by density measurement (determined by flotation or gas displacement pycnometry) is difficult because, as shown in Fig. 6b, differences in the densities of such species seldom exceed 5% and the experimental error of routine measurements is typically 2%. However, the latter can be reduced if special precautions are taken during flotation measurements, especially with regard to eliminating occluded air. Under these conditions, anomalously high or low measured densities may be useful indicators of the presence of pseudopolymorphs. Thus, e.g., a measured density of 1.30(1) g cm^{-3} for a crystal of doxylamine succinate obtained from ethyl acetate was sufficiently different from that of polymorphic Form I (ρ=1.21(1) g cm^{-3}) to indicate the presence of a pseudopolymorph. Subsequent X-ray analysis showed the crystal to have the unexpected composition (doxylamine succinate)$_2$ · (succinic acid) [103].

Among the thermal methods of analysis, thermogravimetric analysis (TGA), differential thermal analysis (DTA) and differential scanning calorimetry (DSC) have been used extensively to quantify thermal events accompanying controlled heating of polymorphs and pseudopolymorphs [18, 23]. Figure 7 shows combined TGA and DSC traces for two pseudopolymorphs of nitrofurantoin [61], containing respectively *N,N*-dimethylformamide (DMF) and dimethylsulphoxide (DMSO).

In TGA, the sample (~5–10 mg) is heated at a predetermined rate and the weight is recorded as a function of temperature. This technique cannot distinguish polymorphs of a given organic compound, but for pseudopolymorphs which lose their included solvents prior to melting or decomposition of the parent ("host") compound, the percentage weight loss may be accurately measured and used to calculate the stoichiometry of the pseudopolymorph. Data recorded from the TGA traces shown in Fig. 7 indicated a nitrofurantoin : DMF stoichiometric ratio of 1:1 and a nitrofurantoin:DMSO ratio of 2:1. A TGA trace may reflect simple one-step weight loss of included solvent or more complex

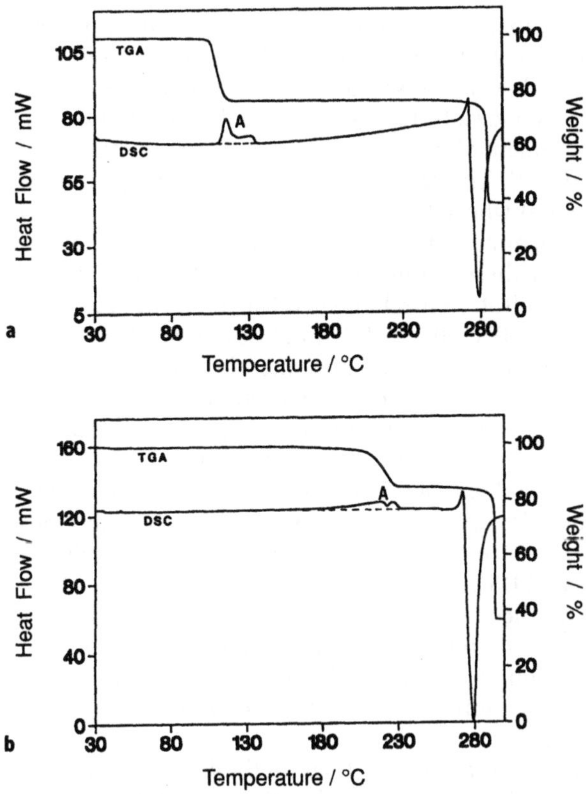

Fig. 7a, b. Combined TGA-DSC traces for pseudopolymorphs of nitrofurantoin containing: **a** DMF; **b** DMSO (Reprnted with permission from [61], copyright 1996, Gordon and Breach Publishers)

multi-step weight losses. Coupled with TGA, evolved gas analysis (EGA) using IR or mass spectrometry is an important means of identifying gaseous products [18] and may be used to quantify both the pyrolysis products from thermal degradation of polymorphs as well as solvents released from desolvation of pseudopolymorphs. Another important application of the TGA method is the determination of the activation energy for desolvation of a pseudopolymorph from traces recorded at varying heating rates [104]. Its application to pseudopolymorphs of succinylsulfathiazole [105] and tenoxicam [106], and to inclusion compounds of synthetic hosts [50] have recently been described.

In DTA, the sample temperature (T_s) is compared with that of a reference compound (T_r) as a function of increasing temperature. The resulting plot of ΔT ($=T_s-T_r$) vs T may display endothermic peaks corresponding to desolvation (for pseudopolymorphs) or fusion (for polymorphs) and exothermic peaks representing recrystallization or decomposition processes. In the related DSC technique, the difference in energy inputs into a compound and a reference substance is plotted against T during a controlled temperature programme. In theory, the area under the curve in a DSC trace is directly proportional to the enthalpy change for the thermal event, but special precautions are necessary to obtain accurate values for this quantity [18]. An alternative method for determining enthalpies of desolvation of pseudopolymorphs, based on direct vapour pressure measurement of the liberated solvent as a function of temperature, has been described [50]. The DSC traces in Fig. 7 display endotherms (peaks A, above the baseline) whose temperature ranges correlate well with those of the TGA traces, confirming that these peaks represent desolvation processes. The bimodal nature of these endotherms is difficult to interpret but evidently reflects complex desolvation mechanisms. The difference in the DSC temperature ranges for loss of DMF (100–125 °C) and loss of DMSO (160–235 °C) is striking. These pseudopolymorphs were subsequently shown by single crystal X-ray analysis to contain their respective solvent guest molecules in constricted channels formed by the nitrofurantoin host framework; each DMF molecule is bound to one drug molecule by a single hydrogen bond, whereas each DMSO molecule forms two hydrogen bonds with neighbouring drug molecules, thus providing a possible explanation for the significantly higher onset temperature for the loss of DMSO. In this connection, the use of the simple parameter $T_{on} - T_b$ (T_{on} = extrapolated DSC desolvation onset temperature, T_b = boiling point of the guest solvent) has been proposed as a possible measure of the relative thermal stabilities of inclusion compounds [50]. For both pseudopolymorphs referred to in Fig. 7, desolvation produced the β-polymorph, as confirmed by the characteristic sharp fusion endotherm at 273 °C, followed by a decomposition exotherm peaking at 279 °C.

Recent reviews of thermoanalytical methods and their application to polymorphic systems [18, 23] include descriptions of modern instrumentation, discussions of the influence of instrumental and sample parameters on measured data, as well as guides to the interpretation of DSC traces. One review [23] includes a summary of the thermodynamic rules established by Burger [107] for distinguishing monotropic and enantiotropic transitions. It is worth emphasising that in the absence of supporting techniques such as hot stage microscopy (HSM), X-ray diffraction (XRD) and Fourier transform infrared spectroscopy (FT-IR), errors in the interpretation of DSC data may result. Reference to HSM was made earlier in connection with its use in screening recrystallized batches of a compound to identify polymorphs and pseudopolymorphs. Apart from its ability to yield melting points and transition temperatures for polymorphic changes or glass transitions, HSM is an invaluable adjunct to the DSC technique in that visual recording of thermal events (e.g. desolvation, recrystallization, sublimation) may corroborate assignments of endothermic or exothermic peaks appearing in DSC traces. As pointed out recently [108], visual observation can sometimes be more sensitive indicators of phase changes than DSC measurements. The report describes the construction and use of the device shown in Fig. 8, which allows simultaneous DSC measurement and microscopic observations with a video-camera. A second camera is used to give a direct readout of the varying sample temperature.

Solid-state infrared (IR) spectroscopy in the spectral range 400–4000 cm^{-1} has been used extensively to distinguish different polymorphic forms as well as

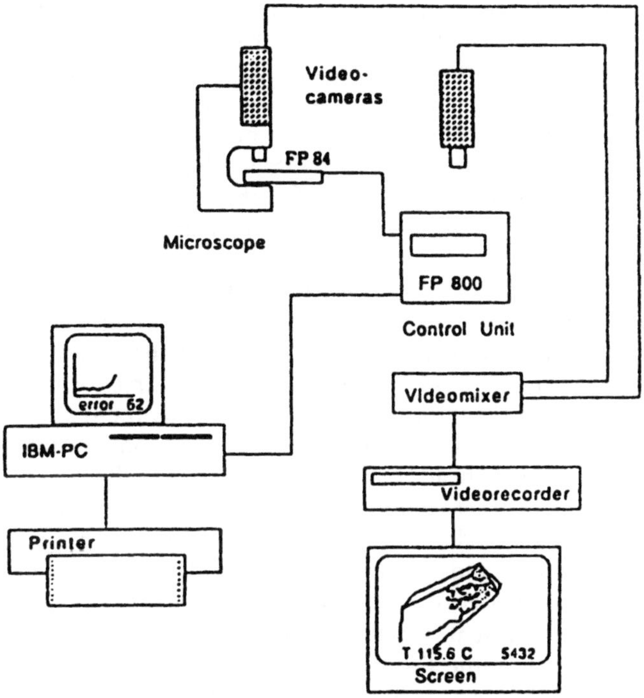

Fig. 8. Schematic description of an apparatus for simultaneous visual and DSC observations using the Mettler Thermosystem FP800 (Reprinted with permission from [108])

different pseudopolymorphs of the same compound. To ensure transparency to IR radiation, samples are usually prepared by trituration of crystalline material and suspension in Nujol or by pressing a co-ground mixture of the compound with KBr into a disk. However, since complete polymorphic transformation can sometimes be effected even by gentle manual grinding, it is advisable to confirm the existence of polymorphs by an independent method prior to a detailed IR analysis. Differences in molecular conformation and crystal packing features for different polymorphs of an organic compound manifest themselves in IR spectra chiefly by frequency shifts and splittings in, e.g., N-H, O-H and C=O absorption peaks (when these functional groups are present) due to differences in hydrogen bonding arrangements. The goal of IR analysis is the deduction of polymorphic structural differences from such spectral effects, but detailed interpretation is often limited and it is somewhat easier to rationalise these effects retrospectively if single crystal X-ray data can be obtained for the various forms.

This is illustrated by the following examples from our laboratory. A study of the literature on the compound sulphamerazine revealed that its polymorphism was controversial. The crystal structure of the drug recrystallized from acetone had been shown to be orthorhombic, Pbca with Z=8 [109]. Centrosymmetric hydrogen bonded (N-H···N) dimers exist in the crystal structure as shown schematically in Fig. 9. We obtained crystals of a second polymorph of sulphamerazine [110] by recrystallization from methanol and established the space group as $Pn2_1a$, Z=8. This form contains the same hydrogen bonded motif as shown in Fig. 9, but it is pseudo-centrosymmetric. The two polymorphs are distinguishable from IR spectra, the form containing the centrosymmetric dimer (with only one unique N-H···N hydrogen bond) yielding only one IR peak (3453 cm^{-1}) assignable to ν_{as}NH whereas that containing the pseudo-centrosymmetric dimer (with two distinct N-H···N hydrogen bonds) displayed a doublet (3493, 3478 cm^{-1}).

We recently reported FT-IR spectral data for seven crystalline forms of the androgen dehydroepiandrosterone [111], showing that differences in both fine structure and in the locations and intensities of major absorption bands can be used to distinguish them. X-ray crystal structures of three forms were determined [112], Form I (a polymorph), Form S1 (a 4:1 hydrate) and Form S4 (a methanol half-solvate). Form I (space group $P2_1$, Z=4) contains infinite chains of steroid molecules linked head-to-tail (shown schematically in Fig. 10) by two crystallographically distinct intermolecular O-H···O=C hydrogen bonds, consistent with two observed O-H stretching bands (3501, 3458 cm^{-1}). Doublets for the O-H absorption bands observed also in both Forms S1 and S4 can be correlated with distinct O-H ···O hydrogen bonds observed by X-ray analysis, and specifically, with unusual "flip-flop" hydrogen bonding in Form S4. In this study, attempts to determine the thermodynamic relationships between the various polymorphic forms were based partly on the application of Burger's IR rule [107] which relates IR band positions to relative stabilities.

The use of IR-spectra to distinguish polymorphs from pseudopolymorphs of the same parent compound, or to distinguish pseudopolymorphs containing different solvents, is relatively straightforward, since solvent molecules which are commonly incorporated in pseudopolymorphic crystals (e.g. water, ketones, alcohols) will exhibit their own characteristic absorption bands. IR spectroscopy was used to distinguish polymorphs of the antidiabetic drug glibenclamide from its solvates with pentanol and toluene [113]. Pseudopolymorphs of the

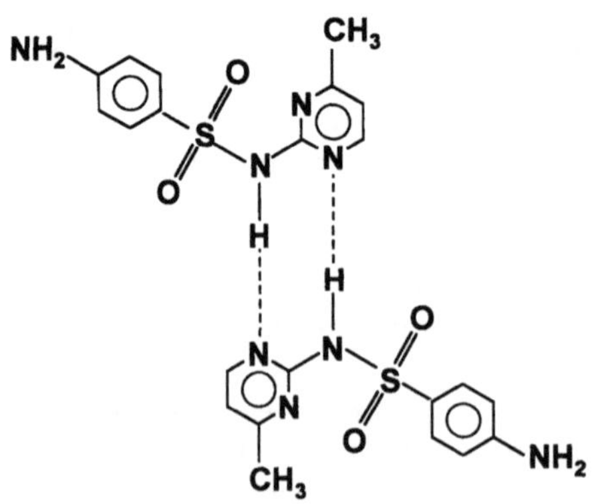

Fig. 9. Schematic representation of the hydrogen bonded dimer occurring in two polymorphs of sulphamerazine

compound gossypol (a toxic constituent of cotton-seed) obtained from three different solvents showed characteristic differences in their IR spectra in the 3500 cm^{-1} region [114]. Pseudopolymorphs of a host containing the same guest in different molar ratios are also known. An alicyclic diol of the helical tubuland family [115] forms two crystallographically distinct lattice inclusion complexes with 1,2-dichlorobenzene having host-guest ratios of 4:1 and 3:1 [116]. These were readily distinguished from their IR spectra in the fingerprint region.

The spectral techniques available for the characterization of polymorphs and solvates have been reviewed [27] and the merits of FT-IR spectroscopy for the acquisition of high-quality spectra have been stressed. The review also lists drug substances whose polymorphism has been investigated by IR-spectroscopy. Another recent review on the use of the IR method to investigate polymorphism is based on spectra for a derivative of sulphadoxine as a model compound [21]. Dynamic aspects of polymorphism are also amenable to study by IR spectroscopy. The effect of hydrogen bonding on the lattice dynamics, polymorphism, mesomorphism and phase tranformations of long-chain carboxylic acids as gauged from IR spectra has recently been reported [117].

A relatively new trend is the application of near-IR spectroscopy for compound identification and quantitation in many industrial environments including chemical processing and food science. The advantages of near-IR over conventional IR spectroscopy for the investigation of polymorphs have been described [118]. A fast, sensitive pattern recognition method for identifying a polymorph and determining its quality using near-IR spectra has been developed [119]. Pattern recognition methods involve training a computer to recognize spectra of acceptable samples of a material and to reject unacceptable ones by multivariate statistical analyses. This study showed that the method described can discriminate between the desired polymorph and other crystalline forms, the statistical results indicating that levels as low as 2% of an adulterating polymorph may be reliably detected. The authors maintain that the near-IR method, which can be automated, provides a reliability comparable to that of XRD for polymorphic identification but is likely to be less costly to implement.

There is a growing interest in the use of Raman spectroscopy for studying polymorphism and pseudopolymorphism [27]. The Raman technique has recently been used to distinguish two triclinic polymorphs of terephthalic acid [120] and, in combination with IR spectroscopy, to study the conformational polymorphism of 1-bromopentane [121].

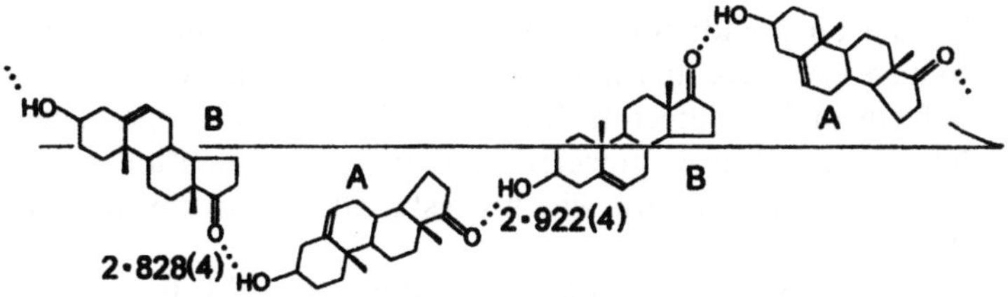

Fig. 10. Head-to-tail hydrogen bonding in Form I of dehydroepiandrosterone. A and B are crystallographically distinct and distances are in Å

The use of white-light microscopy to identify differently coloured polymorphs was mentioned above. Colour polymorphism may be quantified using UV/vis electronic spectroscopy (diffuse reflectance, fluorescence methods). Colour differences for polymorphs of the same compound may originate from differences in charge transfer interactions or different molecular conformations in the crystals. Pseudopolymorphs may also be distinguished using fluorescence spectroscopy since they may display characteristic absorption maxima as well as fluorescence intensities. Systems which exhibit colour polymorphism have been reviewed [27].

Implicit in the examples cited above to illustrate applications of various analytical methods for studying polymorphism is the definitive role of single crystal X-ray diffraction in providing unequivocal evidence for the very existence of these forms. Different polymorphs of a given compound have, by definition, non-equivalent crystal structures and the method of choice for detecting such non-equivalence is single crystal X-ray analysis. In many instances, the differences between polymorphs are so subtle that they can be distinguished only by this method [79]. This technique goes well beyond merely detecting different forms, providing in addition a wealth of structural information for each species at a level of detail and precision currently unsurpassed by other analytical techniques. Furthermore, as illustrated by several of the examples above, detailed knowledge of the molecular parameters and crystal packing features obtained by this method provides a very reassuring basis for the interpretation of results gleaned from other analytical techniques. For these reasons, much effort is expended by researchers in growing single crystals of polymorphs and pseudopolymorphs with adequate quality for complete X-ray structural elucidation.

For routine crystallographic studies of polymorphs or pseudopolymorphs containing small and medium-sized molecules, the methodology of crystal structure determination is well established and the reader is referred to standard monographs on the subject [122, 123]. However, a few points on current routine methodology are worth emphasising in order to place some of the newer developments (to be discussed later) in perspective.

Existing standard methods employ single crystals usually no smaller than 0.1–0.5 mm, automated X-ray intensity data-collection (typically with a four-circle diffractometer) and standard crystallographic software packages for data-processing, structure solution and refinement. Room-temperature data-collection is common, despite the gain in structural definition possible with cryostatic devices for cooling the crystal. Low-temperature analyses are sometimes carried out on pseudopolymorphs in which the included solvent molecules tend to have excessive thermal motion or are disordered at room-temperature. The advent of fast detectors (organic scintillators, image-plates, multiwire area detectors) and more powerful structure determination packages has greatly increased the speed with which crystal structures can be solved and refined [124].

In the absence of single crystals of suspected polymorphic forms, X-ray powder diffraction (XDP) serves as the primary test for non-equivalence of crystal structures, the XDP pattern of such a species being unique [25]. However, since sample preparation involves grinding in a mortar to an average particle size ≤100 μm to minimise preferred orientation effects, the possibility of a pressure-induced phase transformation during sample preparation must be considered. The XRD pattern of an unground sample of the material may be checked as a precautionary measure. Patterns are generally recorded on samples with mass range 200–500 mg using automatic diffractometers with strip chart output. The method is generally non-destructive, allowing recovery of the sample. Film methods (e.g. the Debye-Scherrer technique) require a few mg of material only and are still frequently used when only small amounts of polymorphic forms can be isolated. According to the guidelines specified in the US Pharmacopeia [125], agreement between a sample and a reference standard should be within the calibrated precision of the diffractometer for diffraction angle (typically a reproducibility of ±0.10–0.20° in 2θ) with permissible relative intensity variations of up to 20%.

A recent analytical study stresses the growing need, prompted partly by legislatory requirements, to differentiate polymorphs and to quantify polymorphic mixtures in pharmaceutical production [126]. The compounds benzil and benzoic acid were chosen as a model system for the development of an XRD protocol which could be extended to the quantification of mixtures of drug polymorphs. The study involved the evaluation of sample thickness, the determination of preferred orientation effects, optimum milling conditions and the construction of diffraction intensity-composition calibration curves for mixtures of benzil and benzoic acid. Since the composition of such mixtures can be accurately determined by an independent method, namely HPLC, validation of the quantification of mixtures by the XRD protocol was possible. It was concluded that the protocol is accurate for the model system to within a few percent. It is desirable that the general validity of the approach suggested be tested on a range of real polymorphic systems.

The information contained in an XRD pattern of a powdered material is a significantly condensed version of that obtained by single crystal X-ray analysis and it is a routine matter to reconstruct the XRD pattern of a polymorph by computation, using as input the space group data, refined atomic coordinates, atomic thermal parameters and unit cell data which have been obtained from a single crystal X-ray study of that species. Some years ago, it was recommended [127] that computation of the XRD pattern of a new polymorph should always be carried out if single crystal data become available. The computed pattern, being free of experimental aberrations, serves as the best reference pattern for the polymorph in question. We have followed this practice and can strongly endorse its usefulness, not only for the purpose of identifying the polymorphic form present in a newly crystallized sample, but also for assessing the polymorphic purity of such a sample. Figure 11 shows the experimental XRD pattern of a polymorph of the non-steroidal anti-inflammatory drug tenoxicam and, for comparison, the computed pattern based on the single crystal X-ray data [106]. The level of agreement in peak positions attests to the polymorphic purity of the sample (no extraneous peaks being evident in the experimental pattern) while the discrepancies in corresponding peak intensities indicate some degree of preferred orientation in the sample. Finally, since the sample was a powder derived

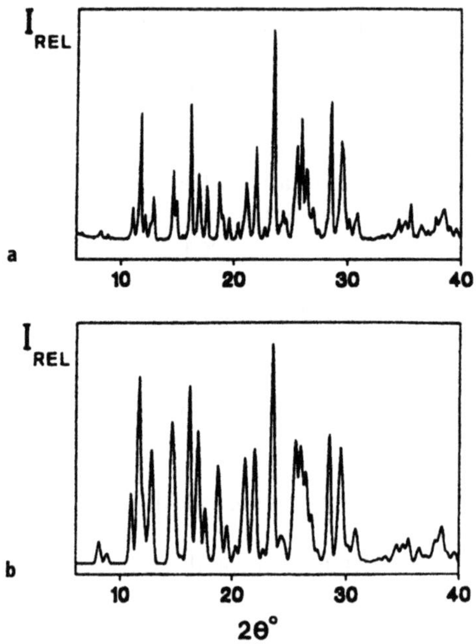

Fig. 11 a, b. XRD patterns for a polymorph of tenoxicam: **a** experimental; **b** calculated from single crystal data. (Reprinted with permission from [106], copyright 1995, American Chemical Society and American Pharmaceutical Association)

from large single crystals previously identified by X-ray analysis, the overall agreement of the patterns reveals the non-trivial result that grinding single crystals of the polymorph does not effect a phase transformation to another polymorph. In the case of the drug probucol (Fig. 3), the experimental XRD patterns for the two polymorphs were identical as a result of transformation of one polymorph into the other during sample preparation [59].

Neutron and electron diffraction techniques have also been used to some extent to study the polymorphism of organic substances. Owing to its ability to locate hydrogen atoms accurately in crystals, neutron diffraction is the method of choice for distinguishing subtle differences in hydrogen bonding for polymorphic crystal structures. However, the technique is costly and the low intensity of neutron beams demands relatively large single crystals (around 1 mm^3 or larger) which are seldom available. Crystallographic studies are therefore usually carried out on polycrystalline samples, yielding powder neutron diffraction patterns analogous to XRD patterns. A fairly recent development is the use of pulsed neutron sources coupled with time-of-flight analysis [100]. As an example of its application to the study of polymorphism, the case of cyclohexanol may be cited [128]. Structural elucidation of three polymorphs (cubic (I), monoclinic (II) and orthorhombic (III)) revealed that I is disordered, that II contains a cyclic dimer of cyclohexanol, and that a polymeric arrangement exists in III.

Because incident electrons interact strongly with crystals, electron diffraction is capable of yielding detailed structural information from very small crystals (typically 0.01–0.02 mm). It has been cited as a more reliable method for determining crystal unit cell and space group data than computer-assisted indexing of XRD patterns [100]. The crystal structure of a high-temperature polymorph of chitosan from electron diffraction data has recently been reported [129]. Structural modelling of this complex molecule involved constrained linked-atom least-squares refinement with stereochemical restraints.

Dilatometric analysis is a less widely used technique for characterising polymorphs, but has, for example, been applied to the study of polymorphic transformations occurring in theobroma oil, methyl stearate and chloramphenicol [25]. Substances which contract as they transform from a metastable (less dense) polymorph to a stable (more dense) polymorph can be studied by measuring their specific volume as a function of temperature. A recent study involved the combined use of dilatometry and neutron scattering to characterise the orthorhombic and monoclinic polymorphs of *m*-nitrophenol [130]. Crystals of both polymorphs showed significant anisotropy of expansion and it was possible to reconcile the direction of lowest expansion with that of the hydrogen bonding interactions. Attempts to correlate these results with those obtained from IR and Raman spectra were subsequently reported [131].

Solubility and dissolution rate analyses are of vital importance for polymorphs and pseudopolymorphs of pharmaceutical relevance. For a given drug, metastable polymorphs tend to have higher solubilities and faster dissolution rates than the stable polymorph. When metastable forms are employed in solid dosage forms (tablets, capsules), they generally yield higher and earlier blood serum levels [25]. Thus, for potent drugs with a narrow therapeutic index (e.g. the cardiotonic digoxin), inadvertent use of a metastable polymorph in a tablet could result in patient death from overdose. In vitro dissolution testing is therefore carried out routinely as part of the quality control of manufactured tablets and capsules. It may also be performed directly on powders or single crystals of the drug polymorphs or pseudopolymorphs comprising the active agent. Essentially, this involves placing the sample in the dissolution fluid, agitating it in a reproducible manner at constant temperature (usually 37 °C), and measuring the drug concentration in solution as a function of time. Compendial testing methods include the rotating basket, rotating paddle, and flow-through cell techniques. Automated systems employ a peristaltic pump which circulates the dissolution fluid through a UV spectrophotometer flow-cell for continuous monitoring of drug concentration. The advantages and disadvantages of modern analytical techniques available for dissolution testing have been reviewed recently [22].

Polymorphs generally dissolve more rapidly than their hydrates, but there have been reports of drug pseudopolymorphs containing, e.g., ethyl acetate or *n*-pentanol, which display enhanced solubilities, both in vitro and in vivo, when compared with their nonsolvated forms [25]. Figure 12 shows powder dissolution rate curves for two polymorphs of nitrofurantoin. The significance of the curves for this system is that the retarded initial dissolution rate for the α-polymorph may render it more favourable for pharmaceutical formulation since there is evidence that adverse side effects may be associated with rapid absorption of the β-polymorph [61].

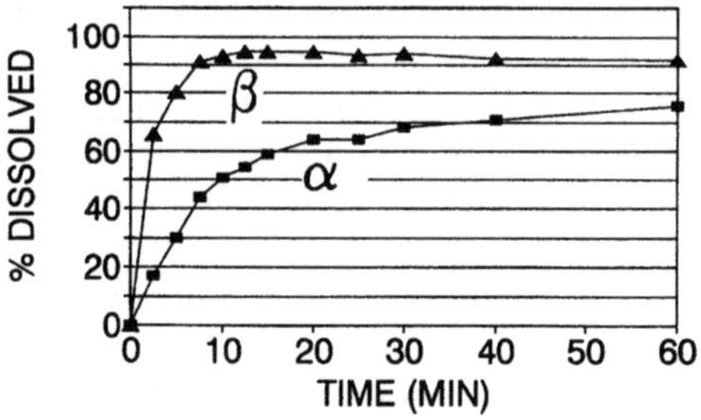

Fig. 12. Powder dissolution rate curves for two nitrofurantoin polymorphs. (Reprinted with permission from [61], copyright 1996, Gordon and Breach Publishers)

Measurements of the equilibrium solubilities of a polymorph as a function of temperature permit evaluation of the enthalpy of dissolution from a van't Hoff plot (i.e. logarithm of solubility vs T^{-1}). The point at which such plots intersect for two different polymorphs of the same compound corresponds to the transition temperature for their interconversion.

It should be stressed that the analytical methods surveyed above achieve their optimum effectiveness and lead to deeper insights into polymorphism and pseudopolymorphism when used in combination. A simple illustration is the combined use of dissolution rate analysis and crystal structure analysis of two polymorphs. Used individually, these techniques would produce merely data, whereas the combined approach could lead to an explanation for the differences in dissolution rates based on differences in crystal packing. Some reported studies utilise a wide range of techniques, permitting a comprehensive analysis of a polymorphic system, while others may use only two, but nevertheless effective, complementary techniques (e.g. DSC and XRD). The following examples selected from the recent literature illustrate this point.

Four polymorphic forms (I-IV) of the drug diflunisal (analgesic, anti-inflammatory) were prepared and characterised by DSC, HSM, IR, XRD, and dissolution studies [132]. The thermoanalytical methods revealed that on heating, all polymorphs recrystallized to the stable one (Form I) prior to melting. Only one transition peak in the DSC traces was observed, for the transformation of Form III to Form I. Intermolecular hydrogen bonding features were deduced from IR spectroscopy and this method was also shown to distinguish all the forms. Measurements of the intrinsic dissolution rates indicated (unexpectedly) that Forms II, III and IV had slightly lower values than that of the stable form. Results of this kind are invaluable for ensuring correct formulation of the drug. Some examples of other important drug polymorphic systems for which similar, fairly comprehensive studies have recently been reported include acyclovir [133], meprobamate [134], prasterone [111], ranitidine hydrochloride [135], and chloramphenicol palmitate [136].

Other compounds whose polymorphism and/or pseudopolymorphism have been investigated in recent years, together with the methods used, include pentachloropyridine (DSC, Raman and IR spectroscopy) [137], cyclohexanol (far-IR, adiabatic calorimetry) [138], nifedipine (HSM, IR, DSC) [139], sulfanilamide (DSC, XRD) [140], paraffins C_{26} – C_{60} (DTA, thermobarometric analysis) [141] and tripalmitin (DSC, thermodiffractometry, microcalorimetry) [142]. This list represents a very small selection, merely serving to indicate the variety of techniques employed.

Several recent developments in the areas of structure determination are having a significant impact on the study of organic crystal polymorphism and these are now surveyed, together with examples of applications to specific systems. In the discussion above, it has been maintained that a knowledge of the crystal structure of a polymorph provides a sound basis for interpretation of other analytical data pertaining to that species and is crucial for even a rudimentary understanding of phase transformations at the molecular level. It is also well known that many polymorphs of technological interest cannot readily be obtained as single crystals suitable for X-ray structural elucidation. However, since the advent of the Rietveld, or "whole-pattern", refinement technique [143] which allows crystal structure refinement using X-ray or neutron powder diffraction data, the status of powder XRD has been elevated from that of a mere "fingerprinting" technique to one of formidable power as a structure-solving tool. This power has been significantly enhanced by recent progress in obtaining trial structural models for Rietveld refinement directly from the diffraction intensities. Since the powder pattern is a one-dimensional projection (on 2θ) of the three-dimensional single crystal diffraction data, a key problem in this challenging research area is unravelling the intensities of severely, or exactly, overlapping reflections. In a recent study, an algorithm for achieving this was reported [144] and, together with software routines for space group determination and a fast iterative Patterson squaring routine, this combination led to the successful ab initio crystal structure determination of an aluminophosphate-based molecular sieve from X-ray powder data exclusively. It is noteworthy that as many as 65% of the reflection data used in this analysis suffered from severe overlap. This approach holds great promise for future structure determination of polymorphs which hitherto have been obtained in the form of powders only.

More recently, a method using powder diffraction for structure solution was described in which structural information is not extracted directly from the intensities; instead, trial structures are generated using a Monte Carlo approach [145] commencing with a collection of atoms initially randomly placed in the crystal unit cell. A series of trial structures is then generated by random movement of the atoms, each trial structure being accepted or rejected depending on the level of agreement between the calculated and experimental powder patterns. The Monte Carlo method employed is based on the Metropolis importance sampling algorithm using the crystallographic R-factor as the basis for constructing trial models. The best model is then used in a conventional Rietveld refinement. The structures of two compounds, with eleven and twelve non-hydrogen atoms in the respective asymmetric units, were solved by this procedure using high-resolution powder X-ray intensity data measured with a position-sensitive detector. The novel feature of this method is the use of the Monte Carlo algorithm for generating trial structures.

Crystal structure prediction based on lattice energy minimisation has already been discussed and attention has been drawn to the serious drawbacks of this approach when used on its own. However, used in combination with the Rietveld method, its chances of success are very significantly improved, as recently demonstrated for a polymorphic system [146]. Of two polymorphs, A and B, of a guanidine derivative (**1**), only form B was available in the form of single crystals and its structure was solved in the conventional manner using X-ray diffraction methods. Polymorph A was available only as a powder and the structural problem was further complicated by molecular features allowing not only for different conformers of the molecule, but also different tautomers. Indexing of 26 lines of an XRD pattern of polymorph A measured on a Guinier camera led to the assignment of the triclinic system. Energy calculations using MOPAC93 were used to search for likely conformers and tautomers suitable for crystal packing calculations. High-precision ab initio quantum mechanical methods were used to confirm the trial models, the four most stable ones being chosen for packing analysis using a corrected Dreiding force-field. Two routes were pursued, one assuming the XRD indexing results, the other ignoring them. In the first case, crystal structures were generated for the space group $P\bar{1}$ only, leading to a most stable structure with unit cell parameters very close to those deduced from the indexing. Subsequent optimisation and Rietveld refinement yielded a final structure with an R-factor of around 10%. When the indexing results were ignored, stable conformers were packed in common space groups (e.g. $P\bar{1}$, $P2_1/c$, $P2_12_12_1$). Of the predicted force-field minimised structures, that which gave the best agreement with the experimental XRD pattern turned out to be one in the space group $P\bar{1}$. Following interactive Rietveld refinement, this structure (gratifyingly) converged to the same one as determined assuming the indexing results. A very important feature of this study was the use of an ordinary inhouse XRD pattern (as opposed to a high-resolution pattern) for the indexing and Rietveld refinement steps. The authors pointed out that both the ab initio quantum mechanical as well as the packing calculations for this problem consumed enormous amounts of computational resources. They nevertheless predicted that structure determination based on an ordinary powder pattern and packing calculations could evolve into a reliable and inexpensive routine procedure in view of the envisaged drop in computational costs which they state is about an order of magnitude every three years.

The structure of a metastable and shortlived polymorph of the drug piracetam (**2**) has been solved from XRD data using the AAP method [147]. This species (Form I) was prepared by heating a mixture of Forms II and III and quenching at room temperature. Form I spontaneously transforms into II at 298 K within a few hours of quenching, but it was possible to capture high-resolution XRD data within two hours using a position-sensitive detector. The same (known) molecular conformation as occurs in Forms II and III

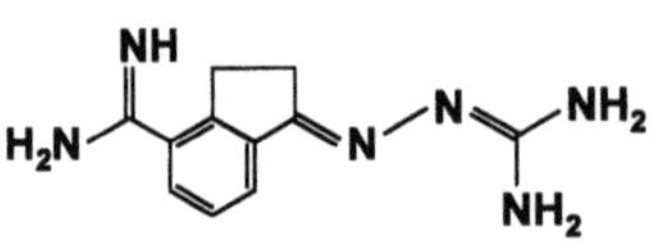

1 guanidine derivative

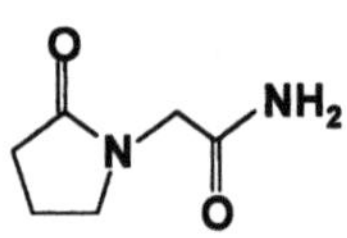

2 piracetam

was assumed in the crystal packing calculations and the AAP method yielded two distinct minima, only the higher energy structure being correct (consistent with the metastability of Form I) as shown by successful Rietveld refinement. This impressive study supports the contention that the need for single crystals for structure solution is becoming less critical [65] and also demonstrates that even shortlived organic polymorphic species can be structurally characterised.

Another major development which continues to have a significant impact on structure determination of polymorphs has been the advent of synchrotron radiation as a source for diffraction studies [148]. This radiation, which is tuneable and whose intensity exceeds that of conventional X-ray sources by orders of magnitude, is bringing about a revolution in structure determination, not only for polymorphs, but for materials in general. In particular, very small crystals (in the micron range) can now be analysed using data generated using synchrotron radiation and collected with ultra-rapid detection devices. Time-resolved studies are also amenable to study with synchrotron radiation owing to its pulsed time structure and the possibility of recording the X-ray diffraction data from picosecond exposures. An on-line system using in situ X-ray diffraction has recently been developed for examination of the polymorphic forms of pharmaceutical materials [149]. It comprises a temperature-controlled stainless steel in situ cell for X-ray measurements attached to a two-circle diffractometer on a beamline at the Daresbury Synchrotron Radiation Source. The pH of a solution of a drug is varied by titration in a glass solution-saturation vessel and crystallization due to change in the solute solubility yields a crystallized slurry which is circulated through the cell for continuous monitoring of its diffraction pattern.

For the (unnamed) antibiotic investigated, Fig. 13 shows a series of diffraction patterns measured at different temperatures. From these, the transformation between two solvates of the drug at around 309 K is clearly evident. Using the system described, enthalpy release following crystallization can also be measured by monitoring the temperature of the crystallization solution. The authors point out the growing awareness of the importance of understanding crystallization processes for many specialty chemicals in process technology and the increasing popularity of on-line characterization techniques. In situ synchrotron X-ray diffraction offers the advantages of non-invasive characterization of polymorphism and polymorphic changes, and the possibility of studying very small crystalline particles.

The investigation of polymorphism and pseudopolymorphism by means of solid-state NMR spectroscopy is also gaining in popularity owing to recent advances in both instrumentation and computer pulse sequence strategies [27]. Most studies employ ^{13}C or ^{15}N NMR since ^{1}H NMR produces broad, uninformative peaks even after removal of proton-proton dipolar interactions. In the MAS ('magic angle spinning') technique, the sample is spun at high velocity at the critical angle of 54.74° with respect to the applied magnetic field. The resulting averaging of the chemical shift anisotropy reduces spectral broadening considerably, yielding sharp peaks with fine structure from which valuable information on individual atomic chemical environments can be deduced. To

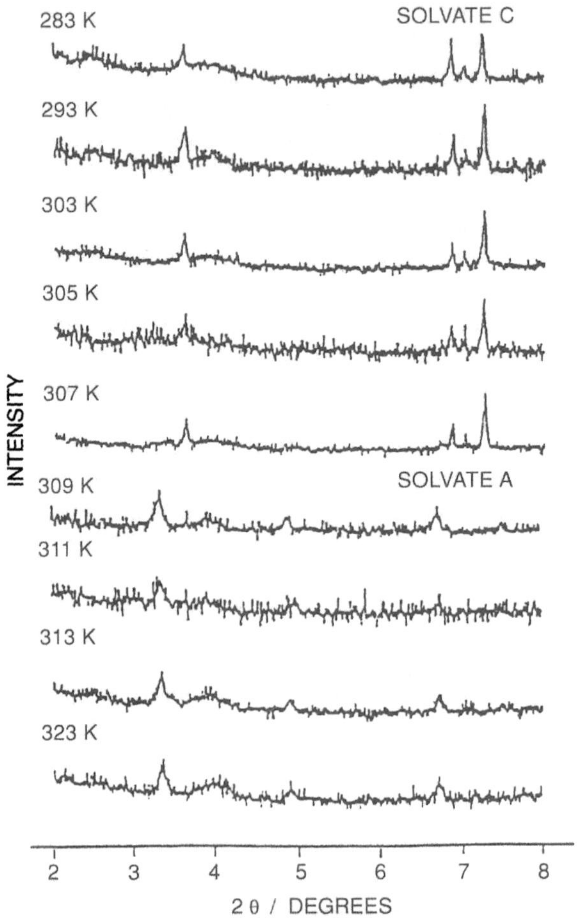

Fig. 13. In situ X-ray diffraction patterns of an antibiotic substance for crystallization range 283–323 K. (Reprinted with permission from [149])

accelerate data-acquisition, enhancement of ^{13}C magnetisation is achieved by the CP ("cross polarization") strategy and the combined method is referred to as CPMAS NMR spectroscopy. With it, polymorphic conformational differences, crystal packing differences and the effects of hydrogen bonding may be characterised. Varying the sample temperature allows dynamic studies of solid-solid phase transformations. Incorporation of solvents in pseudopolymorphs introduces significant changes in local environmental magnetic effects through, e.g., hydrogen bonding, and such changes are therefore also open to study by the CPMAS technique.

Figure 14 shows ^{15}N CPMAS spectra for four polymorphic modifications [150] of the imidazole derivative (**3**), from which it is clear that ^{15}N chemical shifts are good indices for their identification. ^{13}C CPMAS has also recently been used to distinguish three polymorphs of complex triacylglycerols [151]. Spectral

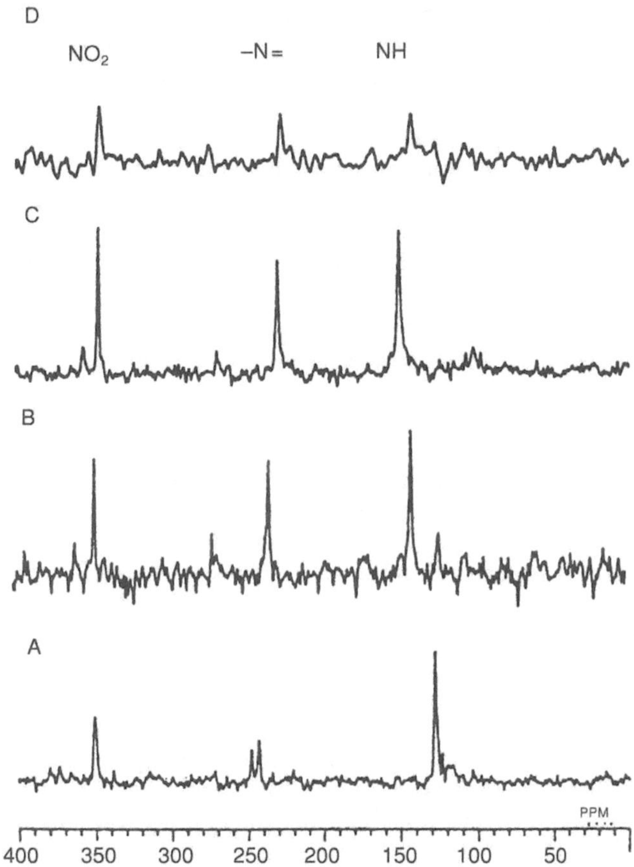

Fig. 14. ^{15}N CPMAS spectra for four polymorphs of the imidazole derivative **3**. (Reprinted with permission from [150], copyright 1996, Gordon and Breach Publishers)

differences were used to characterise conformations and crystal packing differences in the three forms. Examples of applications of CPMAS to the study of polymorphism and pseudopolymorphism of pure drug substances [27] as well as drug-excipient mixtures [152] have been described.

In a recent study of the molecular motion in the crystal of the salt (**4**), the results of ^{13}C CPMAS NMR and high-resolution powder XRD were combined

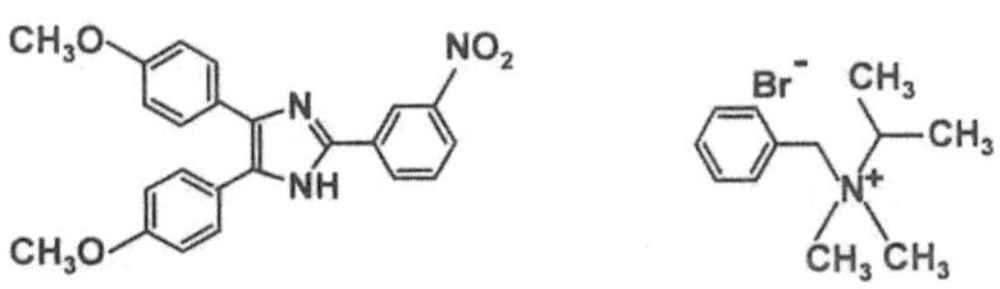

3 imidazole derivative　　4 salt

[153]. An important feature of this study was the determination of the activation energy for rotation of one of the N-methyl groups in the crystal from variable-temperature NMR data. The X-ray crystal structure was solved from high-resolution powder XRD data using direct phasing methods and refined by the Rietveld method. It was possible to reconcile the NMR-derived activation energy with the crystal packing which revealed that one of the N-methyl groups is significantly more hindered than the other. This study illustrates the potential of a combined approach to probing molecular motion in the solid-state. Its application to polymorphic systems is an obvious and desirable extension.

3.3 Polymorphic Systems and Polymorphic Changes – Case Studies

Four selected case studies are cited here to illustrate some of the difficulties encountered in research on organic crystal polymorphism, even when relatively simple molecules are involved. Since space does not permit discussion, précis of these model studies are presented in the hope that the interested reader will consult the original, absorbing accounts for details.

The polymorphism of anthranilic acid has been studied for over a century. Despite this, many properties of crystals of this simple molecule remain obscure and a recent re-examination of the phase transitions was undertaken using X-ray and FT-IR techniques [154]. A special feature of this system is that phase transitions appear to depend subtly on grinding pressure. The main conclusion drawn from the investigation cited is that polymorph I is transformed into polymorph III on heating, not into polymorph II as commonly stated in the literature.

The crystallization behaviour of the high explosive 2,4,6-trinitrotoluene (TNT) has invoked intensive research since the first report on its optical investigation in 1879 [155]. Despite numerous on-going studies, the conditions for crystallizing the monoclinic and orthorhombic forms lack definition. A systematic re-examination using Laue and Weissenberg X-ray methods, DSC, goniometry and computer simulations was undertaken to establish the role of crystallization conditions in determining the nature of polymorphism as well as TNT crystal morphology [156]. Complicating features of this system are extensive twinning in the monoclinic phase and remarkable similarity in the structures of the two polymorphs which results in their XRD patterns being almost indistinguishable. The study established (inter alia) conditions for crystallizing the two forms, finding that their occurrence was determined by growth solvent only and was independent of the rate of crystallization. The orthorhombic polymorph was shown to be metastable relative to the monoclinic polymorph.

Terephthalic acid (TA) is an important industrial chemical used in the production of polyester. Its purification involves multi-stage recrystallization under extreme conditions. The phase characterisation and morphologies of Forms I and II of TA were studied by HSM, thermal analysis, Raman spectroscopy, single crystal Laue X-ray diffraction, lattice energy and morphological calculations [157]. Contrary to earlier reports, this study revealed that Form II (not Form I) is the thermodynamically stable form of TA under ambient conditions. The apparent stability of Form I was shown to be due to twinning in microcrystals leading to "collective stabilisation", which prevents spontaneous transformation to Form II.

Structural, thermodynamic, kinetic and mechanistic aspects of phase transformations among three differently coloured forms of dimethyl 3,6-dichloro-2,5-dihydroxyterephthalate have been studied by various methods [108] including simultaneous DSC and optical microscopy using the apparatus shown in Fig. 8. The three forms, (Y) yellow, (LY) light yellow and (W) white, which exhibit extraordinary polymorphic behaviour, are now known to be conformational polymorphs and the transformations among them appear to occur by nucleation and growth of the new phase outside the domains of the old one, not by single-crystal to single-crystal transformation. Some perspective on research on polymorphism can be gained from the concluding remarks of this account which stress the prevailing lack of understanding of the complexity of phase transitions and the formidable experimental difficulties experienced in obtaining reproducible results.

4 Towards Control of Polymorphism

4.1 The Need for Polymorphic Control

In a review sketching the important developments in synthetic organic chemistry during the past 25 years and predicting future directions [158], it was emphasised that the exciting synthetic targets today are no longer molecules to be prepared "for their own sake", but rather, they are systems possessing specific functions or properties. The systems which are being pursued with this goal in mind are largely those whose properties are governed by non-covalent interactions, as evidenced by the rapid growth of the discipline of supramolecular chemistry in recent years. In the realm of solid-state chemistry, the goals of crystal engineering [3] are, analogously, the design and preparation of materials with specific properties (e.g. metallic conductivity, thermochromism, photoactivity, second-harmonic generation), these properties being strictly dictated by the crystalline assembly or supramolecular organization in the solid-state. Preparation of such assemblies has been severely impeded by the lack of an underlying theoretical framework for understanding the subtle interplay of factors determining crystal packing. There is no guarantee, in the first instance, that a given molecule will form a stable crystal which will survive the treatment to which it will be subjected. The occurrence of polymorphism (in which both thermodynamics and kinetics play a role) is an additional complicating factor, since the three-dimensional array possessing the desired property is but one, unique arrangement among many alternative arrays having similar free energies, but lacking that property. As an example, one property of solid materials currently attracting widespread attention is that of non-linear optical (NLO) behaviour, whose effect is second-harmonic generation (SHG). This property has potential for the design of laser devices, optical communications and information processing. SHG depends on both the molecular hyperpolarizability

(β) of the compound in question and on the orientation of the molecules in the crystal. Specifically, a non-centrosymmetric packing arrangement is required for a solid to exhibit NLO properties and, while it is possible to alter the value of β by synthetic modification, encouraging molecules to adopt a non-centrosymmetric packing arrangement is a formidable challenge.

The question of whether polymorphism is the "nemesis of crystal design" was discussed at length nearly a decade ago [3]. Since then, however, systematic studies in the areas of crystal nucleation, growth and morphology have led to definite proposals for rational stereospecific control of polymorphism as well as some success in its attainment in special cases [36]. Thus, while ab initio crystal structure prediction is still a distant goal [72] and the preparation of specific, desired phases is still elusive, there is at least the possibility of rational intervention during the nucleation or growth stages of crystallization for influencing the outcome, provided that the detailed polymorphic behaviour of the system has been established from systematic study. This again provides strong justification for the application of a wide range of techniques for comprehensive studies of systems exhibiting polymorphism. As implied by case studies described earlier in this report, there are numerous instances in industrial environments where control of both crystal morphology and the polymorphic form of a compound is vital, and here the outlook for success, based on the strategies to be described briefly below, is promising.

4.2 Current Strategies and Prognosis for Crystal Engineering

Practical "control" of the polymorphic outcome of a crystallization process can be interpreted at different levels, ranging from the purely empirical, such as the use of identical crystallization conditions in the hope of obtaining a particular polymorph in a reproducible manner, to a much more ambitious strategy based on, e.g., rational design of additives which could suppress or inhibit nucleation of undesired polymorphs through selective molecular recognition mechanisms, thus allowing unhindered growth of a desired metastable polymorph. Between these extremes, there are different variations, both physical and chemical, that are effective in producing specific polymorphs, as exemplified below. Reports of this kind appear in the recent patent literature, attesting to the technological and commercial importance of the polymorphs concerned. A patent for the preparation of Form 1 of the widely used antiulcerative drug ranitidine hydrochloride describes its crystallization from a mixed solvent in the presence of seed crystals, the preferred process involving in situ reaction of HCl with the free-base [159]. Industrial-scale preparation of the stable polymorphic form of *N*-(4-hydroxyphenyl)retinamide is effected by successive chemical reactions with all-*trans* retinoic acid as the starting material [160]. The desired polymorph is formed by recrystallizing the wet product from 95% EtOH. The β-form of an organic phosphite, reported to be an effective process stabiliser for polyolefins, has been prepared by heating the compound at 125 °C in the absence of solvent [161].

Desolvation of pseudopolymorphs may be considered as a means of polymorphic control. This was discussed earlier in this report in the context of altering the rheological properties of drug polymorphs in particular. However, thermal treatment or vacuum drying has also been applied to more general inclusion compounds containing both natural and synthetic host molecules in order to prepare pure polymorphic forms of the host. Of the seven polymorphic forms of gossypol (**5**) identified thus far [162], only two were obtained by appropriate recrystallizations from solution, while the remainder resulted from desolvation of various gossypol clathrates. The latter could be divided into isostructural groups and it was found that all members of a given group yielded the same polymorph on desolvation. These unusual and intriguing results imply the existence of a "template" effect in this system. Similar experiments with pseudopolymorphs of indomethacin (**6**) also yielded a series of isostructural pseudopolymorphs [83], but in this case there was virtually no correlation between a given series and the identity of the polymorph resulting from desolvation. A systematic study of such systems to elucidate the mechanisms of desolvation is desirable if this process is to be exploited in a predictive way for the control of polymorphism.

Polymorphic inclusion compounds of a host molecule with the same guest can occur if the host-guest interactions are flexible enough to allow nearly equienergetic, but crystallographically distinct host-guest packing arrangements. Reports of the preparations of such polymorphs are rare [163], but would probably increase in frequency if a variety of preparative conditions were to be employed.

As with most organic compounds, inclusion complexes are usually prepared only once and the chances of detecting polymorphism are correspondingly small. The well known host molecule cholic acid (**7**) has been shown to form different polymorphic inclusion complexes with the same guest [164]. The polymorph which crystallizes is determined by the absence or presence of a third chemical species. Crystallization of cholic acid from, e.g., acrylonitrile yielded the 1:1 host-guest complex with space group $P2_12_12_1$, while addition of acrylonitrile to a saturated solution of the host in butan-1-ol yielded a 1:1 cholic acid · acrylonitrile complex with space group $P2_1$. X-ray analyses confirmed that the polymorphs contained different packing arrangements as well as different hydrogen bonding networks. The controlling role of the third component in effecting precipitation of a specific polymorph was not clarified in the report but selective adsorption of solvent molecules on the surfaces of growing crystals may be involved.

During the last ten years, considerable progress has been made in the control and modification of crystal properties using "tailor-made auxiliaries" [35, 36]. These auxiliaries may act either as promoters of crystal growth (making them useful for the study of nucleation processes) or as inhibitors (for use in the control of, e. g., morphology, enantiomeric resolution and polymorphism). The strategy for effecting kinetic resolution of a conglomerate involves the design of chiral-resolved polymeric inhibitors which will, in solution, bind selectively to only one of the growing enantiomeric crystalline phases, thus preventing its growth and permitting unhindered growth of the other enantiomeric phase. Where simultaneous crystallization of the racemic phase might occur, design of the inhibitor would have to take this into account. It must be emphasised that successful design of auxiliaries depends on a detailed knowledge of the packing arrangements, as well as the relationship between packing arrangements and crystal morphology for all species that might appear in the system. An early demonstration of this approach (which was subsequently successfully applied to several other systems) involved resolution of racemic histidine hydrochloride [165]. In the report of this resolution, it was mentioned that this approach was being tested for use in the precipitation of metastable polymorphic crystals by kinetic control.

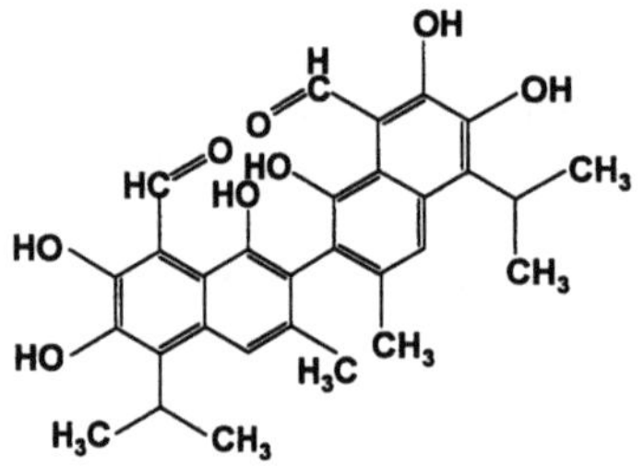

5 gossypol

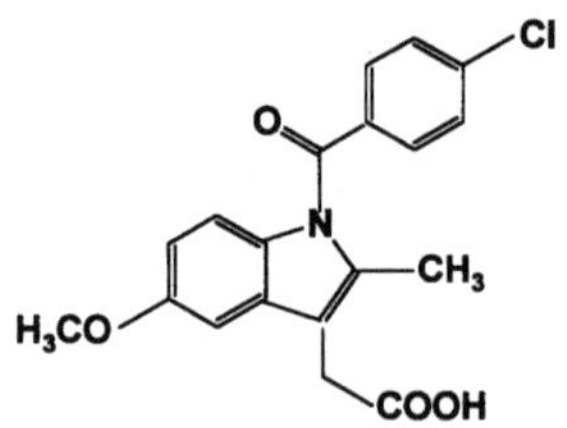

6 indomethacin

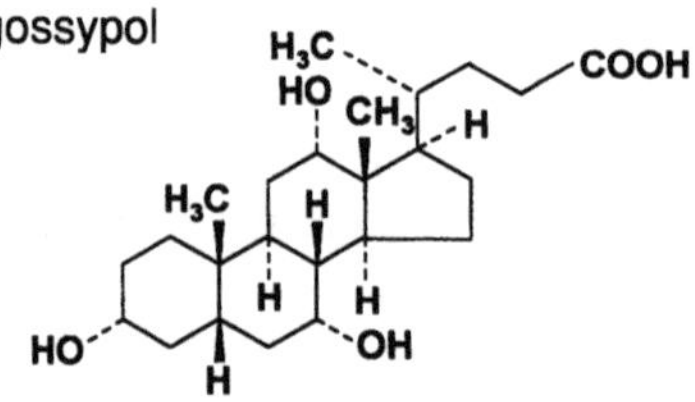

7 cholic acid

Recent application of these principles to polymorphic control is shown schematically in Fig. 15 for a hypothetical dimorphic system in which one polymorph is centrosymmetric and the other crystallizes in a polar space group [36]. In the former crystal, the molecules are arranged in antiparallel orientation whereas in the latter they are aligned along a common direction. A tailor-made auxiliary which binds to both crystals would do so at the two (indistinguishable) ends of the centrosymmetric crystal but only at one end of the polar crystal. The latter would therefore grow at the expense of the former and polymorphic control will have been achieved.

This concept was successfully applied to *N*-(2-acetamido-4-nitrophenyl)pyrrolidene (PAN) where crystallization of the metastable polymorph of PAN could be induced by addition of a remarkably small amount (0.03%) of the designed inhibitor [36]. Generalisation of this strategy is not straightforward, however, since polar crystals may belong to space groups of relatively high symmetry which require the constituent molecules to adopt a variety of orientations with respect to the polar axis [35]. Strategies which have been explored to engineer non-centrosymmetry into crystals for NLO properties have been listed in a recent paper describing the polymorphism of 1,3-bis(*m*-nitrophenyl)urea [166].

Research on the use of tailor-made surfaces for promotion of nucleation has been pursued in parallel with studies of crystal growth inhibition [35, 36]. The objective here is to gain a deeper understanding of the mechanism of molecular

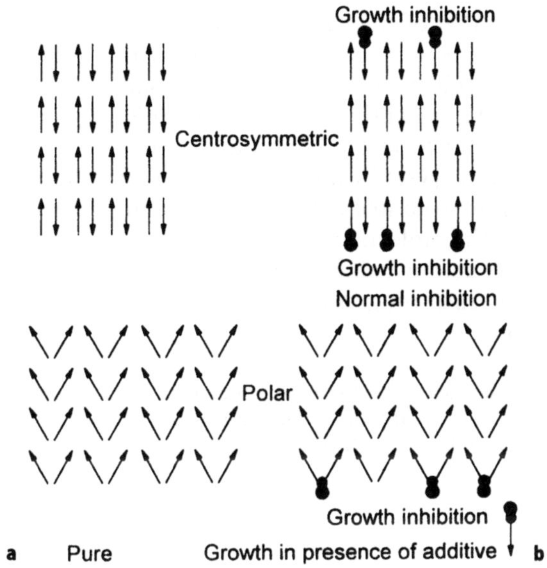

Fig. 15 **a** Schematic illustration of the packing arrangement in a polar and a centrosymmetric crystal. **b** The effect of a designed inhibitor on crystal growth. (Adapted from [36] with permission

clustering associated with the formation of pre-critical nuclei of crystalline phases. The relevance for organic crystal polymorphism is that direct observation of the dynamics of crystal nucleation could elucidate the effects of additives on polymorphic control. Much of this work has been based on the use of Langmuir films to induce nucleation at the air-water interface [36]. To illustrate the level of interpretation which is currently possible from such experiments, a recent study of the self-assembly of crystalline monolayers and multilayers of *n*-alkanes on a water surface is cited [167]. Surface observations were carried out using synchrotron grazing incidence X-ray diffraction and specular X-ray reflectivity. These allowed detailed characterisation of the crystalline films including determination of their space groups. It is noteworthy that these were found to be identical to those of the corresponding three-dimensional crystals, thus demonstrating that the full crystal symmetry already exists even for crystallites whose thickness corresponds to only a few molecular layers.

Related to the above strategy is the concept of engineering solid surfaces for promoting nucleation and epitaxial growth of desired polymorphs. If it is assumed that the pre-nucleation aggregate for a polymorph resembles the mature crystal, it should be possible to design a surface which mimics a particular crystal plane of the desired species, and upon which heterogeneous nucleation will occur and epitaxial polymorphic growth will result. This has been realised in the technique of ledge-directed epitaxy [168] in which the intersection of a terrace plane and a step on a crystal surface (a "ledge") functions as a nucleation site for organic crystals. For polymorphic control, the crucial requirement is that two well-defined, close-packed crystal planes of the polymorph have a dihedral angle which matches that of the substrate ledge site. Successful application of this principle to the selective nucleation and growth of a thermodynamically less favoured polymorph of a complex organic salt on succinic acid single crystals as substrates has recently been reported [169].

The approaches outlined above continue to provide new insights into the fundamental process of crystallization of organic molecules and studies of the type described promise to contribute very significantly in the future to a detailed understanding of the molecular origins of crystal polymorphism and to its effective control. The author believes that the phenomenon of polymorphism will, notwithstanding such progress, retain its aura as a remarkable manifestation of Nature's diversity.

Acknowledgements. My thanks are due to my colleague Prof L R Nassimbeni for constructive comments and to Fiona Jones for assistance with library resources. Research support from the University of Cape Town and the FRD (Pretoria) is gratefully acknowledged.

5
References

1. Mitscherlich E (1822) Ann Chim Phys 19:350
2. Dunitz JD (1996) In: Desiraju GR (ed) The crystal as a supramolecular entity: perspectives in supramolecular chemistry, vol 2. Wiley, New York, p 2
3. Desiraju GR (1989) Crystal engineering. Elsevier, Amsterdam
4. Sato K (1993) J Phys D: Appl Phys 26:B77
5. Dunitz JD, Bernstein J (1995) Acc Chem Res 28:193
6. Herbstein FH (1996) J Mol Struct 374:111
7. Greenberg JH (1996) Mater Sci Eng R16:223
8. Bayard F, Decoret C, Royer J (1990) Stud Phys Theor Chem 69:211
9. Hernqvist L (1990) Food Struct 9:39
10. Kalman A, Argay G (1991) Kem Kozl 73:129
11. Bernstein J (1991) Polymorphism and the investigation of structure-property relations in organic solids. In: Garbarczyk JB, James DW (eds) Organic crystal chemistry. Oxford University Press, Oxford, chap 2
12. Cecchini MAG (1992) Publ. ACIESP 82:65
13. Bernstein J (1993) J Phys D: Appl Phys 26:B66
14. Poole PH, Grande T, Sciortino F, Stanley HE, Angell CA (1995) Comput Mater Sci 4:373
15. Lotz B (1995) Macromol Symp 94:97
16. Bernstein J, Davis RE, Shimoni L, Chang N-L (1995) Angew Chem Int Ed Engl 34:1555
17. Wolff JJ (1996) Angew Chem Int Ed Engl 35:2195
18. Ford JL, Timmins P (1989) Pharmaceutical thermal analysis, 1st edn. Ellis Horwood Limited, Chichester
19. Byrn SR, Tobias B, Kessler D, Frye J, Sutton P, Saindon P, Kozlowski J (1989) Trans Am Crystallogr Assoc 24:41
20. Borka L, Haleblian JK (1990) Acta Pharm Jugosl 40:71
21. Borka L (1990) Spectroscopy (Eugene, Oreg) 5:12
22. Mehta AC (1994) Anal Proc 31:245
23. Giron D (1995) Thermochim Acta 248:1
24. Laird T (1990) In: Kennewell PD (ed) Comprehensive medicinal chemistry, vol 1. Pergamon, Oxford, p 321
25. Curry SH, Thakker KM (1990) In: Taylor JB (ed) Comprehensive medicinal chemistry, vol 5. Pergamon, Oxford, p 545

26. Yasui M (1995) Nippon Kessho Gakkaishi 37:69
27. Brittain HG (1997) J Pharm Sci 86:405
28. Sato K (1989) Zesz Nauk-Politech Lodz, Fiz 9:5
29. Kitamura M (1991) Kagaku Kogaku 55:263
30. Kitamura M (1992) Funtai Kogaku Kaishi 29:118
31. Kocevar J (1993) Farm Vestn (Ljubljana) 44:55
32. Kitamura M (1994) Bunri Gijutsu 24:365
33. Boistelle R, Klein JP, Guyot-Hermann AM (1996) STP Pharma Prat 6:111
34. Yamamoto H, Harano Y (1995) Bunri Gijutsu 25:381
35. Lahav M, Leiserowitz L (1993) J Phys D: Appl Phys 26:B22
36. Weissbuch I, Popovitz-Biro R, Lahav M, Leiserowitz L (1995) Acta Crystallogr B51:115
37. Haleblian JK (1975) J Pharm Sci 64:1269
38. Byrn SR (1982) Solid-state chemistry of drugs. Academic Press, New York
39. Hoelgaard A, Møller N (1983) Int J Pharm 15:213
40. Caira MR, Nassimbeni LR, Van Oudtshoorn B (1993) J Pharm Sci 82:1006
41. Burger A (1995) NATO Advanced research workshop crystals:Supramolecular materials. Sestri Levante, Italy
42. McCrone WC (1965) Polymorphism. In: Fox D, Labes MM, Weissberger (eds) Physics and chemistry of the organic solid state. Interscience, New York p 726
43. Khoshkhoo S, Anwar J (1993) J Phys D: Appl Phys 26:B90
44. Bartell LS (1995) J Phys Chem 99:1080
45. Merlo M, Santos A, Gago Duport L (1993) Mater Res Soc Sympos Proc 290:317
46. Chakraborty D, Bhatia SK (1996) Ind Eng Chem Res 35:1985
47. Morin B, Elder KR, Sutton M, Grant M (1995) Phys Rev Lett 75:2156
48. Alfonso GC, Moretti P, Pallenzona P, Yin J (1995) Opt Eng (Bellingham, Wash) 34:3385
49. Tiller WA (1986) J Cryst Growth 76:607
50. Caira MR, Nassimbeni LR (1996) In: MacNicol DD, Toda F, Bishop R (eds) Comprehensive supramolecular chemistry, vol 6. Pergamon Press, Oxford, chap 25
51. Cardew PT, Davey RJ (1985) Proc R Soc London A 398:415
52. Berkovitch-Yellin Z (1985) J Am Chem Soc 107:8239
53. Kitamura M (1993) J Chem Eng Jpn 26:303
54. Kitamura M, Furukawa H, Asaeda M (1994) J Cryst Growth 141:193
55. Koga K, Kawakami R, Kagara K (1996) Kagaku Kogaku Ronbunshu 22:1174
56. Sudo S, Sato K, Harano Y (1991) J Chem Eng Japan 24:628
57. Janik M, Malarski Z, Mrozinski J, Wajcht J, Zborucki Z (1991) J Crystallogr Spectr Res 21:519
58. Bernstein J (1987) In: Desiraju GR (ed) Organic solid state chemistry. Elsevier, Amsterdam, p 471
59. Gerber JJ, Caira MR, Lötter AP (1993) J Crystallogr Spectr Res 23:863
60. Pienaar EW, Caira MR, Lötter AP (1993) J Crystallogr Spectr Res 23:785
61. Caira MR, Pienaar EW, Lötter AP (1996) Mol Cryst Liq Cryst 279:241
62. Caira MR, Botha SA, Flanagan DR (1994) J Chem Crystallogr 24:95
63. Allen FH, Trotter J (1971) J Chem Soc B 1073
64. Bernstein J, Etter MC, MacDonald JC (1990) J Chem Soc Perkin Trans 2 695
65. Gavezzotti A, Filippini G (1997) In: Gavezzotti A (ed) Theoretical aspects and computer modeling of the molecular solid state, vol 1. Wiley, New York, chap 3
66. Gavezzotti A (1991) J Am Chem Soc 113:4622
67. Gdanitz RJ (1992) Chem Phys Lett 190:391
68. Karfunkel HR, Gdanitz RJ, Leusen FJJ (1994) Comput Aided Mat Des 1:185
69. Perlstein J (1994) J Am Chem Soc 116:11,420
70. Leusen FJJ (1996) J Cryst Growth 166:900
71. Baur WH, Kassner D (1992) Acta Crystallogr B48:356
72. Gavezzotti A, Filippini G (1996) J Am Chem Soc 118:7153
73. Thiery M-M, Rerat C (1996) J Chem Phys 104:9079
74. Toscani S, Dzyabchenko A, Agafonov V, Dugue J, Ceolin R (1996) Pharm Res 13:151

75. Lukasheva NV, Sariban A, Mozel T, Brikman Y (1996) Vysokomol Soedin. Ser A Ser B 38:688
76. Williams DE (1996) Acta Crystallogr A52:326
77. Peeters A, Van Alsenoy C, Lenstra ATH, Geise HJ (1995) J Chem Phys 103:6608
78. Gavezzotti A (1994) Acc Chem Res 27:309
79. Gavezzotti A, Filippini G (1995) J Am Chem Soc 117:12,299
80. Kuhnert-Brandstätter M (1971) Thermomicroscopy in the analysis of pharmaceuticals. Pergamon, New York
81. Haleblian J, McCrone WC (1969) J Pharm Sci 58:911
82. Borka L (1974) Acta Pharm Suecica 11:295
83. Caira MR, Gifford Nash K, Nassimbeni LR (1995) Ann Congr Acad Pharm Sci 16th. Bloemfontein, South Africa. Abstract P28
84. Lerk CF, Andreae AC, De Boer AH, Bolhuis GK, Zuurman K, De Hoog P, Kussendrager K, Van Leverink J (1983) J Pharm Pharmacol 35:747
85. Fachaux JM, Guyot-Hermann A-M, Guyot JC, Conflant P, Drache M, Huvenne JP, Bouche R (1992) Congr Int Technol Pharm 6th 5:213
86. Shakhtshneider TP, Boldyrev VV (1993) Drug Dev Ind Pharm 19:2055
87. Pirttimaki J, Laine E, Ketolainen J, Paronen P (1993) Int J Pharm 95:93
88. Otsuka M, Matsuda Y (1990) Pharm Tech Jpn 6:977
89. Otsuka M, Matsuda Y (1993) Drug Dev Ind Pharm 19:2241
90. Otsuka M, Otsuka K, Kaneniwa N (1994) Drug Dev Ind Pharm 20:1649
91. Gubskaya AV, Chishko KA, Lisnyak YV, Blagoy YP (1995) Drug Dev Ind Pharm 21:1965
92. Mullins JD, Macek TJ (1967) J Pharm Sci 56:847
93. Ogawa K, Yui T (1994) Biosci Biotechnol Biochem 58:968
94. Paukszta D, Garbarczyk J, Sterzynski T (1991) Int Union Crystallogr, Crystallogr Symp 4:192
95. Sayapina OV, Koshkin VM (1990) Pis'ma Zh Tekh Fiz 16:58
96. Hamill GP, Yost EA, Sandman DJ (1992) Mol Cryst Liq Cryst Sci Technol Sect A 211:339
97. Chan HK, Gonda I (1989) J Cryst Growth 94:488
98. Caira MR (1994) J Chem Crystallogr 24:695
99. Bettinetti GP, Giordano F, La Manna A, Giuseppetti G (1974) Il Farmaco-Ed Pr 29:493
100. West AR (1996) Basic solid state chemistry. Wiley, Chichester
101. Himes VL, Mighell AD, De Camp WH (1981) Acta Crystallogr B37:2242
102. Lowes MM, Caira MR, Lötter AP, Van der Watt JG (1987) J Pharm Sci 76:744
103. Van Tonder EC, Caira MR, Botha SA, Lötter AP (1990) Int J Pharm 63:35
104. Flynn JH, Wall LA (1966) Polym Lett 4:323
105. Bourne SA, Caira MR, Nassimbeni LR, Shabalala I (1994) J Pharm Sci 83:887
106. Caira MR, Nassimbeni LR, Timme M (1995) J Pharm Sci 84:884
107. Burger A (1982) Acta Pharm Technol 28:1
108. Richardson MF, Yang Q-C, Novotny-Bregger E, Dunitz JD (1990) Acta Crystallogr B46:653
109. Acharya KR, Kuchela KN, Kartha G (1982) J Crystallogr Spectr Res 12:369
110. Caira MR, Mohamed R (1992) Acta Crystallogr B48:492
111. Chang L-C, Caira MR, Guillory JK (1995) J Pharm Sci 84:1169
112. Caira MR, Guillory JK, Chang L-C (1995) J Chem Crystallogr 25:393
113. Suleiman MS, Najib NM (1989) Int J Pharm 50:103
114. Yuan XB, Jiang DH, Shen HB, Ding HL (1991) Yaoxue Xuebao 26:152
115. Bishop R, Dance IG (1991) In: Atwood JL, Davies JED, MacNicol DD (eds) Inclusion compounds, vol 4. Oxford University Press, Oxford, p 1
116. Ung AT, Bishop R, Craig DC, Dance IG, Scudder ML (1993) Tetrahedron 49:639
117. Babkov LM, Puchkovskaya GA (1993) Khim Fiz 12:944
118. Ciurczak EW (1987) Appl Spectrosc Rev 23:147
119. Aldridge PK, Evans CL, Ward HW II, Colgan ST, Boyer N, Gemperline PJ (1996) Anal Chem 68:997
120. Colombo L, Volovsek V, Furic K, Durig JR (1990) J Raman Spectrosc 21:169

121. Ohno K, Yoshida H, Matsuura H (1996) Spectrochim Acta 52 A:1377
122. Glusker JP, Trueblood KN (1985) Crystal structure analysis, a primer, 2nd edn. Oxford University Press, New York
123. Ladd MFC, Palmer RA (1985) Structure determination by X-ray crystallography, Plenum New York
124. Hope H (1994) X-ray crystallography:a fast, first-resort analytical tool. In: Karlin KD (ed) Progr Inorg Chem. Wiley, New York, p 1
125. The United States Pharmacopeia 23 (1995) The United States Pharmacopeial Convention Inc, Rockville MD p 1843
126. Kidd WC, Varlashkin P, Li C-Y (1993) Powder Diffr 8:180
127. Bar I, Bernstein J (1985) J Pharm Sci 74:255
128. Sciesinska E, Mayer J, Natkaniec I, Sciesinski J (1989) Acta Phys Pol A76:617
129. Mazeau K, Winter WT, Chanzy H (1994) Macromolecules 27:7606
130. Wojcik G, Jakubowski B, Szostak MM, Holderna-Matuszkiewicz K, Mayer J, Natkaniec I (1991) Kryszt Mol '91 Mater Ogolnopol Konf, p 169
131. Wojcik G, Jakubowski B, Szostak MM, Holderna-Natkaniec K, Mayer J, Natkaniec I (1992) Phys Status Solidi A134:139
132. Martinez-Oharriz MC, Martin C, Goni MM, Rodriguez-Espinosa C, Tros de Ilarduya-Apaolaza MC, Sanchez M (1994) J Pharm Sci 83:174
133. Kristl A, Srcic S, Vrecer F, Sustar B, Vojnovic D (1996) Int J Pharm 139:231
134. Lefebvre C, Guillaume F, Bouche R, Bouaziz R, Guyot JC (1989) 5th Congr Int Technol Pharm 1:211
135. Madan T, Kakkar AP (1994) Drug Dev Ind Pharm 20:1571
136. Mitra AK, Ghosh LK, Gupta BK (1993) Drug Dev Ind Pharm 19:971
137. Wojcik G, Giermanska-Kahn J, Marqueton Y, Foulon M (1995) Acta Phys Pol A88:339
138. Mayer J, Rachwalska M, Sciesinska E, Sciesinski J (1990) J Phys (Paris) 51:857
139. Burger A, Koller KT (1996) Sci Pharm 64:293
140. Toscani S, Thoren S, Agafonov V, Ceolin R, Dugue J (1995) Pharm Res 12:1453
141. Lourdin D, Roux AH, Grolier JPE (1991) Calorim Anal Therm 22:151
142. Hongisto V, Lehto V-P, Laine E (1996) Thermochim Acta 276:229
143. Rietveld HM (1969) J Appl Crystallogr 2:65
144. Estermann MA, McCusker LB, Baerlocher C (1992) J Appl Crystallogr 25:539
145. Harris KDM, Tremayne M, Lightfoot P, Bruce PG (1994) J Am Chem Soc 116:3543
146. Karfunkel HR, Wu ZJ, Burkhard A, Rihs G, Sinnreich D, Buerger HM, Stanek J (1996) Acta Crystallogr B52:555
147. Louër D, Louër M, Dzyabchenko VA, Agafonov V, Ceolin R (1995) Acta Crystallogr B51:182
148. Helliwell JR, Helliwell M (1996) J Chem Soc, Chem Commun 1595
149. MacCalman ML, Roberts KJ (1995) J Appl Crystallogr 28:620
150. Yamanobe T, Komoto T, Sakaino Y (1996) Mol Cryst Liq Cryst 276:273
151. Arishima T, Sugimoto K, Kiwata R, Mori H, Sato K (1996) J Am Oil Chem Soc 73:1231
152. Rao RC (1992) Spectra 2000 166:33
153. Riddell FG, Bruce PG, Lightfoot P, Rogerson M (1994) J Chem Soc, Chem Commun 209
154. Ojala WH, Etter MC (1992) J Am Chem Soc 114:10,288
155. Friedlander P (1879) Z Kristallogr 3:169
156. Gallagher HG, Sherwood JN (1996) J Chem Soc Faraday Trans 92:2107
157. Davey RJ, Maginn SJ, Andrews SJ, Black SN, Buckley AM, Cottier D, Dempsey P, Plowman R, Rout JE, Stanley DR, Taylor A (1994) J Chem Soc Faraday Trans 90:1003
158. Seebach D (1990) Angew Chem Int Ed Engl 29:1320
159. Ngooi T-K, McGolrick JD, Antczak C, Tindall JLA (1994) CA 121:263,695e
160. Maryanoff CA (1995) CA 122:240,076 k
161. Shum SP, Pastor SD (1996) CA 124:246,949 k
162. Ibragimov BT, Talipov SA (1994) J Incl Phenom Mol Recognit Chem 17:325
163. Ibragimov BT, Beketov K, Makhkamov K, Weber E (1997) J Chem Soc Perkin Trans 2:1349

164. Nakano K, Sada K, Miyata M (1996) J Chem Soc, Chem Commun 989
165. Weissbuch I, Zbaida D, Addadi L, Leiserowitz L, Lahav M (1987) J Am Chem Soc 109:1869
166. Huang K-S, Britton D, Etter MC, Byrn SR (1995) J Mater Chem 5:379
167. Weinbach SP, Weissbuch I, Kjaer K, Bouwman WG, Nielsen JA, Lahav M, Leiserowitz L (1995) Adv Mater 7:857
168. Carter PW, Ward MD (1993) J Am Chem Soc 115:11,521
169. Bonafede SJ, Ward MD (1995) J Am Chem Soc 117:7853

Author Index Volumes 151–198

Author Index Vols. 26–50 see Vol. 50
Author Index Vols. 51–100 see Vol. 100
Author Index Vols. 101–150 see Vol. 150

The volume numbers are printed in italics

Adam W, Hadjiarapoglou L (1993) Dioxiranes: Oxidation Chemistry Made Easy. *164*:45–62
Alberto R (1996) High- and Low-Valency Organometallic Compounds of Technetium and Rhenium. *176*:149–188
Albini A, Fasani E, Mella M (1993) PET-Reactions of Aromatic Compounds. *168*:143–173
Allan NL, Cooper D (1995) Momentum-Space Electron Densities and Quantum Molecular Similarity. *173*:85–111
Allamandola LJ (1990) Benzenoid Hydrocarbons in Space: The Evidence and Implications. *153*:1–26
Alonso JA, Balbás LC (1996) Density Functional Theory of Clusters of Naontransition Metals Using Simple Models. *182*:119–171
Améduri B, Boutevin B (1997) Telomerisation Reactions of Fluorinated Alkenes. *192*:165–233
Anderson RC, McGall G, Lipshutz RJ (1998) Polynucleotide Arrays for Genetic Sequence Analysis. *194*:117–129
Anwander R (1996) Lanthanide Amides. *179*:33–112
Anwander R (1996) Routes to Monomeric Lanthanide Alkoxides. *179*:149–246
Anwander R, Herrmann WA (1996) Features of Organolanthanide Complexes. *179*:1–32
Aoyama Y (1998) Functional Organic Zeolite Analogues. *198*:131–161
Artymiuk PJ, Poirette AR, Rice DW, Willett P (1995) The Use of Graph Theoretical Methods for the Comparison of the Structures of Biological Macromolecules. *174*:73–104
Astruc D (1991) The Use of p-Organoiron Sandwiches in Aromatic Chemistry. *160*:47–96
Azumi T, Miki H (1997) Spectroscopy of the Spin Sublevels of Transition Metal Complexes. *191*:1–40
Baerends EJ, see van Leeuwen R (1996) *180*:107–168
Baker BJ, Kerr RG (1993) Biosynthesis of Marine Sterols. *167*:1–32
Balbás LC, see Alonso JA (1996) *182*:119–171
Baldas J (1996) The Chemistry of Technetium Nitrido Complexes. *176*:37–76
Balzani V, Barigelletti F, De Cola L (1990) Metal Complexes as Light Absorption and Light Emission Sensitizers. *158*:31–71
Balzani V, see Venturi M (1998) *197*:193–228
Bardin VV, see Petrov VA (1997) *192*:39–95
Barigelletti F, see Balzani V (1990) *158*:31–71
Bassi R, see Jennings RC (1996) *177*:147–182
Battersby AR, Leeper FJ (1998) Biosynthesis of Vitamin B_{12}. *195*:143–193
Baumgarten M, Müllen K (1994) Radical Ions: Where Organic Chemistry Meets Materials Sciences. *169*:1–104
Beau J-M and Gallagher T (1997) Nucleophilic C-Glycosyl Donors for C-Glycoside Synthesis. *187*:1–54
Bechthold A F-W, see Kirschning A (1997) *188*:1–84
Begley TP, Kinsland C, Taylor S, Tandon M, Nicewonger R, Wu M, Chin H-J, Kelleher N, Campobasso N, Zhang Y (1998) Cofactor Biosynthesis: A Mechanistic Perspective. *195*:93–142

Berces A, Ziegler T (1996) Application of Density Functional Theory to the Calculation of Force Fields and Vibrational Frequencies of Transition Metal Complexes. *182*:41–85
Bersier J, see Bersier PM (1994) *170*:113–228
Bersier PM, Carlsson L, Bersier J (1994) Electrochemistry for a Better Environment. *170*:113–228
Besalú E, Carbó R, Mestres J, Solà M (1995) Foundations and Recent Developments on Molecular Quantum Similarity. *173*:31–62
Bignozzi CA, see Scandola F (1990) *158*:73–149
Billing R, Rehorek D, Hennig H (1990) Photoinduced Electron Transfer in Ion Pairs. *158*:151–199
Bissell RA, de Silva AP, Gunaratne HQN, Lynch PLM, Maguire GEM, McCo, CP, Sandanayake KRAS (1993) Fluorescent PET (Photoinduced Electron Transfer) Sensors. *168*:223–264
Blasse B (1994) Vibrational Structure in the Luminescence Spectra of Ions in Solids. *171*:1–26
Bley K, Gruber B, Knauer M, Stein N, Ugi I (1993) New Elements in the Representation of the Logical Structure of Chemistry by Qualitative Mathematical Models and Corresponding Data Structures. *166*:199–233
Boullanger P (1997) Amphiphilic Carbohydrates as a Tool for Molecular Recognition in Organized Systems. *187*:275–312
Boutevin B, see Améduri B (1997) *192*:165–233
Brandi A, see Goti A (1996) *178*:1–99
Brunvoll J, see Chen RS (1990) *153*:227–254
Brunvoll J, Cyvin BN, Cyvin SJ (1992) Benzenoid Chemical Isomers and Their Enumeration. *162*:181–221
Brunvoll J, see Cyvin BN (1992) *162*:65–180
Brunvoll J, see Cyvin SJ (1993) *166*:65–119
Bundle DR (1990) Synthesis of Oligosaccharides Related to Bacterial O-Antigens. *154*:1–37
Buot FA (1996) Generalized Functional Theory of Interacting Coupled Liouvillean Quantum Fields of Condensed Matter. *181*:173–210
Burke K, see Ernzerhof M (1996) *180*:1–30
Burrell AK, see Sessler JL (1991) *161*:177–274
Burton DJ, Lu L (1997) Fluorinated Organometallic Compounds. *193*:45–89
Butz T, see Seebach D (1998) *197*:125–164
Caffrey M (1989) Structural, Mesomorphic and Time-Resolved Studies of Biological Liquid Crystals and Lipid Membranes Using Synchrotron X-Radiation. *151*:75–109
Caira MR (1998) Crystalline Polymorphism of Organic Compounds. *198*:163–208
Caminade A-M, see Majoral J-P (1998) *197*:79–124
Campagna S, see Venturi M (1998) *197*:193–228
Campobasso N, see Begley TP (1998) *195*:93–142
Canceill J, see Collet A (1993) *165*:103–129
Carbó R, see Besalú E (1995) *173*:31–62
Carlson R, Nordhal A (1993) Exploring Organic Synthetic Experimental Procedures. *166*:1–64
Carlsson L, see Bersier PM (1994) *170*:113–228
Carreras CW, Pieper R, Khosla C (1997) The Chemistry and Biology of Fatty Acid, Polyketide, and Nonribosomal Peptide Biosynthesis. *188*:85–126
Ceulemans A (1994) The Doublet States in Chromium (III) Complexes. A Shell-Theoretic View. *171*:27–68
Chambers RD, Vaughan JFS (1995) Nucleophilic Reactions of Fluorinated Alkenes. *192*:1–38
Chambron J-C, Dietrich-Buchecker Ch, Sauvage J-P (1993) From Classical Chirality to Topologically Chiral Catenands and Knots. *165*:131–162.
Chan WH, see Lee AWM (1997) *190*: 101–129
Chang CWJ, Scheuer PJ (1993) Marine Isocyano Compounds. *167*:33–76
Chen RS, Cyvin SJ, Cyvin BN, Brunvoll J, Klein DJ (1990) Methods of Enumerating Kekulé Structures. Exemplified by Applified by Applications of Rectangle-Shaped Benzenoids. *153*:227–254
Chen RS, see Zhang FJ (1990) *153*:181–194
Cheng J, Kricka LJ, Sheldon EL, Wilding P (1998) Sample Preparation in Microstructured Devices. *194*:215–231

Chiorboli C, see Scandola F (1990) *158*:73–149
Chin H-J, see Begley TP (1998) *195*:93–142
Chiu P, Lautens M (1997) Using Ring-Opening Reactions of Oxabicyclic Compounds as a Strategy in Organic Synthesis. *190*:1–85
Cimino G, Sodano G (1993) Biosynthesis of Secondary Metabolites in Marine Molluscs. *167*: 77–116.
Ciolowski J (1990) Scaling Properties of Topological Invariants. *153*:85–100
Clark T (1996) Ab Initio Calculations on Electron-Transfer Catalysis by Metal Ions. *177*:1–24
Cohen MH (1996) Strenghtening the Foundations of Chemical Reactivity Theory. *183*:143–173
Collet A, Dutasta J-P, Lozach B, Canceill J (1993) Cyclotriveratrylenes and Cryptophanes: Their Synthesis and Applications to Host-Guest Chemistry and to the Design of New Materials. *165*:103–129
Colombo M G, Hauser A, Güdel HU (1994) Competition Between Ligand Centered and Charge Transfer Lowest Excited States in bis Cyclometalated Rh^{3+} and Ir^{3+} Complexes. *171*:143–172
Cooper DL, Gerratt J, Raimondi M (1990) The Spin-Coupled Valence Bond Description of Benzenoid Aromatic Molecules. *153*:41–56
Cooper DL, see Allan NL (1995) *173*:85–111
Cordero FM, see Goti A (1996) *178*:1–99
Cyvin BN, see Chen RS (1990) *153*:227–254
Cyvin SJ, see Chen RS (1990) *153*:227–254
Cyvin BN, Brunvoll J, Cyvin SJ (1992) Enumeration of Benzenoid Systems and Other Polyhexes. *162*:65–180
Cyvin SJ, see Cyvin BN (1992) *162*:65–180
Cyvin BN, see Cyvin SJ (1993) *166*:65–119
Cyvin SJ, Cyvin BN, Brunvoll J (1993) Enumeration of Benzenoid Chemical Isomers with a Study of Constant-Isomer Series. *166*:65–119
Dartyge E, see Fontaine A (1989) *151*:179–203
De Cola L, see Balzani V (1990) *158*:31–71
Dear K (1993) Cleaning-up Oxidations with Hydrogen Peroxide. 16
de Meijere A, see Haag R (1998) *196*:137–165
de Mendoza J, see Seel C (1995) *175*:101–132
de Raadt A, Ekhart CW, Ebner M, Stütz AE (1997) Chemical and Chemo-Enzymatic Approaches to Glycosidase Inhibitors with Basic Nitrogen in the Sugar Ring. *187*:157–186
de Silva AP, see Bissell RA (1993) *168*:223–264
Descotes G (1990) Synthetic Saccharide Photochemistry. *154*:39–76
Desiraju GR, see Nangia A (1998) *198*:57–95
Dias JR (1990) A Periodic Table for Benzenoid Hydrocarbons. *153*:123–144
Dietrich-Buchecker Ch, see Chambron J-C (1993) *165*:131–162
Dobson JF (1996) Density Functional Theory of Time-Dependent Phenomena. *181*:81–172
Dohm J, Vögtle, F (1991) Synthesis of (Strained) Macrocycles by Sulfone Pyrolysis. *161*:69–106
Dolbier WR jr. (1997) Fluorinated Free Radicals. *192*:97–163
Drakesmith FG (1997) Electrofluorination of Organic Compounds. *193*:197–242
Dreizler RM (1996) Relativistic Density Functional Theory. *181*:1–80
Driguez H (1997) Thiooligosaccharides in Glycobiology. *187*:85–116
Dutasta J-P, see Collet A (1993) *165*:103–129
Eaton DF (1990) Electron Transfer Processes in Imaging. *156*:199-226
Ebner M, see de Raadt A (1997) *187*:157–186
Edelmann FT (1996) Rare Earth Complexes with Heteroallylic Ligands. *179*:113–148
Edelmann FT (1996) Lanthanide Metallocenes in Homogeneous Catalysis. *179*:247–276
Effenhauser CS (1998) Integrated Chip-Based Microcolumn Separation Systems. *194*:51–82
Ehrfeld W, Hessel V, Lehr H (1998) Microreactors for Chemical Synthesis and Biotechnology – Current Developments and Future Applications. *194*:233–252
Ekhart CW, see de Raadt A (1997) *187*:157–186
El-Basil S (1990) Caterpillar (Gutman) Trees in Chemical Graph Theory. *153*:273–290
Engel E (1996) Relativistic Density Functional Theory. *181*:1–80

Ernzerhof M, Perdew JP, Burke K (1996) Density Functionals: Where Do They Come From, Why Do They Work? *190*:1 – 30
Fasani A, see Albini A (1993) *168*:143 – 173
Fernández-Mayoralas A (1997) Synthesis and Modification of Carbohydrates using Glycosidases and Lipases. *186*:1 – 20
Fessner W-D, Walter C (1997) Enzymatic C–C Bond Formation in Asymmetric Synthesis. *184*:97 – 194
Fessner W-D, see Petersen M (1997) *186*:87 – 117
Feuerbacher N, Vögtle F (1998) Iterative Synthesis in Organic Chemistry. *197*:1 – 18
Fontaine A, Dartyge E, Itie JP, Juchs A, Polian A, Tolentino H, Tourillon G (1989) Time-Resolved X-Ray Absorption Spectroscopy Using an Energy Dispensive Optics: Strengths and Limitations. *151*:179 – 203
Foote CS (1994) Photophysical and Photochemical Properties of Fullerenes. *169*:347 – 364
Fossey J, Sorba J, Lefort D (1993) Peracide and Free Radicals: A Theoretical and Experimental Approach. *164*:99 – 113
Fox MA (1991) Photoinduced Electron Transfer in Arranged Media. *159*:67 – 102
Freeman PK, Hatlevig SA (1993) The Photochemistry of Polyhalocompounds, Dehalogenation by Photoinduced Electron Transfer, New Methods of Toxic Waste Disposal. *168*:47 – 91
Fuchigami T (1994) Electrochemical Reactions of Fluoro Organic Compounds. *170*:1 – 38
Fuhr G, Shirley SG (1998) Biological Application of Microstructures. *194*:83 – 116
Fuller W, see Grenall R (1989) *151*:31 – 59
Galán A, see Seel C (1995) *175*:101 – 132
Gallagher T, see Beau J-M (1997) *187*:1 – 54
Gambert U, Thiem J (1997) Chemical Transformations Employing Glycosyltransferases. *186*:21 – 43
Gehrke R (1989) Research on Synthetic Polymers by Means of Experimental Techniques Employing Synchrotron Radiation. *151*:111 – 159
Geldart DJW (1996) Nonlocal Energy Functionals: Gradient Expansions and Beyond. *190*: 31 – 56
Gerratt J, see Cooper DL (1990) *153*:41 – 56
Gerwick WH, Nagle DG, Proteau, PJ (1993) Oxylipins from Marine Invertebrates. *167*:117 – 180
Gigg J, Gigg R (1990) Synthesis of Glycolipids. *154*:77 – 139
Gislason EA, see Guyon P-M (1989) *151*:161 – 178
Glusker JP (1998) Directional Aspects of Intermolecular Interactions. *198*:1 – 56
Goti A, Cordero FM, Brandi A (1996) Cycloadditions Onto Methylene- and Alkylidene-cyclopropane Derivatives. *178*:1 – 99
Gray HB, see Miskowski VM (1997) *191*: 41 – 57
Greenall R, Fuller W (1989) High Angle Fibre Diffraction Studies on Conformational Transitions DNA Using Synchrotron Radiation. *151*:31 – 59
Greiveldinger G, see Seebach D (1998) *197*:125 – 164
Gritsenko OV, see van Leeuwen R (1996) *180*:107 – 168
Gross EKU (1996) Density Functional Theory of Time-Dependent Phenomena. *181*:81 – 172
Gruber B, see Bley K (1993) *166*:199 – 233
Güdel HU, see Colombo MG (1994) *171*:143 – 172
Gunaratne HQN, see Bissell RA (1993) *168*:223 – 264
Guo XF, see Zhang FJ (1990) *153*:181 – 194
Gust D, Moore TA (1991) Photosynthetic Model Systems. *159*:103 – 152
Gutman I (1992) Topological Properties of Benzenoid Systems. *162*:1 – 28
Gutman I (1992) Total π-Electron Energy of Benzenoid Hydrocarbons. *162*:29 – 64
Guyon P-M, Gislason EA (1989) Use of Synchrotron Radiation to Study-Selected Ion-Molecule Reactions. *151*:161 – 178
Haag R, de Meijere A (1998) Unsaturated Oligoquinanes and Related Systems. *196*:137 – 165
Hadjiarapoglou L, see Adam W (1993) *164*:45 – 62
Hagen S, Hopf H (1998) Modern Routes to Extended Aromatic Compounds. *196*:45 – 89
Hamilton AD, see Meléndez RE (1998) *198*:97 – 129

Hart H, see Vinod TK (1994) *172*:119–178
Harbottle G (1990) Neutron Acitvation Analysis in Archaecological Chemistry. *157*:57–92
Hashimoto K, Yoshihara K (1996) Rhenium Complexes Labeled with $^{186/188}$Re for Nuclear Medicine. *176*:275–292
Hatlevig SA, see Freeman PK (1993) *168*:47–91
Hauser A, see Colombo MG (1994) *171*:143–172
Hayashida O, see Murakami Y (1995) *175*:133–156
He WC, He WJ (1990) Peak-Valley Path Method on Benzenoid and Coronoid Systems. *153*:195–210
He WJ, see He WC (1990) *153*:195–210
Heaney H (1993) Novel Organic Peroxygen Reagents for Use in Organic Synthesis. *164*:1–19
Heidbreder A, see Hintz S (1996) *177*:77–124
Heinze J (1989) Electronically Conducting Polymers. *152*:1–19
Helliwell J, see Moffat JK (1989) *151*:61–74
Hennig H, see Billing R (1990) *158*:151–199
Herrmann WA, see Anwander R (1996) *179*:1–32
Hesse M, see Meng Q (1991) *161*:107–176
Hessel V, see Ehrfeld W (1998) *194*:233–252
Hiberty PC (1990) The Distortive Tendencies of Delocalized π Electronic Systems. Benzene, Cyclobutadiene and Related Heteroannulenes. *153*:27–40
Hintz S, Heidbreder A, Mattay J (1996) Radical Ion Cyclizations. *177*:77–124
Hirao T (1996) Selective Transformations of Small Ring Compounds in Redox Reactions. *178*:99–148
Hladka E, Koca J, Kratochvil M, Kvasnicka V, Matyska L, Pospichal J, Potucek V (1993) The Synthon Model and the Program PEGAS for Computer Assisted Organic Synthesis. *166*:121–197
Ho TL (1990) Trough-Bond Modulation of Reaction Centers by Remote Substituents. *155*:81–158
Holas A, March NH (1996) Exchange and Correlation in Density Functional Theory of Atoms and Molecules. *180*:57–106
Höft E (1993) Enantioselective Epoxidation with Peroxidic Oxygen. *164*:63–77
Hoggard PE (1994) Sharp-Line Electronic Spectra and Metal-Ligand Geometry. *171*:113–142
Holmes KC (1989) Synchrotron Radiation as a source for X-Ray Diffraction – The Beginning. *151*:1–7
Hopf H, see Kostikov RR (1990) *155*:41–80
Hopf H, see Hagen S (1998) *196*:45–89
Houk KN, see Wiest O (1996) *183*:1–24
Humbs W, see Yersin H (1997) *191*:153–249
Hutchinson J, Sandford G (1997) Elemental Fluorine in Organic Chemistry. *193*:1–43
Indelli MT, see Scandola F (1990) *158*:73–149
Inokuma S, Sakai S, Nishimura J (1994) Synthesis and Inophoric Properties of Crownophanes. *172*:87–118
Itie JP, see Fontaine A (1989) *151*:179–203
Ito Y (1990) Chemical Reactions Induced and Probed by Positive Muons. *157*:93–128
Itzstein von M, Thomson RS (1997) The Synthesis of Novel Sialic Acids as Biological Probes. *186*:119–170
Jackman RJ, see Qin D (1998) *194*:1–20
Jennings RC, Zucchelli G, Bassi R (1996) Antenna Structure and Energy Transfer in Higher Plant Photosystems. *177*:147–182
Jobst G, see Urban GA (1998) *194*:189–213
Johannsen B, Spiess H (1996) Technetium(V) Chemistry as Relevant to Nuclear Medicine. *176*:77–122
John P, Sachs H (1990) Calculating the Numbers of Perfect Matchings and of Spanning Tress, Pauling's Bond Orders, the Characteristic Polynomial, and the Eigenvectors of a Benzenoid System. *153*:145–180

Jones RO (1996) Structure and Spectroscopy of Small Atomic Clusters. *182*:87–118
Jucha A, see Fontaine A (1989) *151*:179–203
Juris A, see Venturi M (1998) *197*:193–228
Jurisson S, see Volkert WA (1996) *176*:77–122
Kaim W (1994) Thermal and Light Induced Electron Transfer Reactions of Main Group Metal Hydrides and Organometallics. *169*:231–252
Kappes T, see Sauerbrei B (1997) *186*:65–86
Kavarnos GJ (1990) Fundamental Concepts of Photoinduced Electron Transfer. *156*:21–58
Kelleher N, see Begley TP (1998) *195*:93–142
Kelly JM, see Kirsch-De-Mesmaeker A (1996) *177*:25–76
Kerr RG, see Baker BJ (1993) *167*:1–32
Khairutdinov RF, see Zamaraev KI (1992) *163*:1–94
Khosla C, see Carreras CW (1997); *188*:85–126
Kikuchi J, see Murakami Y (1995) *175*:133–156
Kim JI, Stumpe R, Klenze R (1990) Laser-induced Photoacoustic Spectroscopy for the Speciation of Transuranic Elements in Natural Aquatic Systems. *157*:129–180
Kinsland C, see Begley TP (1998) *195*:93–142
Kirsch-De-Mesmaeker A, Lecomte J-P, Kelly JM (1996) Photoreactions of Metal Complexes with DNA, Especially Those Involving a Primary Photo-Electron Transfer. *177*:25–76
Kirschning A, Bechthold A F-W, Rohr J (1997) Chemical and Biochemical Aspects of Deoxysugars and Deoxysugar Oligosaccharides. *188*:1–84
Kitazume T, Yamazaki T (1997) Enzymatically Controlled Reactions of Organofluorine Compounds. *193*:91–130
Klaffke W, see Thiem J (1990) *154*:285–332
Klein DJ (1990) Semiempirical Valence Bond Views for Benzenoid Hydrocarbons. *153*:57–84
Klein DJ, see Chen RS (1990) *153*:227–254
Klenze R, see Kim JI (1990) *157*:129–180
Knauer M, see Bley K (1993) *166*:199–233
Knops P, Sendhoff N, Mekelburger H-B, Vögtle F (1991) High Dilution Reactions – New Synthetic Applications. *161*:1–36
Koca J, see Hladka E (1993) *166*:121–197
König B (1998) Carbon Rich Cyclophanes with Unusual Properties – an Update. *196*:91–136
Koepp E, see Ostrowicky A (1991) *161*:37–68
Kohnke FH, Mathias JP, Stoddart JF (1993) Substrate-Directed Synthesis: The Rapid Assembly of Novel Macropolycyclic Structures via Stereoregular Diels-Alder Oligomerizations. *165*:1–69
Korchowiec J, see Nalewajski RF (1996) *183*:25–142
Kostikov RR, Molchanov AP, Hopf H (1990) Gem-Dihalocyclopropanos in Organic Synthesis. *155*:41–80
Kratochvil M, see Hladka E (1993) *166*:121–197
Křen V (1997) Enzymatic and Chemical Glycosylations of Ergot Alkaloids and Biological Aspects of New Compounds. *186*:45–64
Kricka LJ, see Cheng J (1998) *194*:215–231
Krogh E, Wan P (1990) Photoinduced Electron Transfer of Carbanions and Carbacations. *156*:93–116
Krohn K, Rohr J (1997) Angucyclines: Total Syntheses, New Structures, and Biosynthetic Studies of an Emerging New Class of Antibiotics. *188*:127–195
Kryutchkov SV (1996) Chemistry of Technetium Cluster Compounds. *176*:189–252
Kuck D (1998) The Centropolyindanes and Related Centro-Fused Polycyclic Organic Compounds. *196*:167–220
Kumar A, see Mishra PC (1995) *174*:27–44
Kunkeley H, see Vogler A (1990) *158*:1–30
Kuwajima I, Nakamura E (1990) Metal Homoenolates from Siloxycyclopropanes. *155*:1–39
Kvasnicka V, see Hladka E (1993) *166*:121–197
Lammerink TS, see van den Berg A (1998) *194*:21–49

Lange F, see Mandelkow E (1989) *151*:9 – 29
Lautens M, see Chiu P (1997) *190*: 1 – 85
Lecomte J-P, see Kirsch-De-Mesmaeker A (1996) *177*:25 – 76
van Leeuwen R, Gritsenko OV, Baerends EJ (1996) Analysis and Modelling of Atomic and Molecular Kohn-Sham Potentials. *180*:107 – 168
Lee AWM, Chan WH (1997) Chiral Acetylenic Sulfoxides and Related Compounds in Organic Synthesis. *190*: 103 – 129
Leeper FJ, see Battersby AR (1998) *195*:143 – 193
Lefort D, see Fossey J (1993) *164*:99 – 113
Lehr H, see Ehrfeld W (1998) *194*:233 – 252
Lipshutz RJ, see Anderson RC (1998) *194*:117 – 129
Little RD, Schwaebe MK (1997) Reductive Cyclizations at the Cathode. *185*:1 – 48
Lopez L (1990) Photoinduced Electron Transfer Oxygenations. *156*:117 – 166
López-Boada R, see Ludena EV (1996) *180*:169 – 224
Lozach B, see Collet A (1993) *165*:103 – 129
Lu L, see Burton DJ (1997) *193*:45 – 89
Ludena EV, López-Boada (1996) Local-Scaling Transformation Version of Density Functional Theory: Generation of Density Functionals. *180*:169 – 224
Lüning U (1995) Concave Acids and Bases. *175*:57 – 100
Lundt I (1997) Aldonolactones as Chiral Synthons *187*:117 – 156
Lymar SV, Parmon VN, Zamarev KI (1991) Photoinduced Electron Transfer Across Membranes. *159*:1 – 66
Lynch PLM, see Bissell RA (1993) *168*:223 – 264
Maguire GEM, see Bissell RA (1993) *168*:223 – 264
Majoral J-P, Caminade A-M (1998) Divergent Approaches to Phosphorus-Containing Dendrimers and their Functionalization. *197*:79 – 124
Mandelkow E, Lange G, Mandelkow E-M (1989) Applications of Synchrotron Radiation to the Study of Biopolymers in Solution: Time-Resolved X-Ray Scattering of Microtubule Self-Assembly and Oscillations. *151*:9 – 29
Mandelkow E-M, see Mandelkow E (1989) *151*:9 – 29
March NH, see Holas A (1996) *180*:57 – 106
Maslak P (1993) Fragmentations by Photoinduced Electron Transfer. Fundamentals and Practical Aspects. *168*:1 – 46
Mathias JP, see Kohnke FH (1993) *165*:1 – 69
Mattay J, Vondenhof M (1991) Contact and Solvent-Separated Radical Ion Pairs in Organic Photochemistry. *159*:219 – 255
Mattay J, see Hintz S (1996) *177*:77 – 124
Matyska L, see Hladka E (1993) *166*:121 – 197
McCoy CP, see Bissell RA (1993) *168*:223 – 264
McGall G, see Anderson RC (1998) *194*:117 – 129
Mekelburger H-B, see Knops P (1991) *161*:1 – 36
Mekelburger H-B, see Schröder A (1994) *172*:179 – 201
Mella M, see Albini A (1993) *168*:143 – 173
Meléndez RE, Hamilton AD (1998) Hydrogen-Bonded Ribbons, Tapes and Sheets as Motifs for Crystal Engineeering. *198*:97 – 129
Memming R (1994) Photoinduced Charge Transfer Processes at Semiconductor Electrodes and Particles. *169*:105 – 182
Meng Q, Hesse M (1991) Ring Closure Methods in the Synthesis of Macrocyclic Natural Products. *161*:107 – 176
Merz A (1989) Chemically Modified Electrodes. *152*:49 – 90
Mestres J, see Besalú, E (1995) *173*:31 – 62
Meyer B (1990) Conformational Aspects of Oligosaccharides. *154*:141 – 208
Meyer J-U, see Stieglitz T (1998) *194*:131 – 162
Mezey PG (1995) Density Domain Bonding Topology and Molecular Similarity Measures. *173*:63 – 83

Michalak A, see Nalewajski RF (1996) *183*:25 - 142
Miki H, see Azumi T (1997) *191*:1 - 40
Mishra PC, Kumar A (1995) Mapping of Molecular Electric Potentials and Fields. *174*:27 - 44
Miskowski VM, Gray HB (1997) Magnetic and Spectroscopic Properties of $Os_2(O_2CR)_4Cl_2$. Evidence for a $^3(\delta^*\pi^*)$ Ground State. *191*:41 - 57
Misumi S (1993) Recognitory Coloration of Cations with Chromoacerands. *165*:163 - 192
Mizuno K, Otsuji Y (1994) Addition and Cycloaddition Reactions via Photoinduced Electron Transfer. *169*:301 - 346
Mock WL (1995) Cucurbituril. *175*:1 - 24
Moeller KD (1997) Intramolecular Carbon - Carbon Bond Forming Reactions at the Anode. *185*:49 - 86
Moffat JK, Helliwell J (1989) The Laue Method and its Use in Time-Resolved Crystallography. *151*:61 - 74
Molchanov AP, see Kostikov RR (1990) *155*:41 - 80
Moore TA, see Gust D (1991) *159*:103 - 152
Müllen K, see Baumgarten M (1994) *169*:1 - 104
Murakami Y, Kikuchi J, Hayashida O (1995) Molecular Recognition by Large Hydrophobic Cavities Embedded in Synthetic Bilayer Membranes. *175*:133 - 156
Nagle DG, see Gerwick WH (1993) *167*:117 - 180
Nakamura E, see Kuwajima I (1990) *155*:1 - 39
Nalewajski RF, Korchowiec J, Michalak A (1996) Reactivity Criteria in Charge Sensitivity Analysis. *183*:25 - 142
Nangia A, Desiraju GR (1998) Supramolecular Synthons and Pattern Recognition. *198*:57 - 95
Narayanan VV, Newkome GR (1998) Supramolecular Chemistry within Dendritic Structures. *197*:19 - 77
Nédélec J-Y, Périchon J, Troupel M (1997) Organic Electroreductive Coupling Reactions Using Transition Metal Complexes as Catalysts. *185*:141 - 174
Newkome GR, see Narayanan VV (1998) *197*:19 - 77
Nicewonger R, see Begley TP (1998) *195*:93 - 142
Nicotra F (1997) Synthesis of C-Glycosides of Biological Interest. *187*:55 - 83
Nishimura J, see Inokuma S (1994) *172*:87 - 118
Nolte RJM, see Sijbesma RP (1995) *175*:25 - 56
Nordahl A, see Carlson R (1993) *166*:1 - 64
Okuda J (1991) Transition Metal Complexes of Sterically Demanding Cyclopentadienyl Ligands. *160*:97 - 146
Omori T (1996) Substitution Reactions of Technetium Compounds. *176*:253 - 274
Oscarson S (1997) Synthesis of Oligosaccharides of Bacterial Origin Containing Heptoses, Uronic Acids and Fructofuranoses as Synthetic Challengers. *186*:171 - 202
Ostrowicky A, Koepp E, Vögtle F (1991) The "Vesium Effect": Synthesis of Medio- and Macrocyclic Compounds. *161*:37 - 68
Otsuji Y, see Mizuno K (1994) *169*:301 - 346
Pálinkó I, see Tasi G (1995) *174*:45 - 72
Pandey G (1993) Photoinduced Electron Transfer (PET) in Organic Synthesis. *168*:175 - 221
Parmon VN, see Lymar SV (1991) *159*:1 - 66
Patterson HH (1997) Luminescence and Absorption Studies of Transition Metal Ions in Host Crystals, Pure Crystals and Surface Environments. *191*:59 - 86
Perdew JP, see Ernzerhof M (1996) *180*:1 - 30
Périchon J, see Nédélec J-Y (1997) *185*:141 - 174
Percy JM (1997) Building Block Approaches to Aliphatic Organofluorine Compounds. *193*:131 - 195
Perlmutter P (1997) The Nucleophilic Addition/Ring Closure (NARC) Sequence for the Stereocontrolled Synthesis of Heterocycles. *190*:87 - 101
Petersen M, Zannetti MT, Fessner W-D (1997) Tandem Asymmetric C–C Bond Formations by Enzyme Catalysis. *186*:87 - 117
Petersilka M (1996) Density Functional Theory of Time-Dependent Phenomena. *181*:81 - 172

Petrov VA, Bardin VV (1997) Reactions of Electophiles with Polyfluorinated Olefins. *192*:39 – 95
Pieper R, see Carreras CW (1997)*188*:85 – 126
Poirette AR, see Artymiuk PJ (1995) *174*:73 – 104
Polian A, see Fontaine A (1989) *151*:179 – 203
Ponec R (1995) Similarity Models in the Theory of Pericyclic Macromolecules. *174*:1 – 26
Pospichal J, see Hladka E (1993) *166*:121 – 197
Potucek V, see Hladka E (1993) *166*:121 – 197
Proteau PJ, see Gerwick WH (1993) *167*:117 – 180
Qin D, Xia Y, Rogers JA, Jackman RJ, Zhao X-M, Whitesides GM (1998) Microfabrication, Microstructures and Microsystems. *194*:1 – 20
Raimondi M, see Copper DL (1990) *153*:41 – 56
Rajagopal AK (1996) Generalized Functional Theory of Interacting Coupled Liouvillean Quantum Fields of Condensed Matter. *181*:173 – 210
Reber C, see Wexler D (1994) *171*:173 – 204
Rettig W (1994) Photoinduced Charge Separation via Twisted Intramolecular Charge Transfer States. *169*:253 – 300
Rheiner PB, see Seebach D (1998) *197*:125 – 164
Rice DW, see Artymiuk PJ (1995) *174*:73 – 104
Riekel C (1989) Experimental Possibilities in Small Angle Scattering at the European Synchrotron Radiation Facility. *151*:205 – 229
Rogers JA, see Qin D (1998) *194*:1 – 20
Rohr J, see Kirschning A (1997) *188*:1 – 83
Rohr J, see Krohn K (1997) *188*:127 – 195
Roth HD (1990) A Brief History of Photoinduced Electron Transfer and Related Reactions. *156*:1 – 20
Roth HD (1992) Structure and Reactivity of Organic Radical Cations. *163*:131 – 245
Rouvray DH (1995) Similarity in Chemistry: Past, Present and Future. *173*:1 – 30
Roy R (1997) Recent Developments in the Rational Design of Multivalent Glycoconjugates. *187*:241 – 274
Rüsch M, see Warwel S (1993) *164*:79 – 98
Sachs H, see John P (1990) *153*:145 – 180
Saeva FD (1990) Photoinduced Electron Transfer (PET) Bond Cleavage Reactions. *156*:59 – 92
Sahni V (1996) Quantum-Mechanical Interpretation of Density Functional Theory. *182*:1 – 39
Sakai S, see Inokuma S (1994) *172*:87 – 118
Sandanayake KRAS, see Bissel RA (1993) *168*:223 – 264
Sandford G, see Hutchinson J (1997) *193*:1 – 43
Sauerbrei B, Kappes T, Waldmann H (1997) Enzymatic Synthesis of Peptide Conjugates – Tools for the Study of Biological Signal Transduction. *186*:65 – 86
Sauvage J-P, see Chambron J-C (1993) *165*:131 – 162
Schäfer H-J (1989) Recent Contributions of Kolbe Electrolysis to Organic Synthesis. *152*: 91 – 151
Scheuer PJ, see Chang CWJ (1993) *167*:33 – 76
Schlüter A-D (1998) Dendrimers with Polymeric Core: Towards Nanocylinders. *197*:165 – 191
Schmidtke H-H (1994) Vibrational Progressions in Electronic Spectra of Complex Compounds Indicating Stron Vibronic Coupling. *171*:69 – 112
Schmittel M (1994) Umpolung of Ketones via Enol Radical Cations. *169*:183 – 230
Schönherr T (1997) Angular Overtap Model Applied to Transition Metal Complexes and d^N-Ions in Oxide Host Lattices. *191*:87 – 152
Schröder A, Mekelburger H-B, Vögtle F (1994) Belt-, Ball-, and Tube-shaped Molecules. *172*: 179 – 201
Schulz J, Vögtle F (1994) Transition Metal Complexes of (Strained) Cyclophanes. *172*:41 – 86
Schwaebe MK, see Little RD (1997) *185*:1 – 48
Seebach D, Rheiner PB, Greiveldinger G, Butz T, Sellner H (1998) Chiral Dendrimers. *197*:125 – 164

Seel C, Galán A, de Mendoza J (1995) Molecular Recognition of Organic Acids and Anions - Receptor Models for Carboxylates, Amino Acids, and Nucleotides. *175*:101-132
Seiders TJ, see Sritana-Anant Y (1998) *196*:1-43
Sellner H, see Seebach D (1998) *197*:125-164
Sendhoff N, see Knops P (1991) *161*:1-36
Serroni S, see Venturi M (1998) *197*:193-228
Sessler JL, Burrell AK (1991) Expanded Porphyrins. *161*:177-274
Sheldon EJ, see Cheng J (1998) *194*:215-231
Sheldon R (1993) Homogeneous and Heterogeneous Catalytic Oxidations with Peroxide Reagents. *164*:21-43
Sheng R (1990) Rapid Ways of Recognize Kekuléan Benzenoid Systems. *153*:211-226
Siegel JS, see Sritana-Anant Y (1998) *196*:1-42
Shirley SG, see Fuhr G (1998) *194*:83-116
Shoji S (1998) Fluids for Sensor Systems. *194*:163-188
Sijbesma RP, Nolte RJM (1995) Molecular Clips and Cages Derived from Glycoluril. *175*: 57-100
Simpson TJ (1998) Application of Isotopic Methods to Secondary Metabolic Pathways. *195*: 1-48
Sodano G, see Cimino G (1993) *167*:77-116
Sojka M, see Warwel S (1993) *164*:79-98
Solà M, see Besalú E (1995) *173*:31-62
Sorba J, see Fossey J (1993) *164*:99-113
Soumillion J-P (1993) Photoinduced Electron Transfer Employing Organic Anions. *168*:93-141
Spiess H, see Johannsen B (1996) *176*:77-122
Sritana-Anant Y, Seiders TJ, Siegel JS (1998) Design of Novel Aromatics Using the Loschmidt Replacement on Graphs. *196*:1-43
Stanek Jr J (1990) Preparation of Selectively Alkylated Saccharides as Synthetic Intermediates. *154*:209-256
Staunton J, Wilkinson B (1998) The Biosynthesis of Aliphatic Polyketides. *195*:49-92
Steckhan E (1994) Electroenzymatic Synthesis. *170*:83-112
Steenken S (1996) One Electron Redox Reactions between Radicals and Organic Molecules. An Addition/Elimination (Inner-Sphere) Path. *177*:125-146
Stein N, see Bley K (1993) *166*:199-233
Stick RV (1997) The Synthesis of Novel Enzyme Inhibitors and Their Use in Defining the Active Sites of Glycan Hydrolases. *187*:187-213
Stieglitz T, Meyer J-U (1998) Microtechnical Interfaces to Neurons. *194*:131-162
Stoddart JF, see Kohnke FH (1993) *165*:1-69
Strasser J, see Yersin H (1997) *191*:153-249
Stütz AE, see de Raadt A (1997) *187*:157-186
Stumpe R, see Kim JI (1990) *157*:129-180
Suami T (1990) Chemistry of Pseudo-sugars. *154*:257-283
Suppan P (1992) The Marcus Inverted Region. *163*:95-130
Suzuki N (1990) Radiometric Determination of Trace Elements. *157*:35-56
Tabakovic I (1997) Anodic Synthesis of Heterocyclic Compounds. *185*:87-140
Takahashi Y (1995) Identification of Structural Similarity of Organic Molecules. *174*: 105-134
Tandon M, see Begley TP (1998) *195*:93-142
Tasi G, Pálinkó I (1995) Using Molecular Electrostatic Potential Maps for Similarity Studies. *174*:45-72
Taylor S, see Begley TP (1998) *195*:93-142
Thiem J, Klaffke W (1990) Synthesis of Deoxy Oligosaccharides. *154*:285-332
Thiem J, see Gambert U (1997) *186*:21-43
Thomson RS, see Itzstein von M (1997) *186*:119-170
Timpe H-J (1990) Photoinduced Electron Transfer Polymerization. *156*:167-198
Tobe Y (1994) Strained [n]Cyclophanes. *172*:1-40

Tolentino H, see Fontaine A (1989) *151*:179–203
Tomalia DA (1993) Genealogically Directed Synthesis: Starbust/Cascade Dendrimers and Hyperbranched Structures. *165*
Tourillon G, see Fontaine A (1989) *151*:179–203
Troupel M, see Nédélec J-Y (1997) *185*:141–174
Ugi I, see Bley K (1993) *166*:199–233
Urban GA, Jobst G (1998) Sensor Systems. *194*:189–213
van den Berg A, Lammerink TSJ (1998) Micro Total Analysis Systems: Microfluidic Aspects, Integration Concept and Applications. *194*:21–49
Vaughan JFS, see Chambers RD (1997) *192*:1–38
Venturi M, Serroni S, Juris A, Campagna S, Balzani V (1998) Electrochemical and Photochemical Properties of Metal-Containing Dendrimers. *197*:193–228
Vinod TK, Hart H (1994) Cuppedo- and Cappedophanes. *172*:119–178
Vögtle F, see Dohm J (1991) *161*:69–106
Vögtle F, see Knops P (1991) *161*:1–36
Vögtle F, see Ostrowicky A (1991) *161*:37–68
Vögtle F, see Schulz J (1994) *172*:41–86
Vögtle F, see Schröder A (1994) *172*:179–201
Vögtle F, see Feuerbacher N (1998) *197*:1–18
Vogler A, Kunkeley H (1990) Photochemistry of Transition Metal Complexes Induced by Outer Sphere Charge Transfer Excitation. *158*:1–30
Volkert WA, Jurisson S (1996) Technetium-99m Chelates as Radiopharmaceuticals. *176*:123–148
Vondenhof M, see Mattay J (1991) *159*:219–255
Voyer N (1997) The Development of Peptide Nanostructures. *184*:1–38
Waldmann H, see Sauerbrei B (1997) *186*:65–86
Walter C, see Fessner W-D (1997) *184*:97–194
Wan P, see Krogh E (1990) *156*:93–116
Warwel S, Sojka M, Rösch M (1993) Synthesis of Dicarboxylic Acids by Transition-Metal Catalyzed Oxidative Cleavage of Terminal-Unsaturated Fatty Acids. *164*:79–98
Weinreb SM (1997) *N*-Sulfonyl Imines – Useful Synthons in Stereoselective Organic Synthesis. *190*:131–184
Wessel HP (1997) Heparinoid Mimetics. *187*:215–239
Wexler D, Zink JI, Reber C (1994) Spectroscopic Manifestations of Potential Surface Coupling Along Normal Coordinates in Transition Metal Complexes. *171*:173–204
Whitesides GM, see Qin D (1998) *194*:1–20
Wiest O, Houk KN (1996) Density Functional Theory Calculations of Pericyclic Reaction Transition Structures. *183*:1–24
Wilding P, see Cheng J (1998) *194*:215–231
Wilkinson B, see Staunton J (1998) *195*:49–92
Willett P, see Artymiuk PJ (1995) *174*:73–104
Willner I, Willner B (1991) Artificial Photosynthetic Model Systems Using Light-Induced Electron Transfer Reactions in Catalytic and Biocatalytic Assemblies. *159*:153–218
Woggon W-D (1997) Cytochrome P450: Significance, Reaction Mechanisms and Active Site Analogues. *184*:39–96
Wu M, see Begley TP (1998) *195*:93–142
Xia Y, see Qin D (1998) *194*:1–20
Yamazaki T, see Kitazume T (1997) *193*:91–130
Yersin H, Humbs W, Strasser J (1997) Characterization of Excited Electronic and Vibronic States of Platinuum Mental Compounds with Chelate Ligands by Highly Frequency-Resolved and Time-Resolved Spectra. *191*:153–249
Yoshida J (1994) Electrochemical Reactions of Organosilicon Compounds. *170*:39–82
Yoshihara K (1990) Chemical Nuclear Probes Using Photon Intensity Ratios. *157*:1–34
Yoshihara K (1996) Recent Studies on the Nuclear Chemistry of Technetium. *176*:1–16
Yoshihara K (1996) Technetium in the Environment. *176*:17–36

Yoshihara K, see Hashimoto K (1996) *176*:275 – 192
Zamaraev KI, see Lymar SV (1991) *159*:1 – 66
Zamaraev KI, Kairutdinov RF (1992) Photoinduced Electron Tunneling Reactions in Chemistry and Biology. *163*:1 – 94
Zander M (1990) Molecular Topology and Chemical Reactivity of Polynuclear Benzenoid Hydrocarbons. *153*:101 – 122
Zannetti MT, see Petersen M (1997) *186*:87 – 117
Zhang FJ, Guo XF, Chen RS (1990) The Existence of Kekulé Structures in a Benzenoid System. *153*:181 – 194
Zhang Y, see Begley TP (1998) *195*:93 – 142
Zhao X-M, see Qin D (1998) *194*:1 – 20
Ziegler T, see Berces A (1996) *182*:41 – 85
Ziegler T (1997) Pyruvated Saccharides - Novel Strategies for Oligosaccharide Synthesis. *186*:203 – 229
Zimmermann SC (1993) Rigid Molecular Tweezers as Hosts for the Complexation of Neutral Guests. *165*:71 – 102
Zink JI, see Wexler D (1994) *171*:173 – 204
Zucchelli G, see Jennings RC (1996) *177*:147 – 182
Zybill Ch (1991) The Coordination Chemistry of Low Valent Silicon. *160*:1 – 46

MIX
Papier aus verantwortungsvollen Quellen
Paper from responsible sources
FSC® C105338

If you have any concerns about our products,
you can contact us on
ProductSafety@springernature.com

In case Publisher is established outside the EU,
the EU authorized representative is:
Springer Nature Customer Service Center GmbH
Europaplatz 3, 69115 Heidelberg, Germany

Printed by Libri Plureos GmbH
in Hamburg, Germany